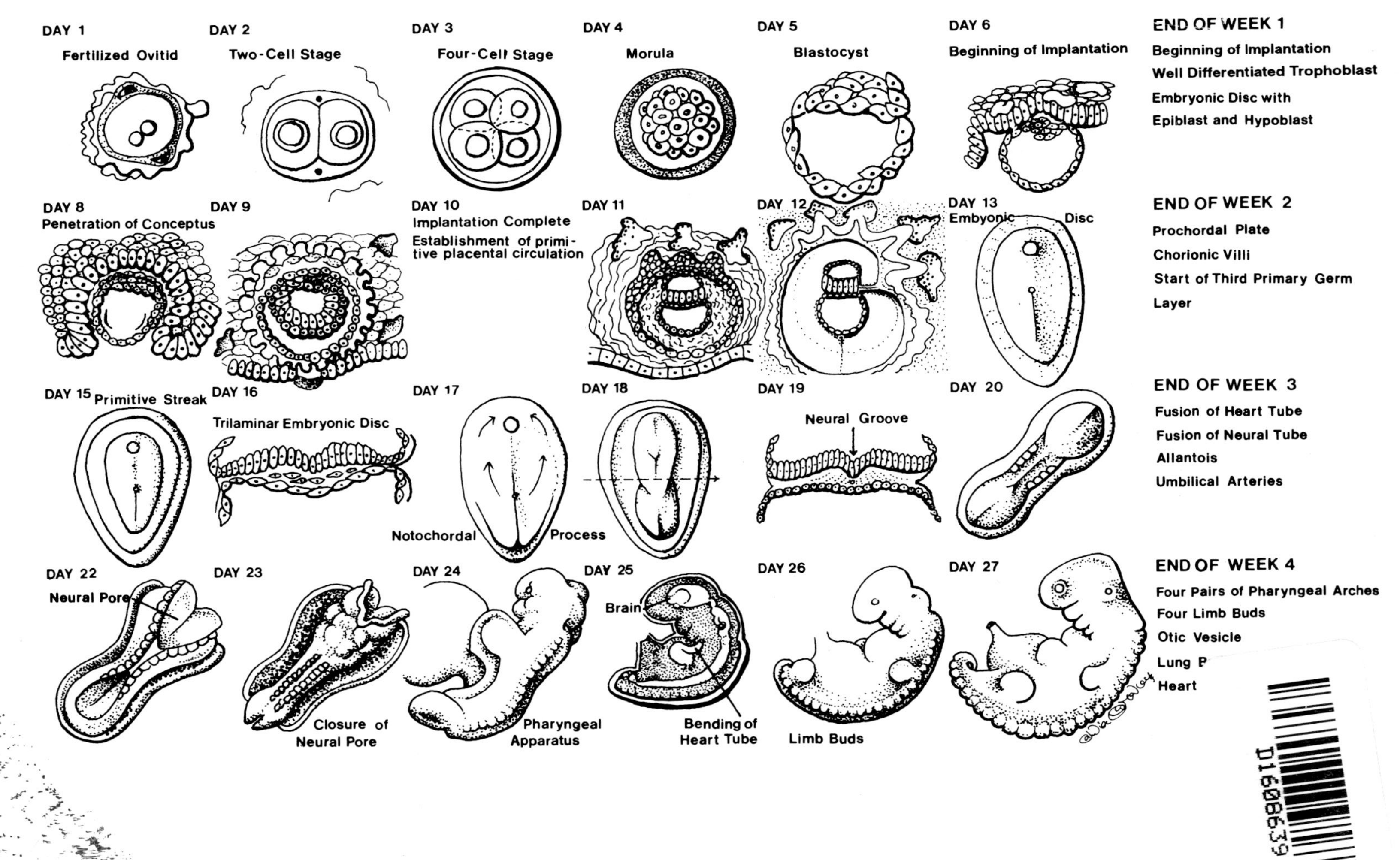
DAY 1
Fertilized Ovitid
DAY 2
Two-Cell Stage
DAY 3
Four-Cell Stage
DAY 4
Morula
DAY 5
Blastocyst
DAY 6
Beginning of Implantation
END OF WEEK 1
Beginning of Implantation
Well Differentiated Trophoblast
Embryonic Disc with
Epiblast and Hypoblast
DAY 8
Penetration of Conceptus
DAY 9
DAY 10
Implantation Complete
Establishment of primi-
tive placental circulation
DAY 11
DAY 12
DAY 13
Embyonic Disc
END OF WEEK 2
Prochordal Plate
Chorionic Villi
Start of Third Primary Germ
Layer
DAY 15
Primitive Streak
DAY 16
Trilaminar Embryonic Disc
DAY 17
Notochordal Process
DAY 18
DAY 19
Neural Groove
DAY 20
END OF WEEK 3
Fusion of Heart Tube
Fusion of Neural Tube
Allantois
Umbilical Arteries
DAY 22
Neural Pore
DAY 23
Closure of
Neural Pore
DAY 24
Pharyngeal
Apparatus
DAY 25
Brain
Bending of
Heart Tube
DAY 26
Limb Buds
DAY 27
END OF WEEK 4
Four Pairs of Pharyngeal Arches
Four Limb Buds
Otic Vesicle
Lung P
Heart

Human Embryology Made Easy

Human Embryology Made Easy

M. Waheed Rana

St. Louis University
School of Medicine
St. Louis, Missouri, USA

harwood academic publishers

Australia • Canada • China • France • Germany • India • Japan • Luxembourg • Malaysia • The Netherlands • Russia • Singapore • Switzerland • Thailand

Amsteldijk 166
1st Floor
1079 LH Amsterdam
The Netherlands

British Library Cataloguing in Publication Data

Rana, M. W.
Human embryology made easy
1.Embryology, Human
I.Title
612.6'4

ISBN 90-5702-545-0

To all developing humans

Verily We [God Almighty] *created human beings from the product of wet earth; then placed him as a small seed* [conceptus] *in a firm resting place. Then We made conceptus a clot, We made the clot a little lump; then We made bones in the lump and clothed the bones with flesh and then produced it as another creation, so blessed by Allah, the best of creators.* (Holy Quran 23:12–14)

He [God Almighty] *created you in the wombs of your mothers, in stages, one after another, in three veils* [membranes] *of darkness.* (39:6)

And We cause what We will to remain in the wombs of your mothers for an appointed time. (22:5)

CONTENTS

PREFACE

Human Embryology Made Easy is a synopsis of the key facts and concepts of human development. The book is intended for use by students who are taking a human embryology course. The book's outline format will help students find up-to-date information quickly, and its inclusion of the underlying mechanisms involved in clinically important congenital anomalies will prove useful to medical, nursing, and allied health students during their clinical training.

Each chapter (except chapter 7) is followed by a set of questions that function as a concise review of that chapter. Students are advised to test themselves to evaluate their understanding of the material, and then check their answers with those provided.

The book makes ample use of illustrations, which are essential for understanding embryology. Some of the illustrations included are from the *Color Atlas of Embryology* (Ulrich Drew, Thieme Medical Publishers, New York, 1995). The terminology of this text is based on Nomina Embryologica adopted by the Tenth International Congress of Anatomists.

I wish to express my gratitude to my students and the faculty of the Department of Anatomy and Neurobiology at St. Louis University School of Medicine for their constructive critique and continuous encouragement.

I am indebted to Kris Sherman for typing the manuscript; to my wife Janice Rana for editing the manuscript; to my editor Sally Cheney for her excellent cooperation during publication of this book; and to Tim Oliver and Pam Robertson, whose hard work during the production stage of the publishing process was invaluable.

CHAPTER

1

Cell Division

In a cell population that is constantly being renewed, individual cells divide periodically. A typical somatic cell division, mitosis, consists of an equal division of nuclear material, so that the two newly formed daughter cells receive exactly the number and kind of chromosomes that the parent cell had. This separation of nuclear material is then followed by division of the cytoplasm.

Before a cell can undergo division, it must increase its mass and contents, and double the mass of its DNA. All of this occurs during the growth period known as interphase. Following this is the M phase, during which nuclear division (mitosis) and cytoplasmic division (cytokinesis) take place. Duplicated DNA must be divided precisely between daughter cells.

Although the cell cycle is continuous, for simplicity and clarity, both interphase and M phase are subdivided into stages. The interphase is composed of the G_1, S and G_2 periods.

I. INTERPHASE

A. G_1 (Gap_1) Period

After completion of cell division, the daughter cells enter the preduplication period, G_1 (gap_1). During this period there is synthesis of RNA and proteins, and total cell mass is increased.

After this, the cell is held at a restriction point. Any cell that passes this point will complete the rest of the stages of the cycle. A trigger or unstable protein (S phase activator) has been proposed. An accumulation of a threshold amount of this

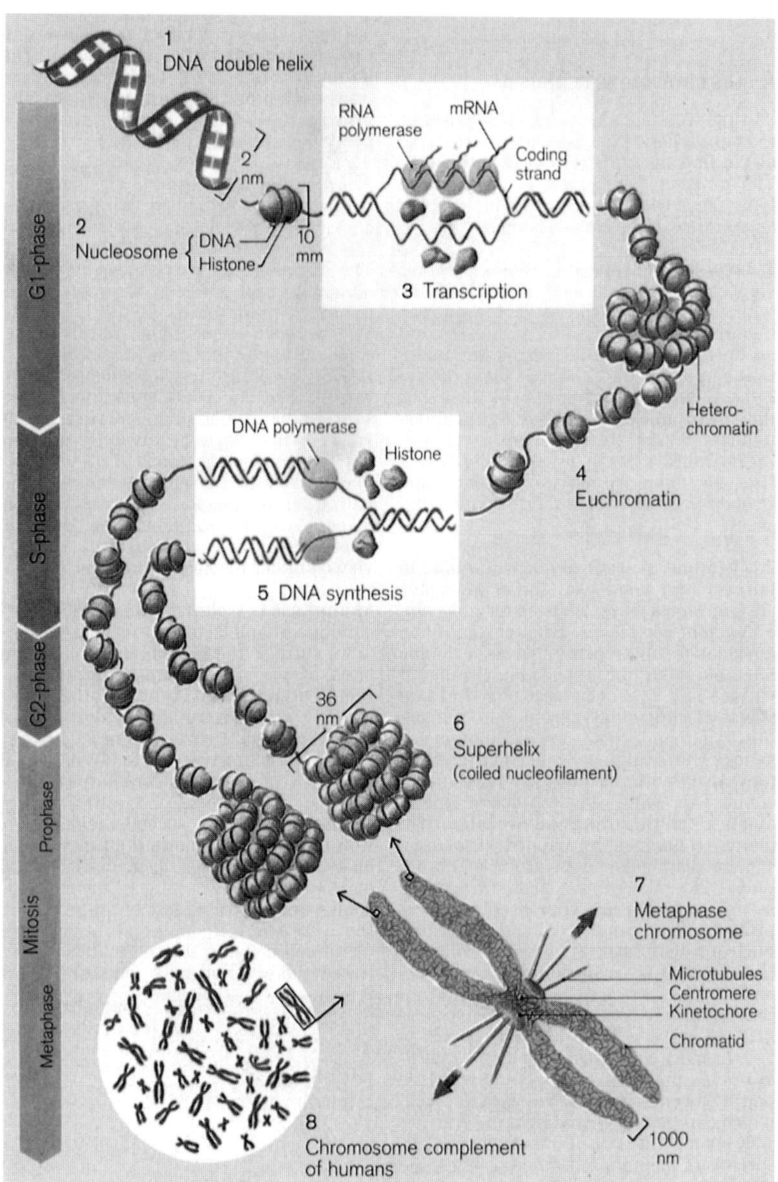

Figure 1.1

Chromosomes (see Color Plate 1.1). Reproduced with permission from Thieme Medical Publishers Inc., New York, 1995, *Color Atlas of Embryology,* Ulrich Drews, Chapter 1: Reproduction.

protein helps the cell to exceed the restriction point. The quiescent cells that do not accumulate this protein are arrested at the restriction point and are considered to be in the G_0 period of interphase. This may be one of the mechanisms by which tissue growth is controlled. Crowding (contact inhibition) and starvation may also inhibit cell division. In many tissues, cells divide only when new cells are needed. Neoplastic cells appear to have lost these growth controls.

B. S (Synthesis) Period

DNA duplication, semiconservative in nature, occurs during period S, a period that is a constant characteristic of the cell type and growth conditions. The DNA is simultaneously replicated in discrete units called replicons. When all replicons have been duplicated, newly formed DNA segments join to complete the daughter molecule (Fig. 1.1).

The centriole pair separate from each other during late G_1. Duplication of each centriole starts during the S period and is completed in G_2. As DNA is replicated, new histones are synthesized during this period. Until DNA replication is complete, the M phase is delayed.

C. G_2 Period

At the end of DNA duplication, the cell enters the preparatory period, G_2. Some proteins essential for cell division are synthesized during this period. A kinase is detected that could be responsible for phosphorylation of proteins of nuclear membrane, which may in turn cause breakdown of nuclear lamins during the M phase. It may also cause phosphorylation of histone H_1 molecules. In this period, components essential to form mitotic spindle are prepared. Most proteins and RNA molecules are synthesized continuously during interphase.

Three diffusible factors that may control events during interphase have been suggested: 1) an S phase activator begins DNA synthesis, 2) an M phase promoting factor (MPF) induces chromosomal condensation, and 3) an M phase delaying factor (MDF) inhibits production of MPF. There is a sequential relationship between the factors that control each successive step. For instance, DNA cannot start replicating unless the DNA re-replication block has been removed during G_1. With appearance of an S phase activator and MDF, DNA synthesis continues until all the DNA has been replicated. MPF cannot be produced until the MDF has disappeared, and cells cannot enter mitosis until an MPF is produced. MPF concentration increases rapidly during early M phase. It is suggested that its surge may be triggered by an increase in the concentration of another protein, cyclin, whose concentration rises steadily. Its increase at threshold level during G_2 activates MPF production. Both reach maximum concentration in the middle of M phase, when cyclin is abruptly destroyed, and MPF disappears. After this, the cyclin concentration again starts to increase steadily.

II. M PHASE

M phase includes **mitosis** (division of nuclear material) and **cytokinesis** (division of cytoplasm).

A. Mitosis: Division of Nuclear Material

1. PROPHASE. The cell becomes spheroid and viscous because of breakdown of cytoskeleton. Dispersed chromatin becomes visible as delicate, longitudinally coiled filaments, known as chromosomes. DNA thread winds around a core that is formed by nucleosomes and appears as chromosomes. These delicate chromosomes undergo further condensation to form metaphase chromosomes. Each daughter centrosome acts as a microtubule organizer, and shows astral rays. As centrosomes begin to move apart, the microtubules in each aster elongate, keeping contact with both centrosomes, thus forming a mitotic spindle. Chromosomes move closer to the nuclear membrane and appear to be composed of two chromatids. They continue to become shorter and thicker. Meanwhile, the nucleolus elongates and disappears among chromosomes. Nuclear membrane disintegrates. Spindle apparatus assembly is initiated.

2. PROMETAPHASE. Prometaphase starts with disintegration of the nuclear membrane. The centrosomes reach opposite poles, and spindle microtubules enter the nuclear region. Each chromosome is seen as two sister chromatids held together at a centromere. A kinetochore, a protein complex, develops on each side of the centromere. Some of the spindle microtubules attach to the kinetochore. These kinetochore microtubules extend in opposite directions from the sister chromatids to one of the poles.

3. METAPHASE. Tension exerted by kinetochore tubules causes chromosomes to move toward the center of the cell and align at the middle of the spindle. The spindle shows polar microtubules extending from opposite poles. Kinetochore microtubules attach sister chromatids to opposite poles and to astral microtubules, which are incorporated in the spindle.

4. ANAPHASE. Anaphase starts abruptly. Shortening of the kinetochore microtubules causes separation of the kinetochore, and the centromere appears split. Cytosolic calcium is increased at this time. The sister chromatids start moving toward opposite poles. The polar microtubules elongate, moving the spindle asters farther apart. Elongated cells show a constriction in the middle.

5. TELOPHASE. Separated chromatids reach the poles. Kinetochore microtubules disappear. New nuclear membranes around each set of chromatids are reconstructed. The chromosomes start unwinding. The nucleoli reappear.

B. Cytokinesis: Division of Cytoplasm

In the middle of the cell, a second cytoskeletal structure, called a contractile ring, appears just beneath the cell membrane. The contractile ring consists of actin and myocin filaments, and it pulls the membrane inward, forming constriction in the middle of the cell perpendicular to the spindle (seen during anaphase). The continuous pull of the ring gradually deepens the constriction, forming a cleavage furrow. The polar microtubules of the spindle are pushed into a small bundle, the mid-body. Finally, the mid-body breaks, and the two daughter cells are separated. Each new cell receives a share of other organelles.

C. Abnormal Mitosis

1. Polyploidy is a condition in which cells contain a multiple of the diploid number of chromosomes. Polyploidy, although common in liver cells, is a result of abnormal mitosis. Neoplastic cells quite often exhibit polyploidy. Actively growing tumors, therefore, show abnormal mitotic figures and giant cells (large cells with multiple nuclei).
2. Aneuploidy is a condition in which cells contain either less than the normal diploid number of chromosomes or one or more extra chromosomes. This type of variation may result in a congenital disease, e.g., Down's syndrome.
3. The term "mitotic figure" refers to any cell undergoing mitosis. The chromatic material, being more condensed, stains more densely than in an interphase nucleus.
4. Colchicine and vinblastine block the formation of the microtubules of a spindle. The cells undergoing mitosis, when treated with these drugs, will be arrested at the metaphase stage.

III. MEIOSIS

Every cell inherits one set of chromosomes (23 in humans) from each parent to constitute a diploid (46) number of chromosomes. Although similar in appearance, two chromosomes, one from each set, somewhat different in their genetic composition, form a homologous pair. The exception to this is the pair of sex chromosomes in the male. Meiosis is the special division during which gametes are formed from the germ cells in the gonads. These gametes contain a haploid number (23) of chromosomes, so that at the time of their union to other gametes (fertilization) the diploid number (46) of chromosomes is restored. This is achieved by two nuclear divisions. During "S" phase, DNA replication occurs. Each chromosome consists of two chromatids. In the first division, homologues are separated; in the second division, the chromatids of each chromosome are separated.

Another characteristic of meiosis is the long prophase of first division, during which an exchange of small portions of DNA occurs randomly between the chromatids of homologues, which permits random independent recombination of genes.

In summary, meiosis causes:

A. a reduction in the number of chromosomes, so that the gametes become haploid cells;

B. random recombination of genes, which makes each chromatid of homologous pairs genetically somewhat different from each other, so that at the union of two gametes from each parent an individual with a new assortment of genes is produced.

A. Meiotic Division I (see Fig. 1.2)

After completion of interphase, germ cells enter prophase I.

1. Prophase I
   Because of its long duration and complexity, prophase I is subdivided into five stages:

   a. Leptotene
      Chromosomes, which were very thin to begin with, become more apparent and individual, and orient themselves toward centrioles.

   b. Zygotene
      Homologous chromosomes begin to pair and fuse at many places along their length. Fusion or synapsis between homologues is initiated by formation of a synaptolemal complex. The complex consists of a central core protein and two lateral elements that are attached to the chromatin of opposite homologues. The alignment of homologues is very precise, by which each gene on opposite chromosomes is brought into juxtaposition. Each homologous pair held thus is called a tetrad.

   c. Pachytene
      At the completion of pairing of the chromosomes, the cells enter the pachytene stage. Large protein complexes, recombination nodules, appear at different points on the core protein of the synaptolemal complex. Recombination nodules bring small regions of DNA of opposing chromatids together. One or more transverse breaks occur in the chromatids at the recombination nodules, followed by crossing over or interchange of DNA segments between chromatids of opposite chromosomes (non-sister chromatids).

   d. Diplotene
      The synaptolemal complexes start to disappear, and homologue chromosomes begin to separate, except at the points of crossing over. The chiasmata become more apparent.

   e. Diakinesis
      Separation of chromosomes continues. The chiasmata appear to move toward the end of the tetrad. Further condensation of chromosomes causes four visibly distinct chromatids.

- **Note**. Female germ cells have two X chromosomes, which pair and separate like other homologue chromosomes. Male germ cells have one X and one Y chromosome. In this case, pairing occurs between the small X and Y segments where there is homology. Crossing over then occurs in this region between non-sister chromatids. This small amount of genetic recombination is sufficient to produce diversity in chromatids.

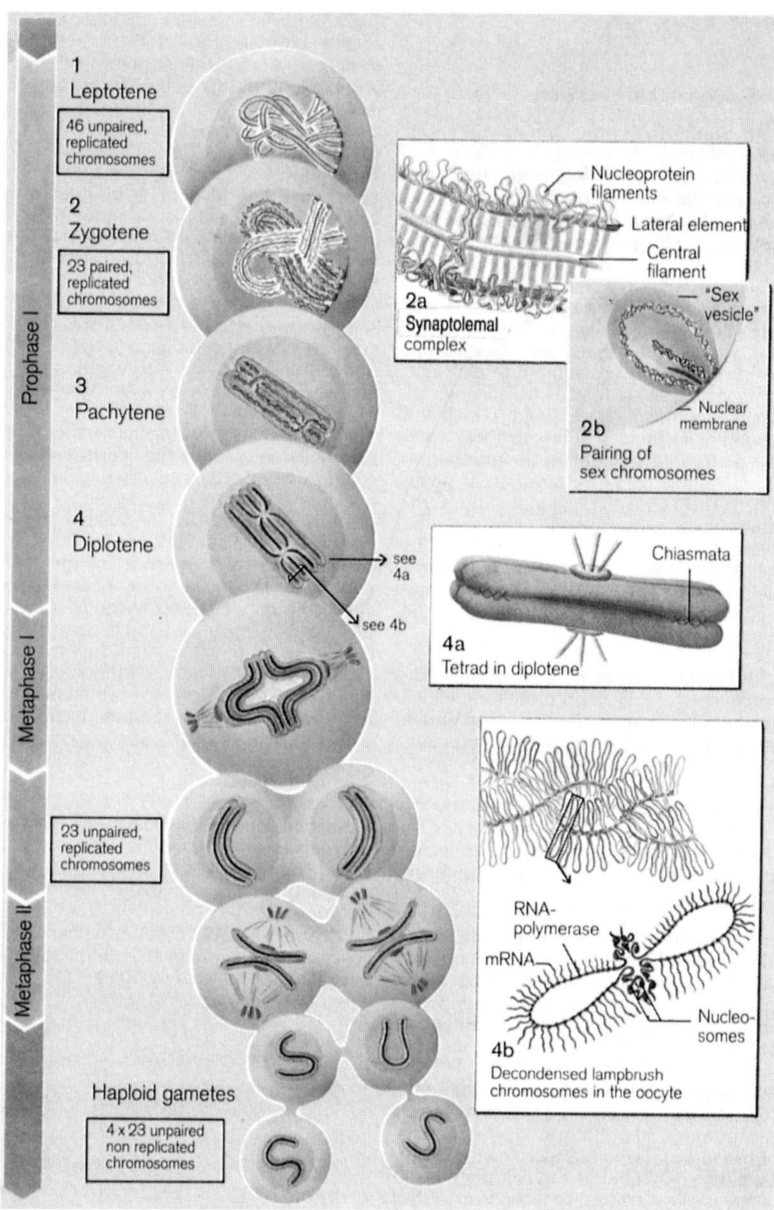

Chromosomes in meiosis

Figure 1.2
Meiosis (see Color Plate 1.2). Reproduced with permission from Thieme Medical Publishers Inc., New York, 1995, *Color Atlas of Embryology*, Ulrich Drews, Chapter 1: Reproduction.

2. Prometaphase I
   The nuclear membrane disintegrates. Kinetochores of sister chromatids remain fused, so that attachment of the chromosomal fibers of the spindle is toward one pole.

3. Metaphase I
   Formation and attachment of the spindle fiber is completed. Maximum condensation of chromosomes is achieved. Homologue pairs are brought to the center of the spindle.

4. Anaphase I
   Chromosomes in homologue pairs begin to separate. Chiasmata dissolve. Chromosomes consisting of two sister chromatids start to move toward opposite poles. One member of each pair starts to move to one pole, and other member starts to move toward opposite pole. Half of the chromosomes move toward one direction and the other half move toward the opposite direction. This segregates the chromosomes into single sets, reducing their number by half. It is important to remember that each of these chromosomes has two sister chromatids, which are a little different in genetic makeup because of crossing over during pachytene.

5. Telophase I
   The chromosome sets complete their migration toward the opposite poles. A nuclear membrane is formed around the chromosomes.

6. Cytokinesis
   A cleavage furrow forms, and the cell is divided into two cells. These cells are now haploid cells. In the male, the cytoplasm of a germ cell (primary spermatocytes) is divided equally, and newly formed cells are known as secondary spermatocytes. In female germ cells (primary oocytes), the cytoplasm during cytokinesis is divided unequally. One cell receives most of the cytoplasm and is known as the secondary oocyte; the other, which is merely a nucleus enclosed in a cell membrane, is known as the first polar body, and does not undergo further division.

7. Interdivision Interphase
   This is relatively short, consisting only of the G_2 period. Secondary spermatocytes and secondary oocytes enter the second part of meiotic division. Secondary oocytes must be fertilized before completing second division.

B. Meiotic Division II

During this second part of meiosis, the chromatids of each chromosome are separated. It resembles mitosis in this regard.

1. Prophase II
   Centrosomes migrate to the opposite poles, and a new spindle is organized. The nuclear membrane disintegrates. Chromosomal fibers from opposite poles attach to the kinetochore of sister chromatids.

2. Metaphase II
   Chromosomes (haploid in number) are arranged in the middle of the spindle.

3. Anaphase II
   Sister chromatids separate and start migrating toward opposite poles.
4. Telophase II
   Chromatids reach their respective poles.
   Kinetochore microtubules disappear.
   New nuclear membrane reforms around the set of chromatids (new chromosomes).
   The chromatin decondenses.
   Nucleoli reappear.
5. Cytokinesis
   Cytoplasm of the cell divides by cleavage. Newly formed gametes stay attached to each other through mid-body.

- **Note**. In the male, the cytoplasm of secondary spermatocytes divides equally, forming two spermatids. Spermatids formed from the same spermatogonium stay connected to each other during their further maturation. In the female, cytoplasm of secondary oocytes divides unequally. The gamete that retains most of the cytoplasm is known as an ootid, and is the only viable gamete. The other gamete, which consists mainly of nuclear material, is known as the second polar body.

C. Abnormal Meiotic Divisions

1. Anaphase lag
   One member of the homologue pair of chromosomes is delayed in moving into one of the daughter cells during anaphase. Consequently, it is not incorporated into the nucleus and is lost from the cell. This may lead to one daughter cell having a normal component of 23 chromosomes and the other having only 22 chromosomes (Fig. 1.3).
2. Nondisjunction
   The homologous pair fails to separate and both members move into one cell. This results in one cell having 22 and the other having 24 chromosomes. After fertilization, these cells could give rise to either one additional or one less chromosome.

D. Fertilization Between Abnormal Gametes

1. Monosomy
   One of the partners of a chromosome pair is missing.
2. Trisomy
   The presence of an extra chromosome. If the additional chromosome is not free but attached to another chromosome, the condition is known as translocation.

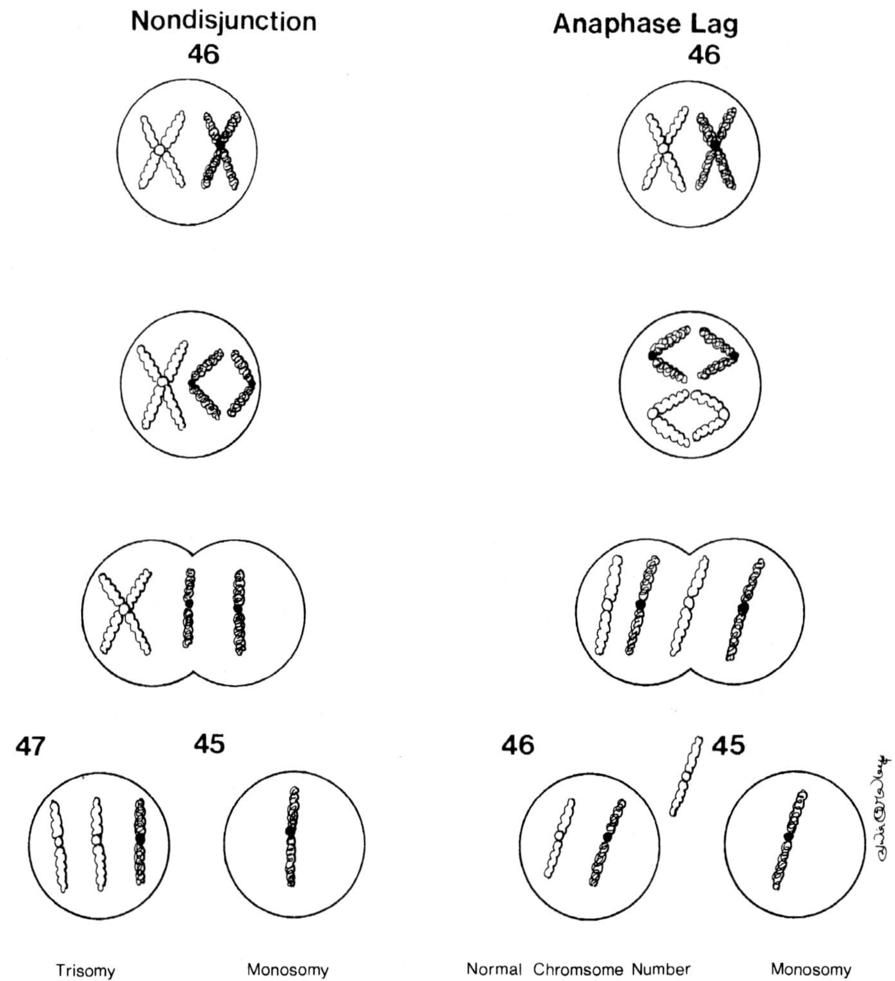

Figure 1.3
Consequences of errors in separation of chromosomes during cell division. (Redrawn from *Clinical Embryology* by Crowley.)

a. Trisomy 21
   Down's syndrome (mongolism) is the most common form of trisomy occurring in Western countries — about 1 in 600 births. The incidence varies with maternal age, being highest after the mother has reached the age of 35 years. In young mothers the incidence is not more than 1 in 1000 births. There are four types of trisomy 21:

   i. Forty-seven chromosomes are present — 90%.

   ii. Forty-eight chromosomes are present, creating double trisomic condition (XXY plus trisomy 21) — 6%.

iii. Forty-six chromosomes are present, and the additional chromosome 21 becomes translocated to either 13, 14, 15, or 21–22. One of the parents, although normal, shows translocation of chromosome 21 — 3%.

iv. Chromosome mosaics, where some normal cells have 46 chromosomes and other cells have 47 chromosomes — 1%.

If parents having normal chromosomes have a Down's syndrome child, the chance of having a second Down's syndrome child is about 2%. If one parent is the carrier of translocated chromosome 21, the chance of having a second Down's syndrome child rises to 33%. Aside from their peculiar physical appearance, Down's syndrome children are mentally defective and show generalized hypotonia and retarded physical development. The incidence of congenital heart disease, umbilical hernia and duodenal atresia is very high. The majority die young.

b. Trisomy 17–18
Less common than Down's syndrome. The infant has a receding chin and a prominent occiput. The fingers are rigidly flexed across the hand and syndactyly may be present. They are also mentally retarded and usually die soon after birth.

c. Trisomy 13–15
The infant has a small head, deformed nose and underdeveloped jaw. Congenital heart disease, cleft lip, cleft palate, deafness and eye defects are common. Most die soon after birth.

d. Other autosomal trisomies have been observed in aborted fetuses.

3. Extra sex chromosomes (Table 1.1).

a. Klinefelter's syndrome 44 + XXY (47) chromosomes
Occurs only in males. At puberty, secondary sexual characteristics develop, but the testicles remain small. Gynecomastia is usually present, and the patients grow very tall. All patients are sterile, and some are mentally retarded.

b. 44 + XXX (47) chromosomes
At times associated with mild mental retardation and/or minor anomalies, rarely with serious gonadal dysgenesis.

c. 44 + XYY (47) chromosomes
Males with unusual height, at times mild mental retardation and/or mental illness, minor somatic abnormalities and aggression.

d. 48 (44 + XXXY or 44 + XXYY) or 49 (44 + XXXXY or 44+ XXXYY) chromosomes
Males have gonadal defect and severe retardation.

TABLE 1.1. Zygotes Produced by Combination of Abnormal Gametes

Abnormal gamete	Normal gamete	Zygote	Anomalies
Female	Male	XXX	Mild mental retardation
XX	X	XXY	Klinefelter's syndrome
O	X	XO	Turner's syndrome
O	Y	YO	Nonviable
male	female		
XY	X	XXY	Klinefelter's syndrome
XX	X	XXX	Superfemale (minor anomalies)
YY	X	XYY	Supermale (unusual height)
O	X	XO	Turner's syndrome

4. Absence of sex chromosomes.
   a. Turner's syndrome (44 + XO)
      Occurs in females.
      i. External genitalia and mammary glands remain infantile after puberty (lack of estrogen).
      ii. Ovaries are hypoplastic (underdeveloped) or missing.
      iii. Other associated anomalies.
         a) Webbing of the neck.
         b) Skeletal deformation.
         c) Congenital heart disease.
         d) Lymphedema of the extremities.
         e) Mental retardation commonly associated with mosaicism due to meiotic chromosomal nondisjunction. XO/XX, XO/XXX, XO/XX/XXX and XO/XY are the most common types.
   b. Absence of X in male (44 + YO) is nonviable.

5. Anomalies involving parts of chromosomes.
   a. Chromosomal loss due to deletions.
      i. Cri-du-chat syndrome (cat-cry syndrome)
      One of the homologues of pair 5 shows a short arm. The infant has a weak, high-pitched cry. There may be congenital heart defect, and some infants are microcephalic.
      ii. Several other deletion syndromes
      18q and 18p syndrome; Dq syndrome, etc.
   b. Partial trisomy
   A "partial trisomy" due to insertions, reciprocal translocations, isochromosomal formation, etc. Although the chromosomal number is 46, one of the altered chromosomes may be detectable in the affected individual.

REVIEW QUESTIONS — Chapter 1

Multiple choice. Select the *most* appropriate answer.

A 16-year-old girl complained of amenorrhea (absence of menses). Physical examination revealed infantile external genitalia, and mammary glands with nipples pointed outward. No estrogens were found in the blood. She was short for her age and showed generalized hypotonia. Four questions pertain:

1. Which of the following karyotypes (chromosomal complements) would be most likely found in this patient:
   a. 45,X
   b. 44,XX
   c. 44,XY
   d. 47,XXY
   e. 46,XX
2. The most appropriate diagnosis is:
   a. Trisomy 21 (Down's syndrome)
   b. Klinefelter syndrome
   c. Turner's syndrome
   d. superfemale
   e. testicular feminization syndrome

3. This condition results from:
   a. anaphase lag
   b. nondisjunction
   c. both
   d. none

4. This chromosomal anomaly occurred during meiosis because:
   a. cyclin was lacking during cell division
   b. M phase promoting factor was missing during gamete formation
   c. M phase delaying factor was missing during S phase
   d. DNA replication was not complete
   e. the cause of this is not known

5. All cells are held at:
   a. S phase
   b. G_2 phase
   c. M phase
   d. restriction point
   e. cytokinesis

6. The feature found *only* in G_2 period is:
   a. nucleosomes
   b. chromosomes
   c. dividing mitochondria
   d. two sets of centrioles
   e. disappearance of nucleolus

7. Normally a cell would not enter M phase (cell division) if:
   a. it is crowding the other cells (contact inhibition)
   b. due to a lack of amino acids, its protein synthesis is restricted (starvation)
   c. the S phase activator does not reach its threshold concentration
   d. its DNA is not replicated
   e. any of the above would inhibit cell division

8. During the prophase I of meiosis:
   a. reduction of the number of chromosomes occurs
   b. random recombination of hereditary material (DNA) occurs
   c. already duplicated DNA is separated
   d. nondisjunction of the chromosomes could occur
   e. all of the above occurs

9. Cat cry (cri-du-chat) syndrome may result from:
   a. anaphase lag
   b. nondisjunction
   c. deletion
   d. translocation
   e. isochromosomal formation

10. In a somatic cell, the restriction point is:
    a. between periods G_1 and S
    b. between periods S and G_2
    c. between G_2 and M phase
    d. just after completion of mitosis
    e. during intermeiotic phase

11. In meiosis, the composition of chromatid is changed during:
    a. prophase I
    b. metaphase I
    c. anaphase I
    d. anaphase II
    e. prophase II

12. Klinefelter's syndrome is a trisomy of chromosome number
    a. 13
    b. 18
    c. 21
    d. X
    e. Y

13. In a dividing cell, the replication of DNA occurs during:
    a. interphase
    b. prophase
    c. metaphase
    d. anaphase
    e. telophase

14. A primary oocyte is arrested during the diplotene phase of meiosis. It completes meiosis:
    a. just before ovulation
    b. during ovulation
    c. at fertilization
    d. before birth
    e. in the neonatal period

15. Kinetochore
    a. is a microtubule organizing center
    b. causes centriole pair replication
    c. is a multiprotein complex with a trilaminar plate-like structure
    d. marks the middle of the microtubule spindle
    e. all of the above are correct

16. Meiotic division is characterized by all of the following except:
    a. replication of DNA
    b. restoration of a diploid number of chromosomes
    c. crossing over
    d. random recombination of hereditary material
    e. formation of polar bodies

17. In meiosis, the centromeres of chromosomes divide during:
    a. prophase I
    b. metaphase I
    c. anaphase I
    d. metaphase II
    e. anaphase II

18. Many factors (proteins) influence the events during cell division. The concentration of which of the following increases steadily?
    a. S phase activator
    b. restriction point trigger
    c. M phase promoting factor
    d. cyclin
    e. M phase delaying factor

19. Deletion of chromosomes may occur due to:
    a. nondisjunction
    b. anaphase lag
    c. translocation of a chromosome
    d. crossing over
    e. all of the above

20. All of the following occur during the G_2 period except:
    a. end of DNA duplication
    b. breakdown of nuclear membrane laminins
    c. phosphorylation of the histone H_1 molecule
    d. synthesis of M phase promoting factor
    e. synthesis microtubules for spindle apparatus

ANSWERS TO REVIEW QUESTIONS

---

1. a	6. d	11. a	16. b
2. c	7. e	12. d	17. d
3. c	8. b	13. a	18. d
4. e	9. c	14. a	19. d
5. d	10. a	15. e	20. b

CHAPTER

2

GAMETOGENESIS

Gametogenesis is the formation of gametes: spermatozoa in males and oocytes in females. This process occurs in gonads, where primordial germ cells undergo meiosis, through which is achieved reduction in the number of chromosomes to haploid (half that in the primordial cell) and random exchange of DNA segments. After their production by meiosis, the gametes undergo further differentiation and alteration in their shape (Fig. 2.1, Table 2.1).

I. OOGENESIS

This is the sequence of events in ovaries by which oogonia are transformed into secondary oocytes. The secondary oocytes then complete meiosis at fertilization, resulting in a large nonmotile gamete, the ootid.

During early development, primordial cells differentiate in the yolk sac and then migrate to the developing ovaries. Here they multiply by mitosis and mature to primary oocytes. The primary oocytes become surrounded by flat epithelial cells, forming primordial follicles. Oocytes in the follicles enter prophase I of meiosis and are arrested at diplotene stage. The cyclic AMP secreted by follicular cells blocks the completion of meiosis by the oocyte. At puberty, under the surge of pituitary hormones, many follicles in the ovaries start growing every month. The flat follicular cells proliferate and form many layers of cuboidal cells and are referred to as primary follicles. Primary follicles acquire a fluid-filled cavity (antrum) and become secondary follicles.

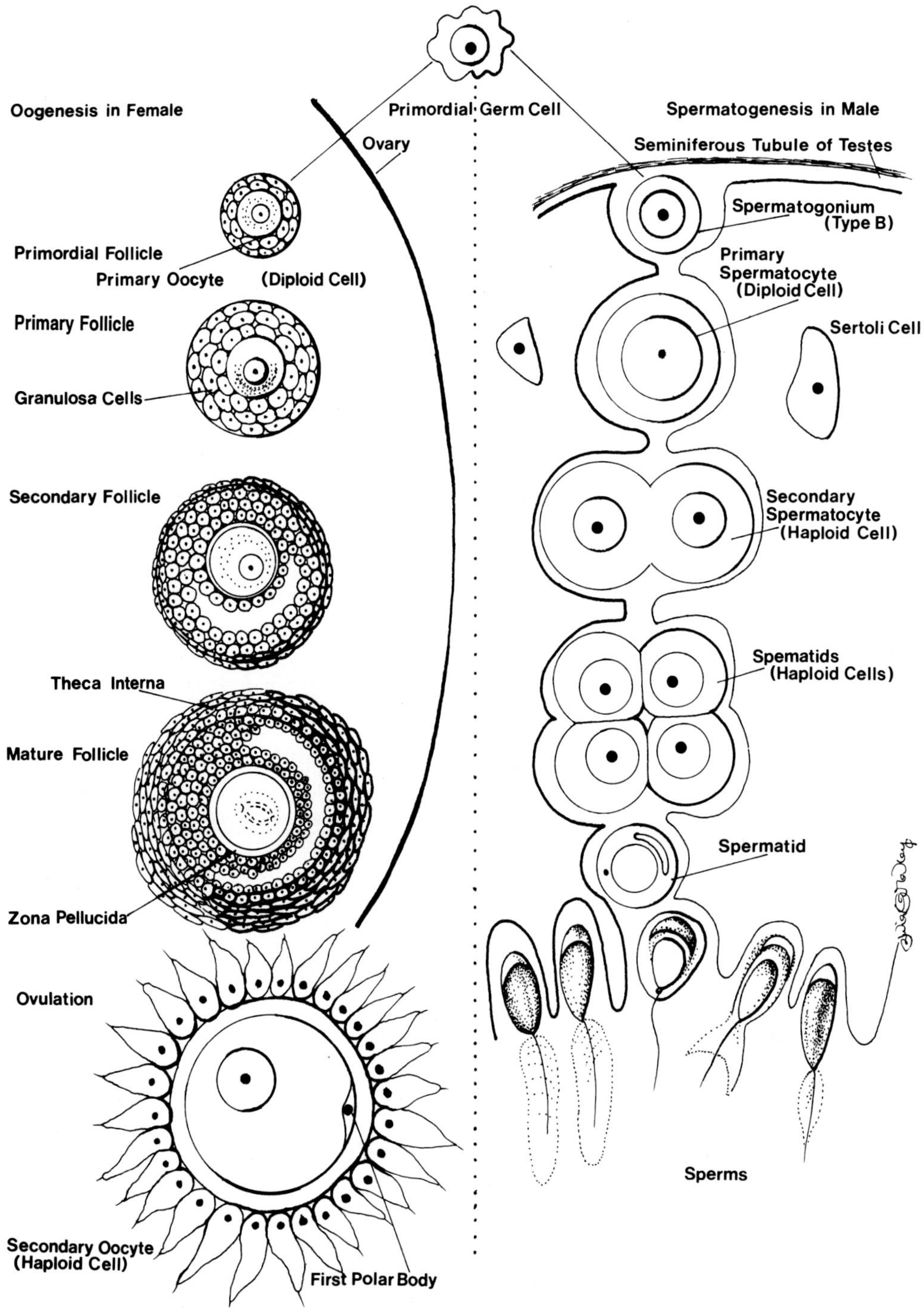

Figure 2.1
Gametogenesis.

TABLE 2.1. Endocrine Cells in Gonads Affecting Gametogenesis

Cells	Major receptors	Secretory products	Major functions
Granulosa cell of ovary	FSH	estrogen, progesterone, inhibin	affect the metabolism and growth/maturation of oocytes; arrest meiosis
Theca cells	LH	androgens	testosterone enters granulosa cells and is converted to estrogens
Oocytes		zona pellucida glycoproteins, glycosaminoglycans	proliferation and differentiation of granulosa cells and follicular organization
Sertoli cells of testes	FSH	estrogen, inhibins, androgen-binding proteins (also growth factor transferrin, retinal binding proteins)	maintenance and coordination of spermatogenesis, maintenance of the blood/testes barrier; inhibitory influence of Leydig cells
Leydig cells	LH	testosterone	stimulate Sertoli cells and spermatocytes
Spermatids		protamines	displace histones for compaction of chromatin

In the growing follicle, the oocytes also undergo changes, increasing the size of the follicle. At first, the organelles and fat globules form a crescent-shaped mass at one side of the nucleus; then, with further growth, these organelles become scattered throughout the cytoplasm. A translucent zone (layer) consisting of glycoproteins, known as the zona pellucida, appears between the oocytes and follicular cells. The cytoplasmic processes from the innermost layer of follicular cells penetrate the zona.

Both the oocyte and the follicular cells probably secrete zona pellucida material. Just prior to ovulation, only one of the growing follicles gains ascendancy and reaches maturity. In this mature follicle, the follicular cells of the innermost layer retract their processes and start separating themselves from the rest of the follicular cell layers. By this separation, the inhibition by cyclic AMP from the follicular cells is removed and meiosis resumes. The first part of the meiotic division of primary oocytes, which began during the third month of fetal life, is completed a few hours before ovulation. This results in a large secondary oocyte and a small first polar body. Both of these cells are haploid. The chromosomes of the first polar body become separated from most of the cytoplasm. A secondary oocyte with the first polar body within the zona pellucida surrounded by a layer of cells (corona radiata) is ovulated. A surge of luteinizing hormone (LH) and prostaglandins influences ovulation. The remaining mature follicle transforms into corpus luteum. All the other growing follicles will undergo atresia and become atretic follicles.

II. SPERMATOGENESIS

This is a sequence of events in the testes by which spermatogonia are transformed into primary spermatocytes. The primary spermatocytes then complete meiotic division to give rise to four equal-sized haploid cells, the spermatids. The spermatids then undergo morphological changes to form motile spermatozoa.

During early development, primordial germ cells differentiate in the yolk sac and then migrate to seed the testes. Here they multiply by mitosis and differentiate into spermatogonia. Spermatogonia are incorporated into developing seminiferous tubules of the testes and remain relatively inactive. At puberty, under the surge of pituitary hormones and stimulation by interstitial cells, spermatogonia begin proliferation. The spermatogonia divide by mitosis to give rise to more spermatogonia. Some of these cells differentiate into primary spermatocytes. The primary spermatocytes enter meiosis. After the first division, they give rise to secondary spermatocytes that undergo second division (completion of meiosis) to result in four equal-sized haploid cells, the spermatids.

III. SPERMIOGENESIS

Transformation of spermatid to spermatozoon. The spermatid is a small, nonmotile gamete and cannot fertilize the secondary oocyte. It must become motile to pass through the female genital tract to reach the site of fertilization. The spermatids become embedded in the apical folding of Sertoli cells and undergo a series of morphological changes.

Several small membrane bound granules appear in the vesicles of the Golgi apparatus. Later vesicles coalesce to form a large acrosomal granule. The acrosomal granule adheres to the nuclear envelope and spreads over half of the nucleus, forming a head cap. The nucleus becomes elongated, flattened and filamentous. The nuclear filaments subsequently shorten and thicken into coarse dense chromatin. The centrosomes migrate to the pole opposite the acrosomal granule. One centriole grows out as a slender flagellum. The cytoplasmic microtubules arrange into a manchette. With the appearance of the manchette, the spermatid elongates and the bulk of the cytoplasm becomes displaced to the flagellar pole. The mitochondria gather around the proximal segment of the flagellum, arranging helically to complete the midpiece. The excess cytoplasm is detached, with the manchette leaving a fibrous sheath around the principal piece of the sperm. The fully formed spermatozoon now leaves the Sertoli cell and becomes free within the lumen of the seminiferous tubule. The total duration of spermatogenesis is estimated at about 64 days.

IV. DIFFERENCES BETWEEN SPERMATOGENESIS AND OOGENESIS

A. In spermatogenesis, each parent cell produces four mature spermatozoa that are equal in size and functional capability. In oogenesis, only one mature gamete (ootid) is produced. The other daughter cells are discarded as polar bodies.

B. Following the onset of puberty, spermatogenesis proceeds continually, and complete maturation of spermatozoa is completed within about two months. In contrast, the oocyte remains in a prolonged prophase of first meiotic division, lasting from fetal life well into adult life. Some of the germ cells may be suspended in prophase for as long as 45 years. This prolonged prophase appears to be the reason for the relatively high incidence of chromosomal abnormalities in late-age pregnancies. The ootids that are discharged in older women have been subjected for a prolonged period of time to various adverse environmental factors, which may predispose them to abnormal separation of chromosomes.

C. Throughout their life after puberty, males are capable of producing sperm. However, after the age of 45 years the spermatogenesis is reduced, and the number of nonviable and abnormal sperm increases. At the age of 70 years, 25% become impotent. Late in their fourth or early in their fifth decade, females undergo menopause. The ovaries no longer contain any follicles, which causes estrogen deficiency. In the absence of estrogen, breast, uterus and vagina show senile changes. This is also reflected in the skin, bony skeleton and vascular system.

REVIEW QUESTIONS — Chapter 2

---

1. Which of the following organelles in a spermatid will accumulate the enzymes to penetrate the zona pellucida?
   a. mitochondria
   b. nucleolus
   c. Golgi apparatus
   d. endoplasmic reticulum
   e. lysosomes

2. The structure(s) which are ovulated with the secondary oocyte is (are):
   a. second polar body
   b. corona radiata cells
   c. mature follicles
   d. corpus luteum
   e. all of the above

3. The secondary follicles contain:
   a. primary oocytes
   b. several layers of granulosa cells
   c. an antrum
   d. zona pellucida
   e. all of the above

4. Regarding the zona pellucida, all of the following statements are correct except:
   a. it is a glycoprotein layer synthesized mainly by the primary oocyte
   b. it contains species-specific receptor molecules
   c. it becomes modified at fertilization to block polyspermy
   d. it protects the embryo during early development
   e. it penetrates the endometrium to facilitate implantation
5. Oogonia divide by mitosis:
   a. during the fetal period
   b. after birth
   c. after puberty
   d. during reproductive period
   e. all of the above
6. During spermiogenesis spermatids secrete:
   a. testosterone
   b. inhibins
   c. protamines
   d. androgen-binding proteins
   e. all of the above
7. Which of the following cells undergo mitosis after birth?
   a. primary spermatocyte
   b. secondary oocyte
   c. spermatid
   d. first polar body
   e. spermatogonia
8. Which of the following cells contain 23 strands of DNA?
   a. primary spermatocyte
   b. spermatogonia
   c. secondary spermatocyte
   d. spermatid
   e. primary oocyte
9. The primary spermatocyte completes its meiotic division:
   a. at fertilization
   b. in the testes
   c. in the epididymis
   d. during ejaculation
   e. in the female genital tract
10. Oocytes in follicles enter prophase I of meiosis and are arrested at the diplotene stage. Which of the following block the further stages of meiosis?
    a. follicular stimulating hormone (FSH)
    b. luteinizing hormone (LH)
    c. cyclic AMP secreted by follicles
    d. androgens
    e. estrogens

11. All of the following are secreted by granulosa cells of ovarian follicle except for:
    a. estrogens
    b. progesterone
    c. inhibins
    d. zona pellucida glycoproteins
    e. androgens

12. The ovulatin is influenced by a surge of luteinizing hormones and:
    a. protamines
    b. prostaglandins
    c. inhibins
    d. progesterone
    e. all of the above

13. The primary spermatocytes:
    a. are formed before birth and are arrested at prophase I of meiosis
    b. are the only cells formed in the testes after birth
    c. are released from the testes to fertilize ootid
    d. differentiate at puberty
    e. are haploid cells

ANSWERS TO REVIEW QUESTIONS

1. c	2. b	3. e	4. e
5. a	6. c	7. e	8. d
9. b	10. c	11. e	12. b
13. d			

CHAPTER

3

FERTILIZATION AND CLEAVAGE

Embryonic life commences with fertilization, which is the progression of events that begins when a spermatozoon makes contact with an oocyte or its investments, and ends with the intermingling of maternal and paternal chromosomes at metaphase of the first mitotic division of the zygote. Successful initiation of fertilization requires that the ovum descending in the uterine tube meet the ascending viable sperm at the right time, in the right place and under the right circumstances (Fig. 3.1).

I. REQUIREMENTS PRELIMINARY TO FERTILIZATION

A. Seminal Fluid

During a normal ejaculation, this fluid amounts to about 3 ml and contains 200–300 million spermatozoa. In the seminiferous tubules the sperm are nonmotile; as they pass through the ductus deferens they become mature. After ejaculation the sperm become motile. Seminal plasma is a composite of secretions from various accessory genital glands. About 30% originates from the prostate and contains acid phosphatase, neuraminidase, citric acid, aminotransferases, dehydrogenases, zinc and magnesium. Approximately 60% originates from seminal vesicles and contains fructose and prostaglandins. About 10% comes from the epididymis, bulbourethral glands and urethral glands. The seminal plasma provides a carrier medium and helps to activate the sperm.

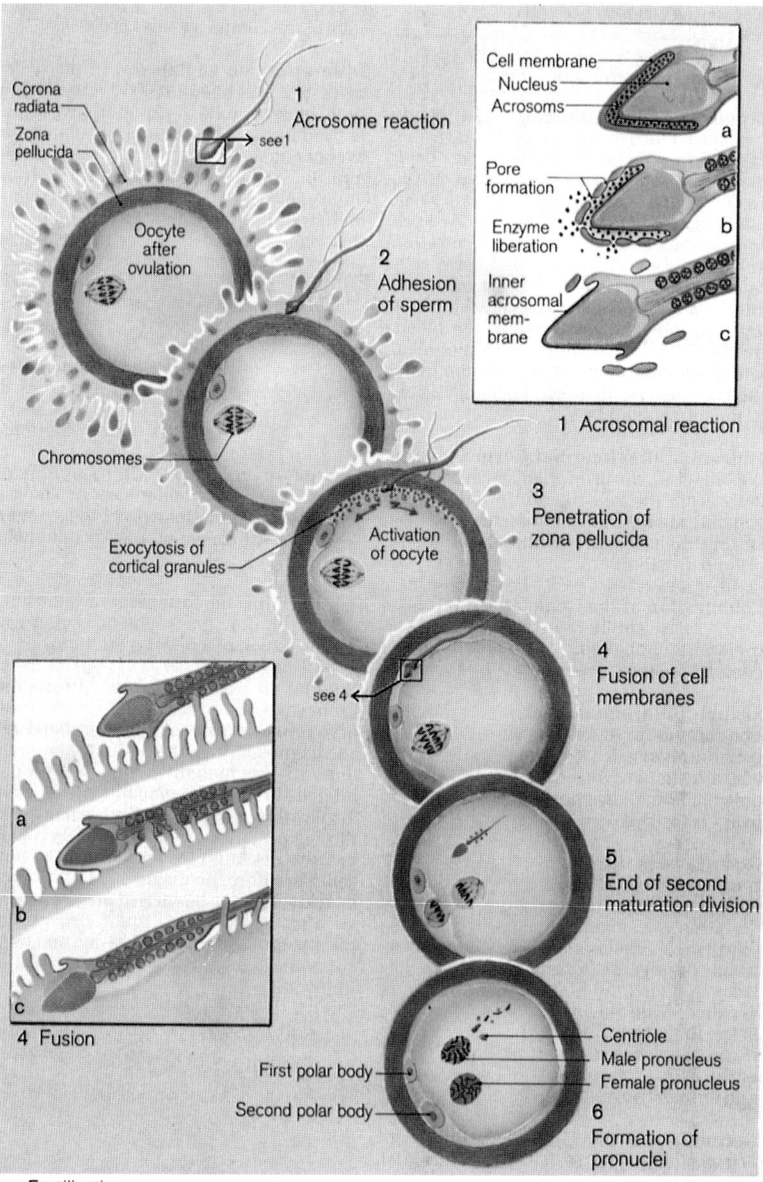

Figure 3.1

Fertilization (see Color Plate 3.1). Reproduced with permission from Thieme Medical Publishers Inc., New York, 1995, *Color Atlas of Embryology*, Ulrich Drews, Chapter 1: Reproduction.

B. Transport and Viability of Sperm

1. Vaginal pool

   The enzyme vesiculase coagulates the ejaculated semen and forms a vaginal plug, which probably prevents backflow of semen from the vagina. Coagulation of semen is followed by reliquification.

2. Transport through the cervix
   Various mechanisms, such as the contractile activity of vagina and cervix, and the properties of cervical mucus and orgasm of the female are involved. Seminal protease, foreign proteins, sex hormones and pH also affect this transport. At midcycle the cervical mucus is arranged into parallel glycoprotein micelles that assist in unidirectional transport of spermatozoa.

3. Transport in the uterus

   a. Rapid phase
      The contractile activity of the myometrium plays a major role in this phase of the transport. Neurohumoral pathways may activate the rapid transport of spermatozoa.

   b. Reserve phase
      The uterotubal junction acts as a barrier. The micelles guide a majority of motile sperm to this barrier, which establishes a concentration gradient of the sperm.

   c. Slow phase
      The contractile activity of myometrium and uterine tube continues for a prolonged period, which causes a sequential slow release of sperm to the site of fertilization.

4. Transport in the oviduct
   The oviduct can move spermatozoa and a secondary oocyte in opposite directions almost simultaneously. The uterotubal junction appears to act as a mechanical barrier that maintains a graded concentration of sperm.

5. Factors inhibitory to fertilization:

   a. The acidity of vaginal secretion, particularly at midcycle, is extremely destructive to the sperm.

   b. Sperm are rapidly separated from the seminal fluid, which is richer in enzyme for utilization of sugars.

   c. The introduction of semen initiates the leukocytic response in the uterus.

   d. Many sperm invade the uterine glands and are lost there.

   e. Spermatozoa reaching the fimbriae are released into the peritoneal cavity.

   f. Complex mucal folds of the uterine tube tend to impede and deflect sperm.

C. Capacitation of the Sperm

There is evidence that sperm are not capable of fertilizing ova immediately upon reaching the female genital tract. They must undergo capacitation before they can penetrate the zona pellucida. Capacitation is a conditioning of the spermatozoon surface that occurs within the female genital tract, and permits the acrosomal reaction in the vicinity of the zona pellucida. Capacitation involves alteration of the lipid and glycoproteins composition in the sperm plasma membrane. The acrosomal reaction involves exposing the inner acrosomal membrane, which comes in contact with filamentous material of the zona pellucida.

1. Estrogen appears to be necessary for the process of capacitation. Estrogen withdrawal considerably depresses the capacitating ability of both the uterus and the oviduct. The oviduct seems to be less dependent upon estrogen than the uterus for this mechanism.
2. Capacitation is somewhat inhibited in the uterus by a high level of progesterone. This effect of progesterone is minimal in the uterine tube.
3. The decapacitation factor, which is brought in by the semen and is at a high level in the cervix, has to be removed.

D. The Fertilizing Power of Sperm

1. The more active the sperm, the greater their ability to fertilize an oocyte.
2. Within narrow limits, the more acidic the vaginal secretion, the less motile the sperm. Alkalinity, on the other hand, increases their motility.
3. Proteolytic enzymes present in semen liquidize the ejaculation coagulum in order to facilitate sperm entry into the uterine cavity. Their hydrolytic action considerably reduces the high viscosity of the cervical mucus mucoid.
4. Abnormal spermatozoa are seen frequently. Up to 10% of the spermatozoa may be abnormal without any loss of fertility.

E. Transport of Oocytes Through the Uterine Tube

After ovulation, along with its corona radiata, the oocyte is sucked into the uterine tube by the current in the peritoneal fluid caused by beating of the cilia of the mucous membrane of the tube. The oocyte is then passed along the tube by rhythmic contraction of the muscular wall and by activity of the cilia.

F. Oocyte Viability

It is not known exactly how long the human oocyte remains fertilizable, but this period is probably less than 1 day. Its protoplasm becomes progressively more coarsely granular, and the oocyte loses its vigor and can no longer be fertilized. In an ovulated oocyte, the continuous release of cortical granules changes the zona pellucida, which in the older oocyte blocks entry of sperm.

II. FERTILIZATION

Fertilization consists of contact of sperm with the zona pellucida, engulfment of sperm by the secondary oocyte, extraction of the second polar body, formation of male and female pronuclei, and beginning of the first (pronuclei) mitotic division of the zygote (embryo). Fertilization usually takes place in the lateral third of the uterine tube.

Only one spermatozoon is required to fertilize an oocyte. The spermatozoon penetrates through the zona pellucida and becomes anchored to the microvilli of the oocyte. This is accomplished by four enzymes associated with the sperm head:

A. Hyaluronidase assists sperm in passing through corona cells.

B. Corona penetrating enzyme inhibits decapacitating factor and disperses corona cells.

C. Neuraminidase-like factor acts on the zona pellucida.

D. Acrosine digests zona and decapacitating factor.

The zona pellucida consists of three glycoproteins: ZP_1, ZP_2 and ZP_3. ZP_3 binds with sperm surface membranes and causes the acrosomal reaction. Specific carbohydrate determinants on ZP_3 recognize the binding site of sperm plasma membrane.

The spermatozoon, including its tail, is engulfed by the secondary oocyte. The cortical granules of the oocyte are liberated, initiating the zona reaction to block polyspermy. Enzymes released from cortical granules alter the ZP_3 in the zona and change the oocyte membrane.

The penetration of a spermatozoon triggers a second meiotic division, which results in the release of the second polar body. The remaining chromosomes reconstitute into the female pronucleus.

The sperm head swells, the chromatin becomes less compact and starts to migrate toward the center of the ootid, during which time it completes its transformation into the male pronucleus. The middle piece and tail of the sperm disintegrate and disappear.

The two pronuclei meet in the center of the ootid. A pair of centrioles appear and a spindle is formed. The nuclear membranes of the pronuclei break down and become confluent with each other, and a set of chromosomes from each pronucleus becomes organized on the spindle. All the chromosomes split longitudinally at the centromere, and the first cleavage (mitotic division) occurs.

Fertilization results in the association of male and female chromosomes and restoration of diploid number.

III. CLEAVAGE

Cleavage consists of a rapid succession of mitotic divisions that produces large numbers of small cells called blastomeres (Fig. 3.2).

Soon the embryo consists of a solid ball of cells enclosed within the zona pellucida, and it is called the morula.

The human zygote reaches the two-cell stage about 30 hours after fertilization, and the four-cell stage approximately 40 to 50 hours after fertilization. The morula, which is at the 12- to 16-cell stage, reaches the endometrial cavity about the third

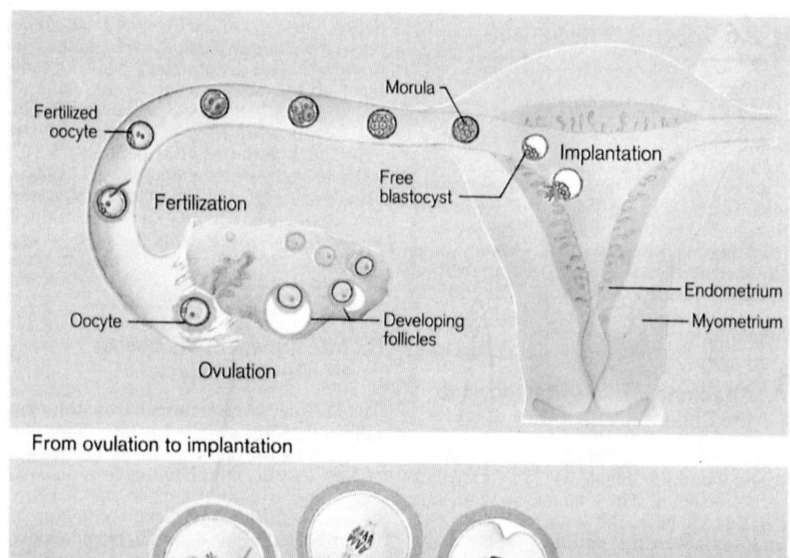

From ovulation to implantation

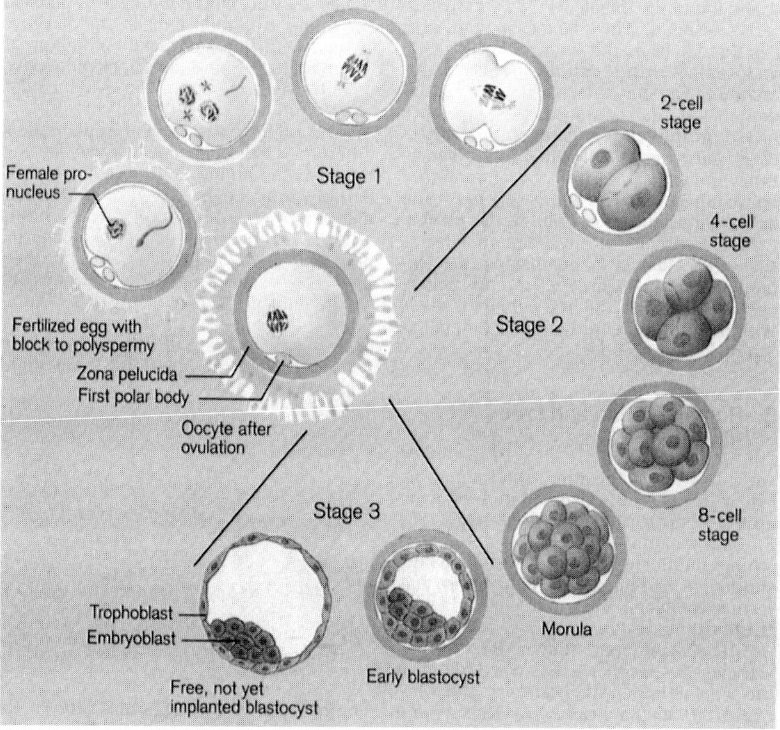

Development of oocyte to blastocyst

Figure 3.2
Overview: "first week" (see Color Plate 3.2). Reproduced with permission from Thieme Medical Publishers Inc., New York, 1995, *Color Atlas of Embryology,* Ulrich Drews, Chapter 2: Human Development.

to fifth day after fertilization and obtains nutrition from the secretion of the endometrium.

Centrally located in the morula, the inner cell mass is completely surrounded by a layer of cells, the outer cell mass. The inner cell mass develops into the embryo; the outer cell mass gives rise to the trophoblast, which later develops into the placenta. The cells at this stage are not determined. Their position in the morula determines their fate. The larger outer cells divide early and precociously to form the trophoblast. Thus, the segregation of blastomere into embryoblast and trophoblast occurs at this stage.

After about five divisions, the cells start secreting intercellular fluid, and small spaces appear among the inner cells. These intercellular spaces on one side of the inner cell mass become confluent. Finally, a single cavity, the blastocele, is formed. The morula is now known as the blastocyst. The zona pellucida now disappears.

With further accumulation of fluid within the blastocele, the trophoblast cells become flattened and the young embryo is now ready to attach itself to the uterine wall.

REVIEW QUESTIONS — Chapter 3

---

1. The acrosomal granule forms during:
   a. capacitation of sperm
   b. mitosis of spermatogonia
   c. fertilization
   d. ovulation
   e. spermiogenesis
2. While in the female genital tract, many sperm are destroyed by:
   a. acidity of vaginal secretion
   b. leukocytosis in uterine cavity
   c. entrance into uterine gland
   d. mucal folds of uterine tube
   e. all of the above
3. The capacitation of sperm is affected by:
   a. estrogens
   b. progesterone
   c. decapacitating factor
   d. all of the above
   e. none of the above
4. Acrosomal enzymes of sperm:
   a. help sperm to penetrate the zona pellucida
   b. initiate division of the secondary oocyte
   c. cause release of cortical granules in the ootid
   d. help sperm to penetrate the plasma membrane of the secondary oocyte
   e. all of the above

5. At fertilization:
   a. male and female pronuclei complete their meiotic division
   b. male and female pronuclei enter interphase
   c. cyclic AMP is secreted by the zygote to block polyspermy
   d. the zona pellucida is discarded
6. For in vitro fertilization the cells recovered from ovarian follicles *most likely* are:
   a. ootids
   b. secondary oocytes
   c. primary oocytes
   d. oogonia
   e. zygotes (cells with two pronuclei)
7. The majority of children born with trisomy 21 have:
   a. cleft lip and palate
   b. rock bottom feet
   c. congenital heart disease
   d. clenched fingers
   e. deafness
8. All of the following are essential for successful fertilization except
   a. capacitation of the sperm
   b. transport of sperm through the female genital tract
   c. transport of the zygote to the uterine cavity
   d. separation of sperm from semen
   e. bidirectional movements of gametes in the uterine tube
9. Factor(s) that are inhibitory to fertilization include:
   a. the zona pellucida
   b. the leukocytic response in the uterus
   c. oral estrogens and progestin
   d. an intrauterine device (IUD)
   e. all of the above
10. Fertilization commences when a sperm makes contact with an oocyte, and ends:
    a. with the intermingling of maternal and paternal chromosomes at metaphase of the first cleavage division
    b. when a sperm is completely taken into the ootid
    c. when a sperm penetrates the zona pellucida
    d. with the formation of a second polar body
    e. with all of the above
11. Transport of sperm through the cervix is assisted by all of the following except:
    a. capacitation of the sperm
    b. the time of the female cycle
    c. properties of the cervical mucus
    d. the contractile activity of the uterus
    e. pH

12. The seminal fluid:
    a. provides a transporting medium for sperm
    b. helps to activate the sperm
    c. liquefies the cervical mucus mucoid
    d. reduces the acidity of vaginal secretions
    e. all of the above

13. During fertilization, the association of an ootid with karyotype 22+0 and a sperm with karyotype of 22+Y results in a:
    a. female with a webbed neck
    b. a nonviable embryo
    c. a female with hypoplastic ovaries
    d. a male with hypoplastic testes
    e. a normal-appearing male until puberty

> The following diagrams are the *abnormal portions* of human karyotype.
> Match the diagram with the numbered statements on p. 34.

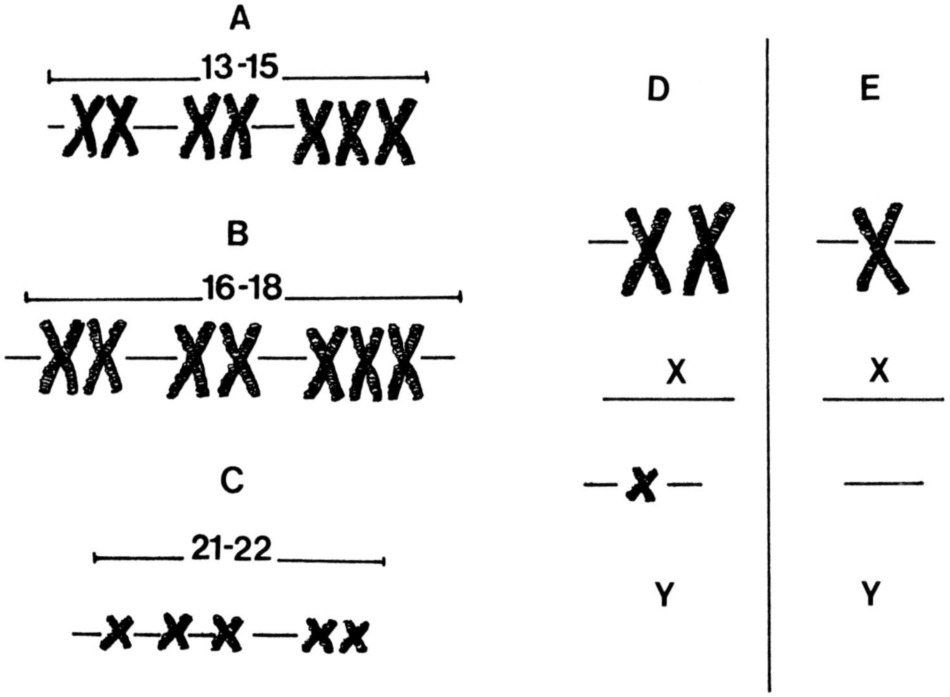

A

13-15

B

16-18

C

21-22

D

X

Y

E

X

Y

14. The individual with this karyotype is a female, the genitalia and mammary glands remain infantile after puberty. _____

15. The most common form of trisomy associated with mental retardation, generalized hypotonia and congenital heart disease. _____

16. Very tall male with small testicles and gynecomastia. _____

17. Form of trisomy associated with rock bottom feet and rigidly flexed fingers. _____

18. Form of trisomy associated with cleft lip and cleft palate. _____

ANSWERS TO REVIEW QUESTIONS

---

1. e 2. e 3. c 4. a
5. b 6. c 7. c 8. c
9. b 10. a 11. a 12. e
13. b 14. e 15. c 16. d
17. b 18. a

CHAPTER

4

IMPLANTATION

Implantation is a highly specialized process that is achieved by intimate contact between the embryonic trophoblast cells and the endometrial cells of the uterus. Although many details of this phenomenon are known, it is still not fully understood. Through this process the embryo (blastocyst) is oriented properly, brought to the site and attached there. The proliferation, penetration and spread of the polar trophoblast and envelopment of maternal vessels within the trophoblast leads to the establishment of placenta.

I. NORMAL IMPLANTATION

The blastocyst begins to implant on about the sixth day (Fig. 4.1).

The combined action of progesterone and estrogen prepares the endometrium of the uterus to undergo a decidual reaction (Fig. 4.2).

A. The endometrium increases in thickness.

B. The endometrial glands become long and tortuous.

C. The endometrial stromal cells become swollen and filled with glycogen.

D. The endometrial arteries become coiled and congested. Decidual changes also occur in response to local stimulation.

The blastocyst induces local morphological changes in endometrium before and after implantation.

A. At the time of implantation, the blood vessels are dilated, and the subendothelial capillary plexus increases in thickness and length.

B. The surface blood vessels increase in size and form blood sinuses.

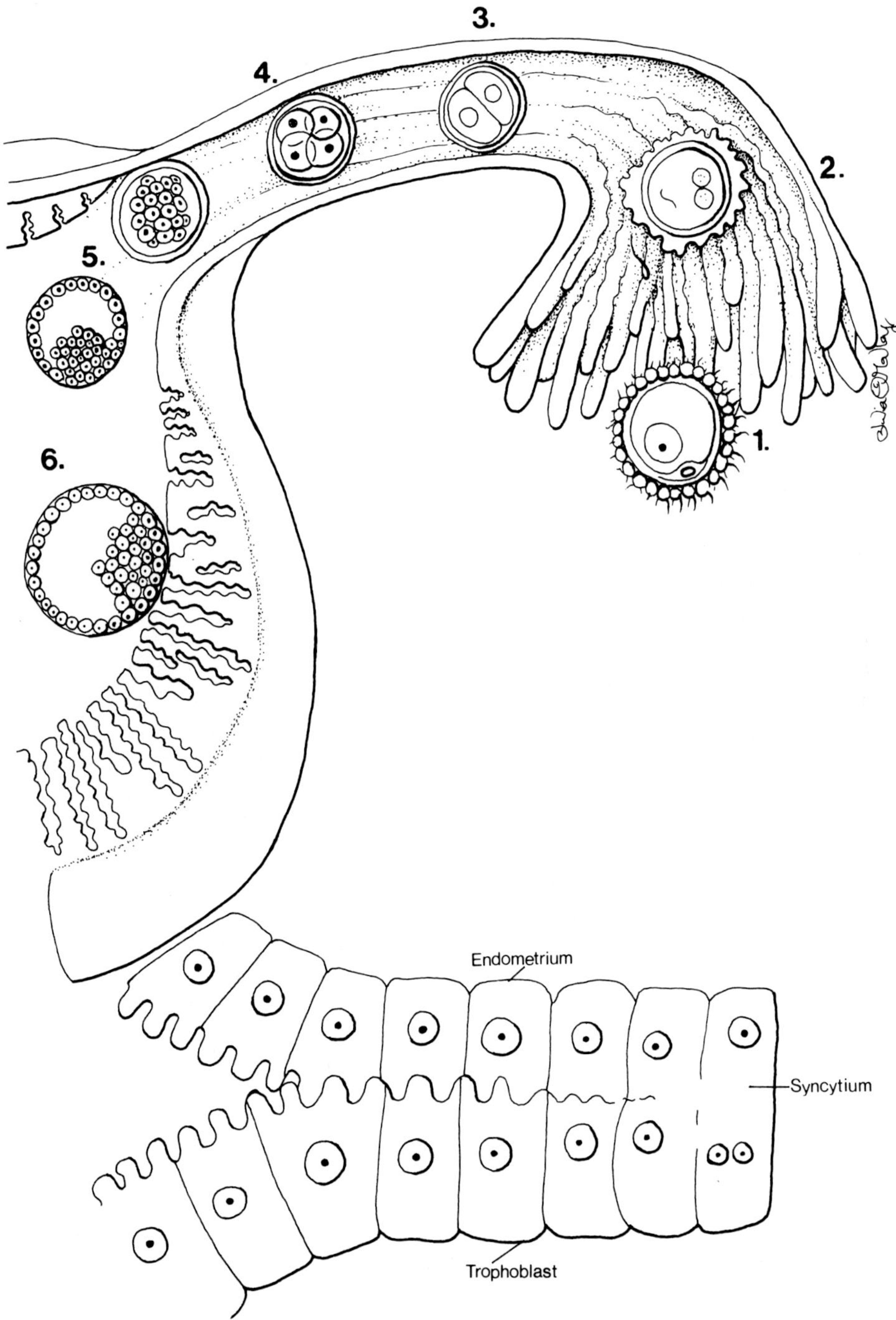

Figure 4.1

The first week of development. A secondary oocyte (1) becomes fertilized (2) in the ampullary region of the uterine tube. The first cleavage division occurs 30 hours after fertilization. At 3 days the morula stage (4) and at 4½ days the blastocyst stage (5) are reached. At about 6 days, the embryo (6) starts to implant itself. Implantation is achieved by fusion of trophoblast cells with endometrial cells.

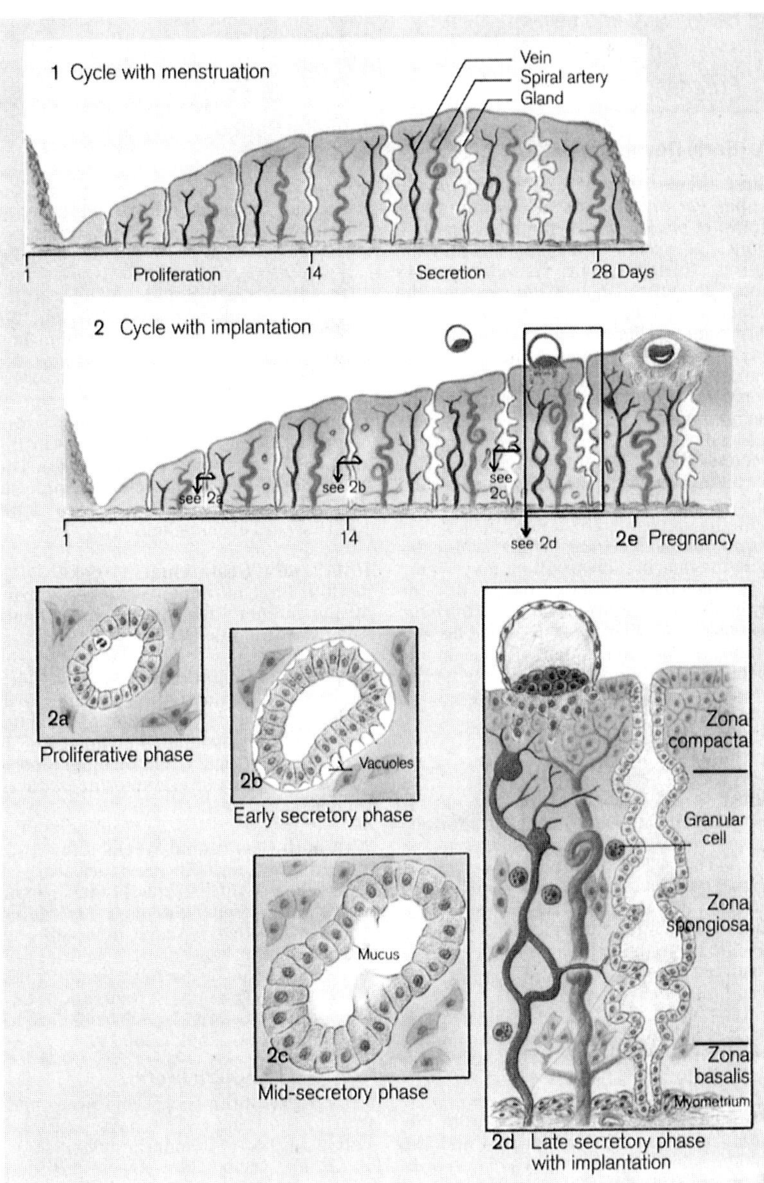

Figure 4.2

Endometrium (see Color Plate 4.2). Reproduced with permission from Thieme Medical Publishers Inc., New York, 1995, *Color Atlas of Embryology*, Ulrich Drews, Chapter 1: Reproduction.

C. The blastocysts secrete bicarbonate, which locally increases the pH of uterine secretion and makes it very sticky.

The zona pellucida is lost shortly before implantation. This is brought about by the following factors:

A. A change in the uterine environment.

B. An intrinsic factor of the trophoblast.

C. An implantation-initiating factor, a proteolytic enzyme secreted by the uterus.

D. The developmental stage of embryo.

E. Vigorous contraction of the trophoblast.

The polar trophoblast cells adjacent to the inner cell mass come into direct contact with the endometrium and become attached to it. The nature of the stimulus for this orientation is not known. The changes in the endometrium at the site of implantation and in the trophoblast cells associated with it may play a role in this. The cell membranes of trophoblast cells and endometrial cells form microvilli, which interdigitate with each other and become intimately related. Cell junctions occur between them. Shortly afterwards, the membranes of microvilli disappear and trophoblast cells fuse with endometrial cells. This coalescence of the polar trophoblast is known as syncytiotrophoblast. All the other trophoblast is known as cytotrophoblast (Fig. 4.1). The blastocyst gradually sinks deep into the endometrium; the site of penetration is sealed over and the endometrium becomes continuous. In some cases a coagulation plug is seen at the site of implant.

Implantation normally takes place in the endometrium of the body of the uterus, most frequently on the upper part of the posterior or anterior wall near the midline, usually between the openings of the endometrial glands. Chorionic gonadotropin is excreted in the urine from approximately the middle of the first month of pregnancy, reaching its peak during the third month. It then decreases sharply in the fourth and fifth months, gradually leveling off until the end of gestation. The excretion of this gonadotropin serves as the basis for diagnostic tests for pregnancy. Human implantation is interstitial in type. The blastocyst is buried entirely within the substance of the endometrium.

II. ABNORMALITIES OF IMPLANTATION

A. Implantation outside its normal location is called an ectopic pregnancy. Possible abnormal sites of implantation are shown in Fig. 4.3.

1. Ovarian pregnancy is quite rare. Ovarian tissue ruptures early, but cases have been recorded in which the pregnancy continued to term.

2. Cervical pregnancy is also rare. Implantation near the internal os results in placenta previa.

3. Abdominal pregnancy usually occurs in the rectouterine cavity, but the blastocyst can attach itself anywhere to the peritoneal lining.

4. Most ectopic pregnancies occur within the uterine tubes and may result from a delay in the transport of the blastocyst through the tube. This delay may be the result of damaged tubal mucosa due to previous infection. In tubal pregnancies, abortion may take place between the sixth and tenth weeks of pregnancy.

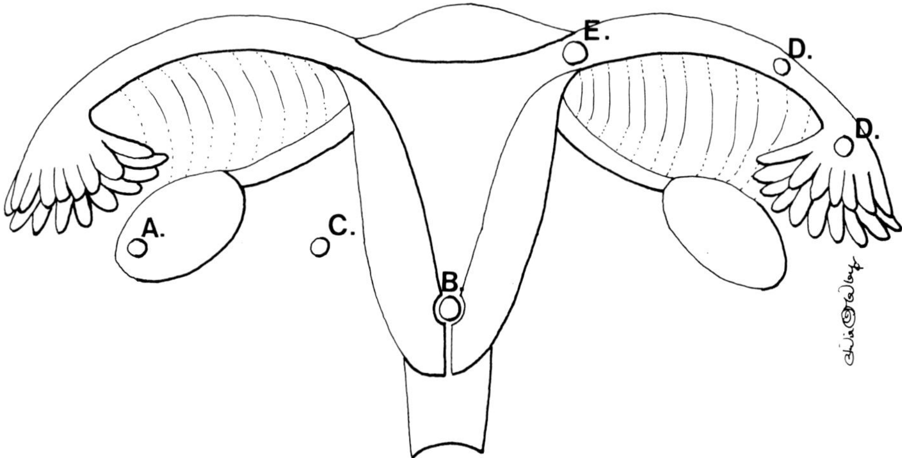

Figure 4.3
Abnormalities of implantation: A) ovarian, B) cervical, C) abdominal, D) uterine tube and E) interstitial portion of the uterine tube.

5. Interstitial pregnancy ruptures soon after implantation because the narrow proximal portion of the tube is incapable of much distention.
6. An ectopic pregnancy may occur in a woman who is using an intrauterine device, which prevents implantation of the blastocyst within the endometrium. The blastocyst can then develop in another location.
7. Rarely does an extrauterine embryo come to full term.

B. Abnormal implantation may result from defective blastomeres and abnormal development of the trophoblast.

C. The embryo is susceptible to maternal infections, such as German measles, which may affect its development.

D. Excessive proliferation of the trophoblast may lead to the formation of a hydatidiform mole. The embryo in this condition dies, and excessively large quantities of chorionic gonadotropins are excreted. The mole can become a chorion epithelioma, a malignant tumor of the chorionic villi. (This is most common in certain parts of Southeast Asia.)

E. Progesterone and estrogen imbalances produce several types of blastocyst degeneration.

F. Chromosomal abnormalities result in spontaneous abortions. It is suggested that about half of all conceptions abort spontaneously before the pregnancy is even detected. Chromosomal anomalies account for half of these.

G. Contraceptive devices interfere with the sperm, oocytes and early embryonic development, and prevent implantation. These include birth control pills, injected or implanted sources of progestin (Depo-Provera), IUDs (medicated or nonmedicated) and antiprogesterone compound RU-486.

REVIEW QUESTIONS — Chapter 4

Select the most appropriate answer.

1. A woman in her 10th week of pregnancy complained of discomfort in her pelvic region and of shortness of breath. Her hemoglobin was found to be low. She most likely had:
   a. a hydatidiform mole
   b. a ruptured tubal pregnancy
   c. an abdominal pregnancy
   d. placenta praevia
   e. a ruptured placenta

2. Implantation normally takes place in the
   a. body of the uterus, most frequently on the upper posterior wall
   b. body of the uterus on the lower posterior wall
   c. cervical region
   d. two-thirds lateral uterine tube
   e. peritoneal cavity

3. At the time of implantation, the zona pellucida is removed by the
   a. action of the uterine enzymes
   b. influence of the intrinsic factor of the trophoblast
   c. vigorous contractions of the blastocyst
   d. penetration of trophoblastic cell processes at the abembryonic pole
   e. all of the above

4. During implantation the blastocyst:
   a. usually implants in the posterior wall of the body of the uterus
   b. releases substances that cause local changes in the endometrial tissue
   c. attaches itself to endometrium at its embryonic pole
   d. all of the above are correct

5. Which of the following events are directly related to implantation?
   a. attachment of blastocyst to the endometrial wall
   b. interdigitation between microvilli of trophoblast cells and endometrial cells
   c. formation of cell junctions between the trophoblast cell membrane and the endometrial cell membrane
   d. formation of syncytium of trophoblast cells and endometrial cells
   e. all of the above

6. The zona pellucida is present around:
   a. a primary oocyte
   b. a secondary oocyte
   c. the morula
   d. an early blastocyst
   e. all of the above

7. All of the following are involved in the process of implantation except
   a. a decidual reaction
   b. localized morphological changes induced by the blastocyst
   c. shedding of the zona pellucida
   d. separation of the inner cell mass
   e. formation of the syncytiotrophoblast

8. Which of the following is essential for successful implantation?
   a. a decidual reaction
   b. transport of the blastocyst to the uterine cavity at the right time
   c. removal of the zona pellucida just before attachment of the blastocyst
   d. release of implantation-initiating factor (IIF) by the endometrium
   e. all of the above are needed

9. The usual site of ectopic pregnancy is the:
   a. abdominal cavity
   b. uterine tube
   c. ovary
   d. cervix of uterus

10. Diagnostic tests for pregnancy are based mainly on:
    a. secretion of human chorionic gonadotropins (HCG)
    b. levels of estrogens
    c. levels of progesterone
    d. levels of thyroxin
    e. secretion of human chorionic somatomammotropins (HCS)

11. The decidual reaction in the endometrium:
    a. occurs only during pregnancy
    b. is brought by the hormones secreted by the ovaries
    c. is brought by the hormones secreted by the cytotrophoblast
    d. is brought by the hormones secreted by the syncytiotrophoblast
    e. occurs after implantation

12. Degeneration of the corpus luteum of pregnancy before week 20 may be caused by insufficiency of:
    a. estrogens
    b. progesterone
    c. human chorionic gonadotropin
    d. luteinizing hormone
    e. follicular stimulating hormone

13. The presence of gestational trophoblastic neoplasm is best confirmed by:
    a. serial HCG secretion
    b. evaluation of the menstrual cycle
    c. the presence of lactation
    d. serial progesterone measurement
    e. serial X-ray films of the chest

ANSWERS TO REVIEW QUESTIONS

---

1. b 2. b 3. e 4. d
5. e 6. e 7. d 8. e
9. b 10. a 11. b 12. c
13. a

CHAPTER

5

PLACENTATION

Placental membranes and the embryo proper develop simultaneously. Only the development related to establishment of placenta is discussed in the present chapter. Early development of the embryo is discussed in the chapter that follows.

I. FIRST WEEK

The blastocyst starts implantation.

II. SECOND WEEK

A. Formation of Syncytium

Coalescence of polar trophoblast with endometrial cells gives rise to multinucleated syncytial cells of the syncytiotrophoblast, which assists in implantation of the embryo (Fig. 5.1). The embryo itself produces immunosuppressive factors that directly suppress the endometrial immune response. Although the endometrium is adequate for nourishment of the conceptus at this stage, for further development this histiotype nourishment must be replaced by hemotype nourishment. This is achieved by establishment of the placenta.

B. Formation of Trophoblastic Lacunae

Numerous vesicles within the syncytium have been observed by electron microscopy. Their confluence gives rise to larger spaces, the lacunae. In humans it

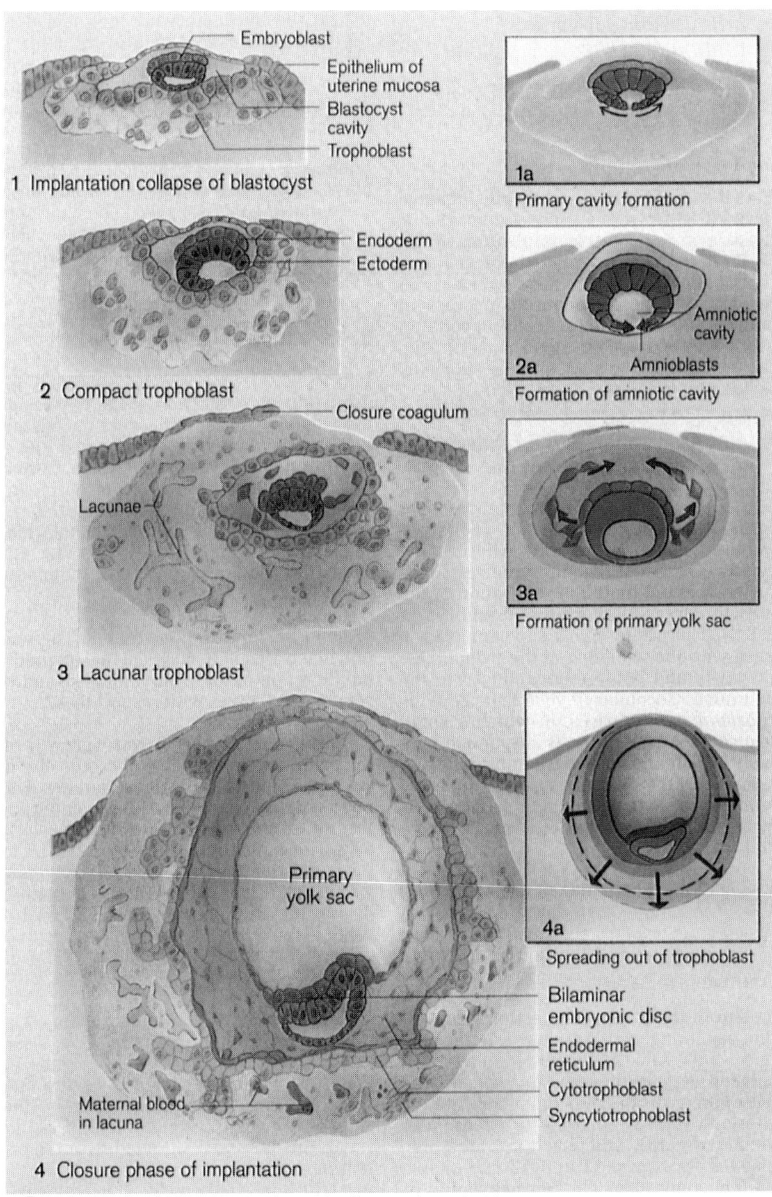

Figure 5.1

Stage 5: bilaminar embryonic disk (see Color Plate 5.1). Reproduced with permission from Thieme Medical Publishers Inc., New York, 1995, *Color Atlas of Embryology,* Ulrich Drews, Chapter 2: Human Development.

appears that syncytiotrophoblastic invasion engulfs the endometrial capillary plexus. A part of this plexus forms a series of spaces that subsequently enlarge to join lacunae. This assures continuity with maternal circulation. The trophoblast, which is not involved in the formation of syncytium, organizes into an epithelial layer, the cytotrophoblast, which lines the enlarged blastocele (the extraembryonic coelom). The syncytiotrophoblast is more active enzymatically. It secretes many substances. One of these, human chorionic gonadotropin, starts at the second week and peaks at 10 weeks of pregnancy, when it can be detected by pregnancy test kits.

C. Formation of the Primary Yolk Sac

On the opposite side of the inner cell mass, a thin membrane, the exocoelomic or Heuser's membrane, of flattened cells appears. (This is also referred to as the intrachorionic mesothelial membrane.) It appears to be formed by delamination of the inner layer of the cytotrophoblast. As it is attached to the embryonic disk, it is suggested that in humans it is formed by the peripheral spread of extraembryonic endoderm (arising from the lower layer of the inner cell mass). The formation of this membrane gives rise to two cavities within the blastocele. The inner cavity lined by this exocoelomic membrane is now called the primary yolk sac; the outer space between this membrane and the cytotrophoblast epithelium is the extraembryonic or chorionic coelom.

D. Formation of the Extraembryonic Mesoblast (Mesoderm)

With the development of the extracoelomic membrane, a meshwork of cells appears within the chorionic coelom. These cells are known as extraembryonic mesoblasts. In addition to their own proliferation, mesoblasts are continuously added by delamination of the cytotrophoblast and by migration of the extraembryonic endoderm. Soon after, the extraembryonic mesoblasts organize into two distinct layers. One of these lines the cytotrophoblast (extraembryonic somatic pleura) and the other lines the extracoelomic membrane (extraembryonic splanchnic pleura) and the amniotic cavity, leaving a connection between the cytotrophoblast and the embryonic disk as a connecting stalk. The combination of the extraembryonic mesodermal layer and the inner layer of the cytotrophoblast is known as the chorion. In humans, the chorion is vascular.

E. Conversion of the Primary Yolk Sac to the Yolk Sac Proper

Late in the second week, the endodermal cells from the edges of the embryonic disk migrate along the interior of the proximal part of the primary yolk sac. The distal part of the yolk sac, which is devoid of endodermal cells, becomes detached, giving rise to a smaller cavity, the yolk sac proper (secondary yolk sac). The yolk sac proper is entirely lined by an endodermal layer, which is now covered by the extraembryonic mesoderm (extraembryonic splanchnic pleura). By formation of the yolk sac proper, the chorionic coelom (cavity) becomes enlarged. The unincorporated distal part of the yolk sac collapses (Fig. 5.2). The yolk sac proper functions as a specialized nutritional membrane, an early hemopoietic organ, and gives rise to primordial germ cells. Later, during folding of the embryo, it becomes incorporated in the embryo as primitive gut.

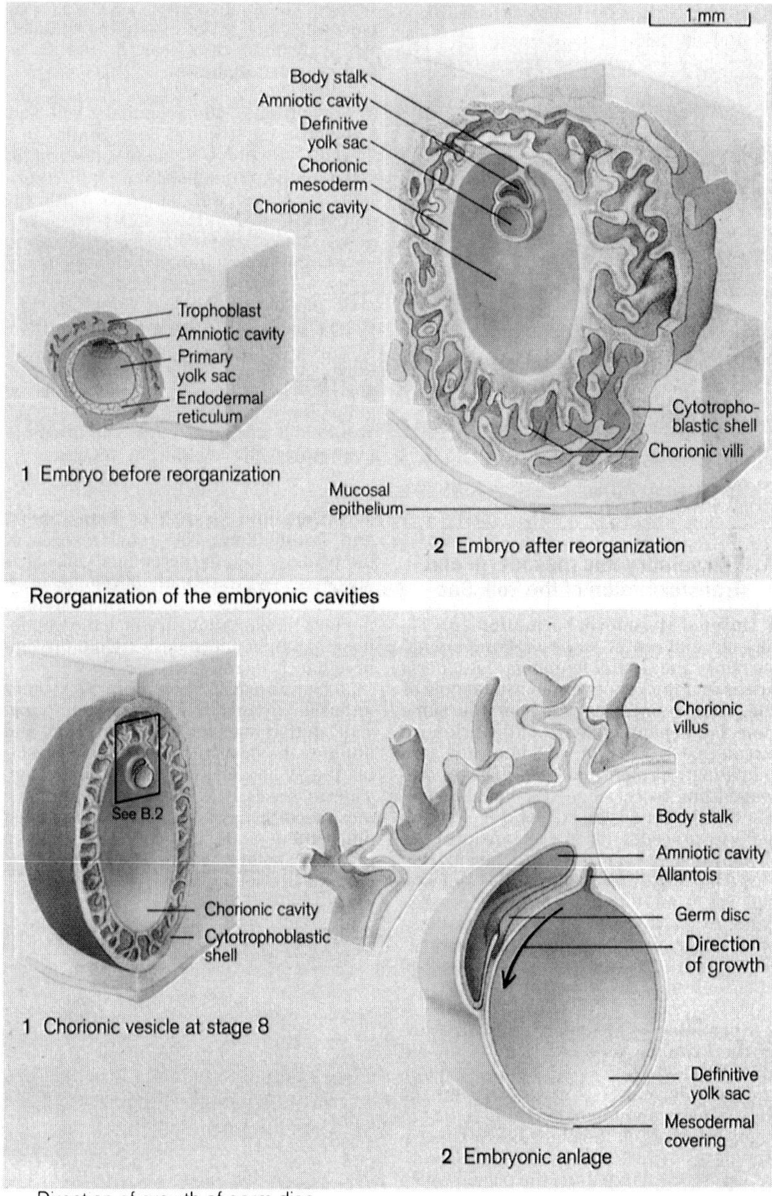

Figure 5.2

Overview: third week (see Color Plate 5.2). Reproduced with permission from Thieme Medical Publishers Inc., New York, 1995, *Color Atlas of Embryology*, Ulrich Drews, Chapter 2: Human Development.

F. Intercommunication of Trophoblastic Lacunae

The syncytiotrophoblast continues to penetrate deep into the endometrium, and trophoblastic lacunae become continuous with endometrial sinusoids. The syncytial cells move in and line these spaces, and the maternal blood enters the enlarged lacunae. The lacunae intercommunicate, forming a labyrinth of blood-filled spaces.

G. Formation of Chorionic (Stem) Villi

Solid cords of cytotrophoblast follow the syncytiotrophoblast into the endometrium and extend into blood-filled lacunar spaces as primary chorionic (stem) villi. Although the core of these villi is cytotrophoblast, they are completely covered by syncytial cells. The extraembryonic mesoblasts penetrate the center of cytotrophoblast cords, forming secondary chorionic (stem) villi. Thus, the mesoblastic crest forms the cores, the cytotrophoblast forms the caps, and syncytial cells cover the secondary chorionic villi. Some of the mesoblasts differentiate into angioblasts and start to form chorionic capillaries in the villi.

H. Allantois

The allantois arises as a diverticulum from the caudal part of the yolk sac. Early blood formation occurs in its walls. It is later incorporated into the umbilical cord as the urachus, which continues with the cloaca.

III. THIRD WEEK

A. Intervillous Space

With further enlargement and coalescence of lacunae, the trophoblastic labyrinth becomes a large, blood-filled space (Fig. 5.2).

B. Cytotrophoblastic Shell

Some of the chorionic villi grow faster. The cytotrophoblast cap penetrates through the syncytium and contacts the maternal tissue. These villi are called the anchoring villi. The cytotrophoblast layer of these villi then spreads on each side to come into contact with the cytotrophoblast layer of the other anchoring villi, forming a continuous layer, the cytotrophoblastic shell. This shell attaches the chorionic sac (placenta) firmly to the endometrium. After formation of the shell, the cytotrophoblast layer from the villi disappears, leaving mesoblast core covered by syncytial cells. Other villi remain free in the intervillous space (Fig. 5.3).

C. Angiogenesis

This involves formation of primitive blood vessels. The extraembryonic mesoblasts in villi, connecting stalk, chorion and yolk sac aggregate to form isolated

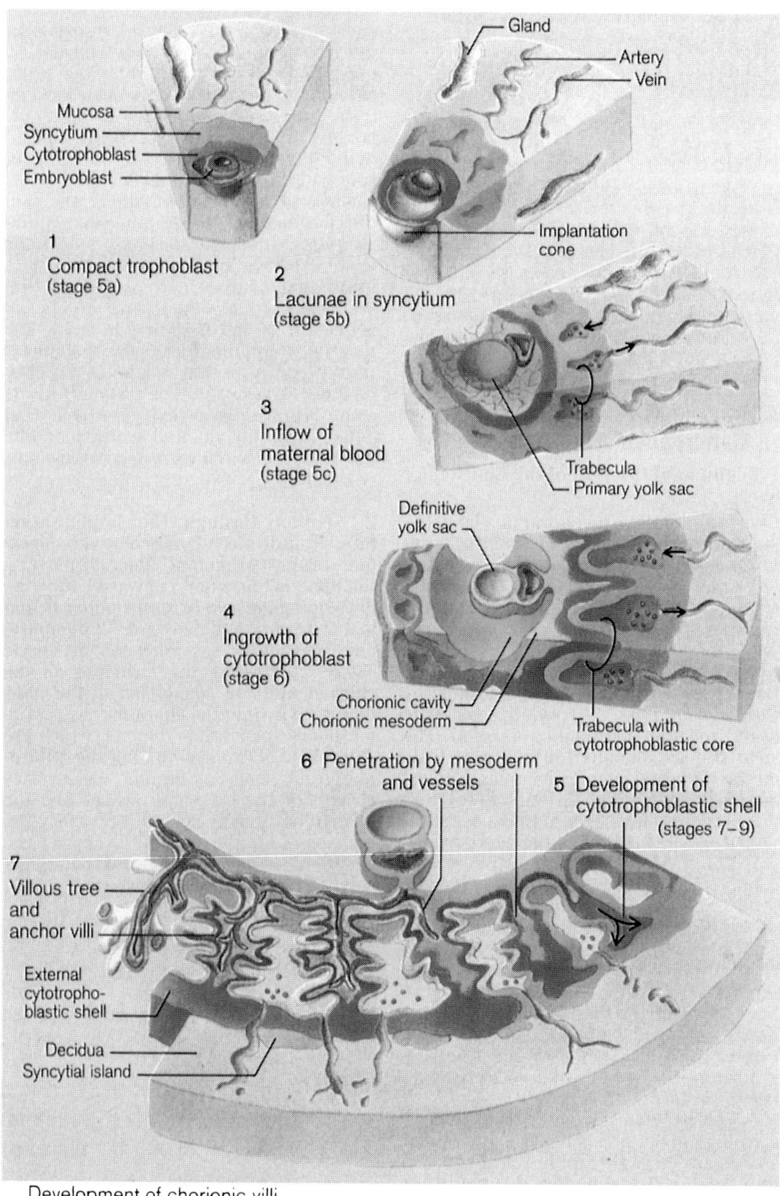

Figure 5.3

Overview: chorion and chorionic villi (see Color Plate 5.3). Reproduced with permission from Thieme Medical Publishers Inc., New York, 1995, *Color Atlas of Embryology*, Ulrich Drews, Chapter 2: Human Development.

masses, the blood islands. The cells in these islands are angioblasts. The angioblasts situated at the periphery differentiate to form the endothelial channels, and those located centrally differentiate into primitive blood cells. The endothelial channels extend into adjacent areas and fuse with each other to form a network. By the end of the third week, extraembryonic blood channels become linked to the developing intraembryonic blood channels. The villi containing blood vessels are known as tertiary villi.

D. Decidua

Under the influence of progesterone and estrogen, the endometrium prepares itself to receive the blastocyst. The stromal cells of the functional layer contain large amounts of glycogen and lipids, and become large decidual cells. The secretory activity of the endometrial glands is also increased. This endometrial layer of the gravid uterus is known as the decidua (meaning "cast off"). Three different regions of the decidua can be recognized (see Chapter 4, Fig. 4.2).

1. The decidua basalis, the part beneath the implanted conceptus, forming the maternal component of placenta.
2. The decidua capsularis, the part that covers the blastocyst and separates it from the uterine cavity.
3. The decidua parietalis, all the remaining uterine mucosa.

The decidua provides nourishment for the embryo and protects the maternal tissue from excessive invasion by the trophoblast.

IV. DEVELOPMENT OF FETAL MEMBRANES AFTER WEEK 3

A. Chorion

The chorionic cavity enlarges, and the villi, which arise adjacent to the decidua capsularis, become compressed and undergo degeneration leaving the chorion laeve, or smooth chorion. The chorionic villi associated with the decidua basalis, on the other hand, rapidly increase in number and size. The chorion in this region, known as the chorion frondosum or villous chorion, forms the fetal component of the placenta. The villi, which were initially straight tubular structures, branch profusely and fill the intervillous spaces. The basal portion of the chorion, from which the villi arise, is called the chorionic plate. The adjacent decidua basalis, together with the cytotrophoblast shell anchoring the villi, is called the basal plate. With further growth of the embryo, the decidua capsularis becomes stretched and thinned, and comes into contact with the decidua parietalis on the opposite wall of the uterus. It fuses with the decidua parietalis and eventually degenerates, leaving the chorion in direct contact with the decidua parietalis (Fig. 5.4).

B. Placenta

The placenta is developed by the mother and the child in symbiosis, and therefore consists of fetal and maternal components.

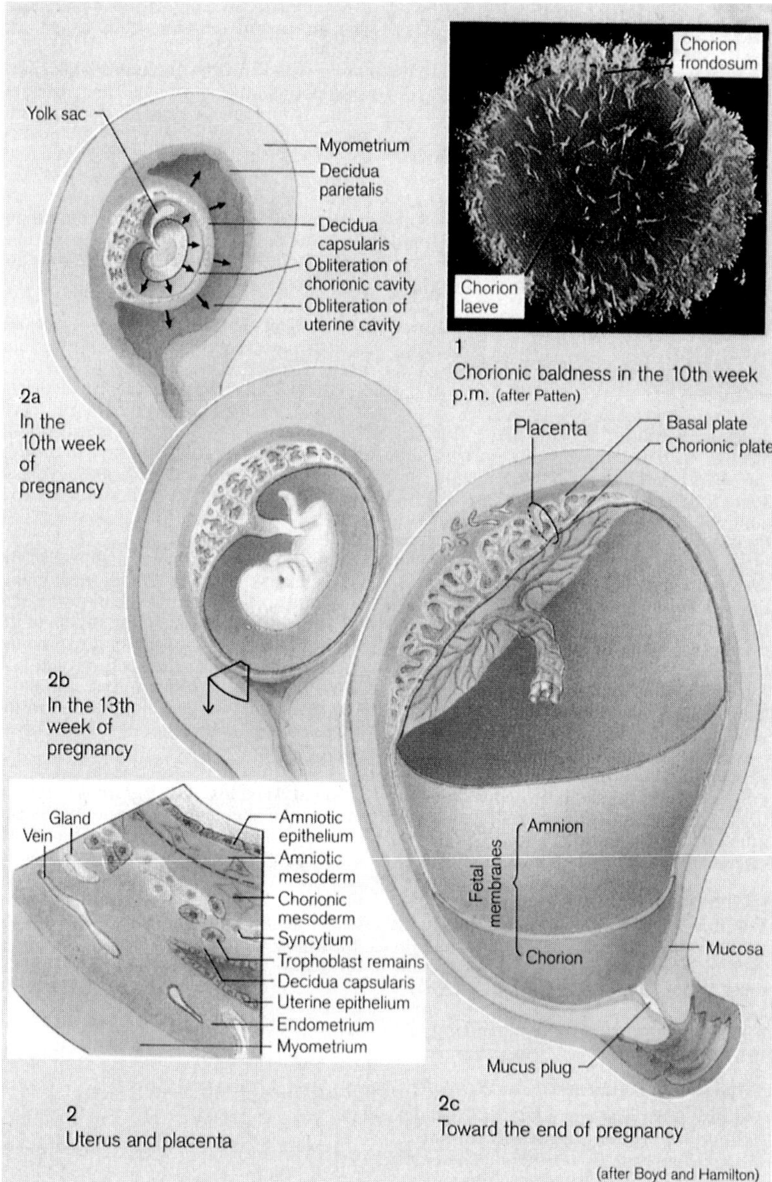

Development of placenta and fetal membranes

Figure 5.4

Overview: placenta and fetal membranes (see Color Plate 5.4). Reproduced with permission from Thieme Medical Publishers Inc., New York, 1995, *Color Atlas of Embryology*, Ulrich Drews, Chapter 2: Human Development.

1. Fetal components of placenta
   The fetal component is formed by the villous chorion and is anchored to the decidua by the cytotrophoblastic shell. Some villi (anchoring villi) that are directly attached to the decidua also participate in anchoring. Other villi enlarge,

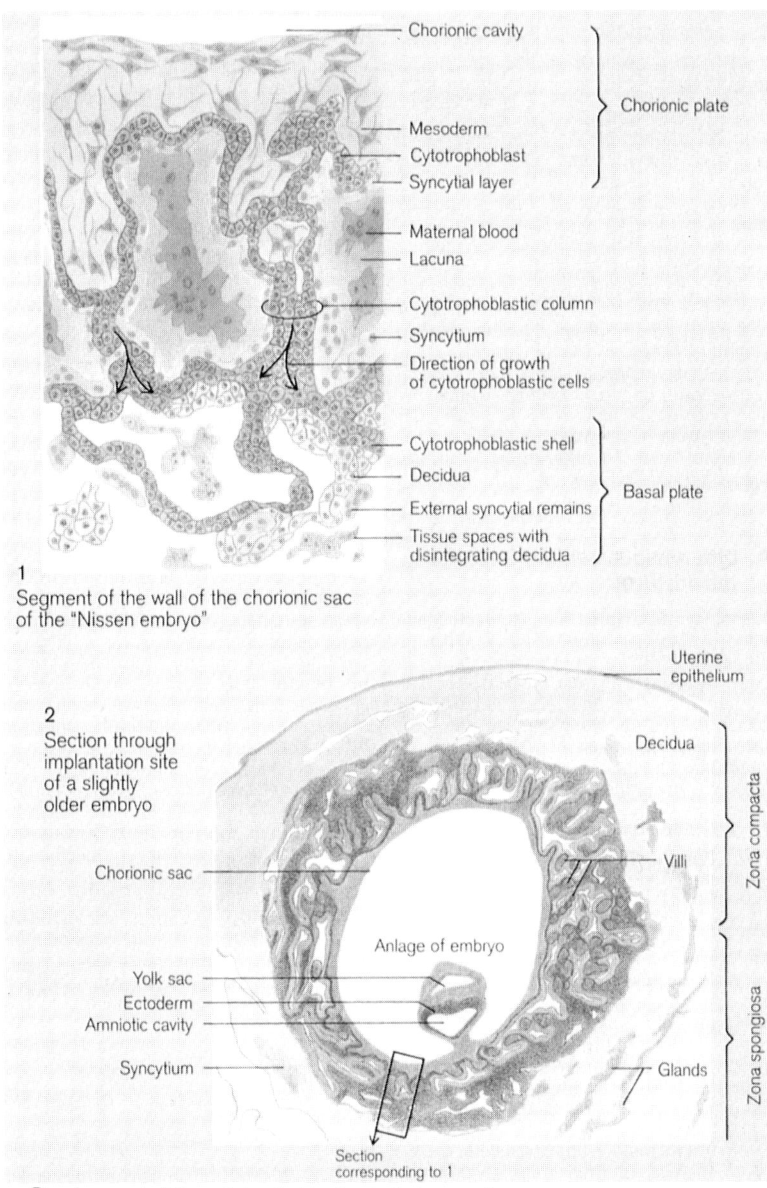

Figure 5.5

Cytotrophoblastic shell (see Color Plate 5.5). Reproduced with permission from Thieme Medical Publishers Inc., New York, 1995, *Color Atlas of Embryology,* Ulrich Drews, Chapter 2: Human Development.

elaborately branch and project freely into the intervillous space of one compartment in communication with others (Fig. 5.5).

2. Formation of placental membrane
   During the early part of pregnancy, villi are covered with both syncytiotrophoblast (syncytium) and cytotrophoblast. After about 20 weeks, the cytotro-

phoblast mostly disappears, leaving only a syncytial layer covering the villi. The amount of connective tissue is greatly reduced; the capillaries become large and numerous. This new covering over the villi is termed the placental membrane. As the pregnancy advances, the placental membrane becomes progressively thinner, and many fetal blood vessels lie closer to the surface syncytium (Fig. 5.6).

3. Maternal component of placenta
   The maternal component of the placenta is the decidua basalis. The intervillous spaces, which are formed by opening of the syncytiotrophoblast lacuna into maternal sinusoids, soon become lined by trophoblast, as described previously. Incomplete partitions of decidua (placental septa) extend from the basal plate and divide the fetal part of the placenta into from 15 to 30 irregular lobules called cotyledons.

4. Placental circulation

   a. Maternal placental circulation
      Multiple spiral endometrial arteries penetrate the basal plate and open into intervillous spaces. The lumen of spiral arteries become narrow, while passing through the basal plate, resulting in increased blood pressure. Blood, which is at a considerably higher pressure, flows around and over the villi and then returns to the floor of the intervillous space, where it enters the endometrial veins found on the entire surface of the decidua basalis. Cotyledons do not form complete partitions, so that blood can flow between adjacent compartments (Figs. 5.6 and 5.7).

   b. Fetal placental circulation
      The umbilical arteries divide freely, and the small branches enter the villi, bringing fetal blood very close to the maternal blood. The fetal blood, after exchanging metabolic and gaseous products, passes into the thin-walled veins, which follow the placental arteries and converge to form the umbilical vein, which carries oxygenated blood to the fetus. There is no actual intermingling of fetal and maternal blood.

5. Functions of the placenta
   The functions of the placenta include a) respiration, b) nutrition, c) excretion, d) protection and e) endocrine secretion (Fig. 5.7).

   a. Respiration
      Oxygen from maternal blood readily diffuses across the placental membrane. Carbon dioxide from the fetal blood passes readily to the opposite side. Carbon monoxide also diffuses readily through the placental membrane.

   b. Nutrition
      Water, electrolytes and glucose are transferred readily. Water-soluble vitamins pass more quickly than fat-soluble vitamins. Free fatty acids are transported, but there is little or no transport of cholesterol, triglycerides or phospholipids. Unconjugated steroids pass freely; testosterone and synthetic progestins cross the placenta freely and may cause external

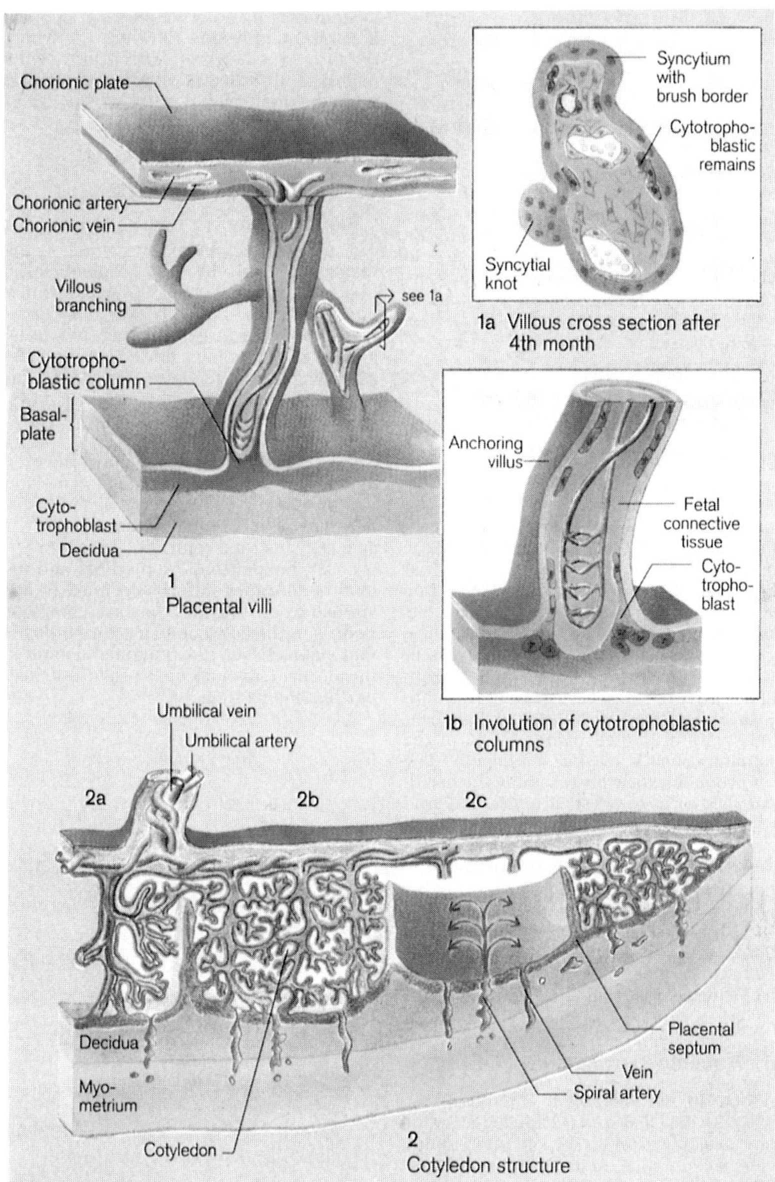

Figure 5.6

Structure of the placenta (see Color Plate 5.6). Reproduced with permission from Thieme Medical Publishers Inc., New York, 1995, *Color Atlas of Embryology*, Ulrich Drews, Chapter 2: Human Development.

masculinization of a female fetus. Protein hormones do not pass significantly; thyroxin and triiodothyroxin pass very slowly.

c. Excretion

Major waste products (urea, uric acid and bilirubin) cross the placental membrane freely.

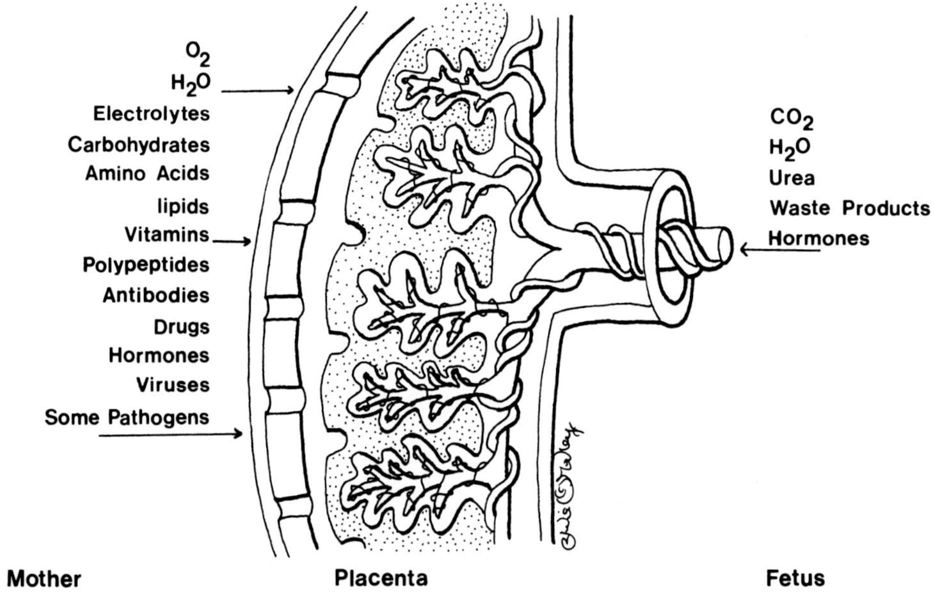

Figure 5.7
Exchange through the placenta.

d. Protection
Most foreign particulate matter such as bacteria are unable to pass through the placental membrane. However, the following harmful substances cross the placenta. Infectious agents such as toxoplasma gondii, rubella and Coxsackie virus, and those associated with variola, varicella, measles, encephalitis and poliomyelitis may pass through the placenta. Treponema pallidum (syphilis-causing organism), as well as some bacteria and protozoa may infect the placenta and then gain entry to fetal blood. Most drugs pass the placental membrane freely. Antibiotics (such as penicillin and sulfonamide) and alcohol and nicotine pass in small amounts. Morphine, barbiturates and general anesthetics when given to the mother pass across and may depress fetal respiratory centers. The psychotropic drugs chlorpromazine, reserpine and meprobamate cross the placenta. Fetal drug addiction may occur following maternal use of drugs such as heroin and cocaine. Maternal antibodies for diseases such as diphtheria, scarlet fever, tetanus, small pox and measles may pass across placenta by pinocytosis and confer some passive immunity to such infections for a variable period after birth. However, no immunity is acquired to pertussis (whooping cough) or chicken pox.

e. Endocrine secretion
The syncytiotrophoblast synthesizes the following hormones:

i. Human chorionic gonadotropin (HCG)
Production of HCG begins very early in pregnancy, the hormone being detectable in the blood and urine as early as the fifth or sixth

week of pregnancy. Its production increases rapidly, reaching its peak at about the 10th week. After this period it declines to relatively low levels. The measurement of HCG levels for pregnancy tests is quite reliable in the first trimester. The production of HCG in twin or multiple pregnancies is very high. It is also increased in case of a hydatidiform mole and choriocarcinoma.

ii. Human chorionic somatomammotropin (HCS) [also called human placental lactogen (HPL)]
The production of HCS rises progressively throughout pregnancy. Its effects are similar to those of pituitary growth hormones. It stimulates maternal metabolic processes to produce adequate amounts of amino acids and fatty acids for both the mother and the fetus. It also stimulates growth of the breasts in preparation for lactation.

iii. Thyrotropin
Production of thyroxin has recently been documented.

iv. Estrogen
Estrogens produced by the placenta: estrone, estradiol, and estriol. The production of estriol increases throughout pregnancy. The steroid precursors produced by fetal adrenal glands are converted to estrone and estradiol by enzymes in the placenta. Estrone and estradiol are circulated back to the fetus, where they are converted to estriol by enzymes in the fetal liver. Estriol is then transferred back to the placenta. Since neither the fetus nor the placenta contains all the enzymes for complete synthesis of these hormones, their production during pregnancy reflects the functional integrity of both the fetus and the placenta. Some maternal diseases, e.g., diabetes and high blood pressure, are associated with incidence of fetal mortality (pregnant women with these conditions are considered to be at high risk). The estrogen output in these patients should be carefully monitored. A decline in estrogen output indicates a threat to the life of the fetus, and in this case delivery should be accomplished as soon as possible. Reduced production of estrogen is also associated with a failure of the fetal adrenal gland to develop, or with anencephaly. In case of congenital adrenal hyperplasia, the output of estrogens could be markedly increased.

v. Progesterone
Progesterone production increases throughout pregnancy; it is excreted in the urine as pregnanediol. For the synthesis of progesterone, cholesterol and other steroid precursors are supplied to the placenta by the mother. As the fetus does not play any role in the production of progesterone, the level of this hormone does not reflect the condition of the fetus. Other hormones — HC corticotropin, insulin-like growth factor, relaxin, corticotropin releasing hormone and endothelin — have recently been reported.

V. ABNORMALITIES OF THE PLACENTA

A. Placenta Membranous

A thin placenta with persistent villi on the smooth chorion so that both the smooth chorion and villous chorion take part in formation of the placenta.

B. Placenta Bipartita or Tripartita

A placenta with two or three incomplete lobes. After delivery, one or two lobes may be retained in the uterus and may cause postpartum uterine hemorrhage or uterine infection (Fig. 5.8).

C. Placenta Duplex, Triplex or Multiplex

A placenta with two, three, or more completely separate lobes.

D. Placenta Circumvallata

The fetal surface of the placenta is partially overlapped by a circular rim of decidua.

E. Battledore Placenta (Marginal Insertion)

The umbilical cord is attached to the margin of the placenta.

F. Velamentous Placenta

The umbilical cord is attached to the membrane at some distance from the margin of the placenta. This may be hazardous to the fetus. The extended vessels may be compressed or ruptured during the course of delivery.

G. Very Small Placenta

Found in women suffering from chronic hypertension.

H. Very Large Placenta

Found in fetal hydrops, a condition of the fetus with severe hemolytic disease resulting from serological incompatibility between the mother and baby.

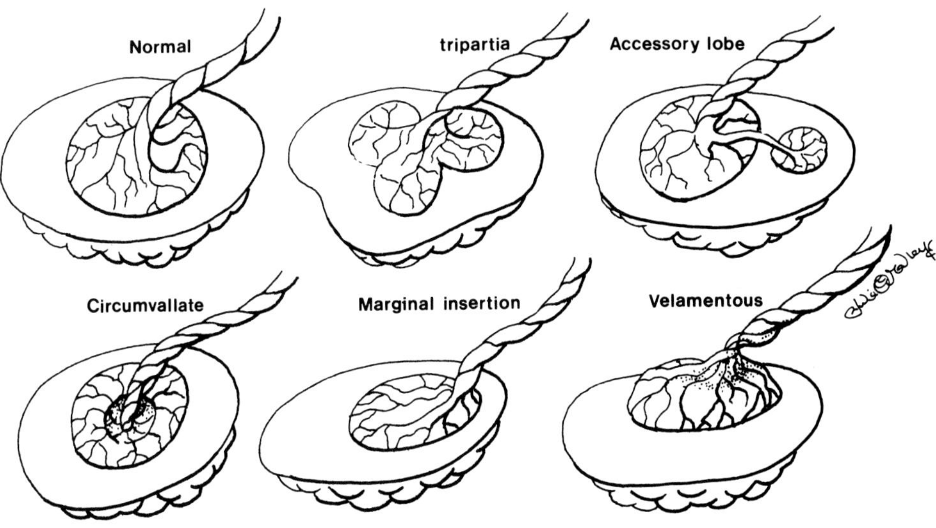

Figure 5.8
Placental anomalies.

VI. PLACENTA PRAEVIA

This involves abnormal attachment of the placenta within the uterus. The placental attachment may completely or partially cover the cervix. In this situation, the placenta fails to stretch to conform to the contour of the expanding lower part of the uterus. For this reason, parts of the placenta tear loose and cause bleeding in the latter part of pregnancy. Placenta praevia is hazardous to both mother and child. Due to the position of the placenta, vaginal delivery is a great risk. Vaginal examination is avoided in these women. Severe painless hemorrhage occurring from the 20th week and onward is a clinical sign of placenta praevia.

VII. CORD ACCIDENTS

An abnormally short cord rarely becomes a problem, but an excessively long cord tends to entwine around the neck or extremity of the fetus.

VIII. ERYTHROBLASTOSIS FETALIS

Rh factors are groups of surface molecules that are present on erythrocyte membranes. Individuals whose erythrocytes carry these factors are Rh^+, and individuals whose blood lacks these factors are Rh^-. If an Rh^- mother is carrying an Rh^+ fetus, and fetal erythrocytes leak into the maternal circulation, the mother will start making antibodies against the fetal erythrocytes, causing dissolution (hemolysis) of fetal erythrocytes. In general, significant leaking does not occur during pregnancy. However, at delivery fetal blood mixes with maternal blood, and mother will produce antibodies against the fetal blood that remain in the maternal circulation and will harm the second Rh^+ fetus and cause destruction of fetal blood cells — erythroblastosis fetalis. This can be prevented by injecting anti-Rh^+ antibodies into an Rh^- mother immediately after the birth of each Rh^+ child. Many cases have been reported of a sufficient amount of Rh^+ erythrocytes passing to the mother's circulation during pregnancy and provoking the mother's immune response. In these cases, transfusions of Rh^- blood are given to the fetus in utero or to the newborn.

IX. HUMAN IMMUNODEFICIENCY SYNDROME (AIDS)

Human immunodeficiency virus (HIV) can be transmitted across the placenta to the fetus, or during childbirth, or through breast milk to the baby. HIV-positive babies may look healthy at birth, but they develop the syndrome by age 3 or 4 years. The disease then appears as in adults, destroying the immune system and leaving babies prone to repeated infections like pneumonia, diarrhea and bronchitis. Babies having AIDS do not survive long if they contract pneumonia.

X. INTRAUTERINE GROWTH RETARDATION (IUGR)

Fetal growth is affected by genetic, infectious, nutritional and teratogenic factors. These factors cause limitation to normal fetal growth. They exert their effects by uteroplacental dysfunction and maternal nutrition. IUGR is most commonly considered when the birth weight is 10% less than it should be for a given gestational age.

A. The Genetic Causes of IUGR

Chromosomal anomalies such as trisomies 9, 13, 18, 21 and 22 can lead to growth retardation. It is also found among monosomy X and triploid infants. Many other genetic syndromes are associated with IUGR. Some inborn errors of metabolism, such as intrauterine anemias, immune deficiencies, mucolipidoses and sphingolipidoses, can lead to retarded intrauterine growth.

B. Infections

Intrauterine infections can be caused by viruses, protozoa, spirochetes and some bacteria. These can produce IUGR. The placenta is involved in most of these. The degree of IUGR is related to severity of infection, stage of gestation and type and virulence of infecting agent.

C. Nutritional Causes

Malnourished mothers not only supply fewer nutrients but also have reduced placental blood flow, which by itself can lead to IUGR. Intense physical activity, smoking, bacterial infections and diarrhea in malnourished mothers further retard intrauterine growth. Any disease of the mother that reduces uteroplacental blood flow or impairs mother's oxygenation can cause IUGR. Pregnancy-induced hypertension, chronic renal disease, collagen disease or anemia are included in some of these disorders.

D. Teratogenic Factors

Antibiotic and therapeutic agents like anticonvulsant trimethadione, phenytoin, anticoagulant warfarin, retinoids (vitamin A analogues) and chemotherapeutic agents can lead to IUGR. Most of these exert their effects during the embryonic period. Heroin, cocaine, alcohol and tobacco are also teratogens. Cigarette smoking is a powerful determinant of IUGR. Addiction to recreational drugs is commonly associated with IUGR.

1. Fetal Alcohol Syndrome

It has been suggested that heavy drinking even once during the first trimester of pregnancy is teratogenic. Chronic consumption of even quite small amounts of alcohol may result in fetal growth retardation. The spectrum of fetal alcohol syndrome ranges from mild to severe. A disturbance in early induction of the forebrain (holoprosencephalon) is the most damaging manifestation of fetal alcohol syndrome. Other features of this syndrome include microcephaly, low birth weight, hirsutism, characteristic facial appearance with short palpebral fissures, a depressed nasal bridge, a thin upper lip and smooth philtrum, cardiac defects, hydrocephaly, CNS defects and dysplasia of the nails. Many other features like cleft lips and palate have also been observed.

2. Heroin or Cocaine Addiction

Heroin and cocaine ingested by pregnant women readily cross the placenta and may cause addiction in the developing fetus. Cocaine is a potent vasoconstrictor and may cause a decrease in placental blood flow. It also affects myometrial contraction, making it more sensitive to labor-initiating signals. Cocaine-using mothers therefore have a higher frequency of fetal mortality and morbidity. These women also have a high frequency of preterm labor. Babies are born premature and addicted to cocaine. Aside from IUGR, infarction of the cerebral cortex and various cardiovascular anomalies have also been reported.

REVIEW QUESTIONS — Chapter 5

1. After implantation (2nd week of development), the blastocele becomes the:
   a. amniotic cavity
   b. extraembryonic coelom
   c. primary yolk sac
   d. intervillous space
   e. placenta

2. The trophoblast differentiates into:
   a. chorion
   b. epiblast
   c. hypoblast
   d. outer cell mass
   e. none of the above

3. The placental membrane (barrier) is formed by all of the following except:
   a. the basement membrane of placental (fetal) capillaries
   b. the endothelium of placental (fetal) capillaries
   c. the cytotrophoblastic shell
   d. the syncytiotrophoblast
   e. the extraembryonic mesoderm

4. The human placenta secretes all of the following except:
   a. estrogen
   b. progesterone
   c. thyroxin
   d. human chorionic gonadotropin
   e. follicular stimulating hormone

5. Cells from the cytotrophoblast participate in formation of all of the following except:
   a. amnioblasts
   b. extraembryonic mesoderm
   c. exocoelomic (Heuser's) membrane
   d. cytotrophoblastic shell
   e. primary chorionic villi

6. Which of the following layers forms an outer covering on the umbilical cord?
   a. extraembryonic mesoderm
   b. yolk sac
   c. amnion
   d. chorion
   e. mesenchyme

7. The factors that play a role in nonrecognition of the embryo by the mother's immune system are:
   a. low major histocompatibility antigens (classes I and II) in the trophoblast
   b. local decidual barriers (reaction) in the endometrium
   c. selective suppression of the mother's immune system
   d. all of the above
   e. none of the above

8. Heuser's or the exocoelomic membrane lines:
   a. the amniotic cavity
   b. the blastocele
   c. the primitive yolk sac
   d. the yolk sac proper
   e. the chorionic cavity

9. The yolk sac proper:
   a. is limited by the exocoelomic membrane
   b. is formed by separation of the cytotrophoblastic layers
   c. disappears after the third week
   d. is lined by endodermal cells

10. During late pregnancy, which of the following layers of placental membrane is in direct contact with maternal blood?
    a. the cytotrophoblastic shell
    b. the syncytiotrophoblast
    c. the extraembryonic mesoderm
    d. the intraembryonic mesoderm
    e. the endothelium of placental blood vessels

11. The extraembryonic tissue derived from the outer cell mass is the:
    a. amnion
    b. yolk sac
    c. allantois
    d. hypoblast
    e. placenta

12. During implantation of the embryo, lacunae merge with endometrial sinuses to form the intervillous space, which is lined by:
    a. endothelium
    b. endometrial epithelium
    c. syncytiotrophoblast
    d. cytotrophoblast
    e. all of the above

13. The decidua that takes part in the formation of the maternal component of the placenta is:
    a. the decidua basalis
    b. the decidua capsularis
    c. the decidua parietalis
    d. the villous chorion
    e. the smooth chorion

14. Paternal imprinting selectively turns off some genes involved in development of the embryo, whereas maternal imprinting turns off some genes involved in the formation of extraembryonic structures. The formation of a hydatidiform mole is an example of:
    a. maternal imprinting
    b. paternal imprinting
    c. both
    d. neither

15. During placentation, delamination of the innermost layer of the cytotrophoblast results in formation of the:
    a. exocoelomic membrane
    b. chorionic cavity
    c. primary yolk sac
    d. extraembryonic mesoderm
    e. all of the above

16. All of the chorionic villi are covered by the:
    a. syncytiotrophoblast
    b. cytotrophoblast
    c. extraembryonic mesoderm
    d. maternal endothelium
    e. all of the above

17. The decidua parietalis
    a. forms the maternal part of the placenta
    b. forms the fetal part of the placenta
    c. does not take part in the formation of the placenta
    d. covers the embryo after implantation
    e. fuses with the chorion when the fetus grows larger

18. In a normal pregnancy, the urinary concentration of HCG
    a. parallels fetal development
    b. parallels placental development
    c. peaks in the 1st trimester
    d. peaks in the 2nd trimester
    e. peaks in the 3rd trimester

19. Abnormal attachment of placenta within the uterus is termed:
    a. velamentous placenta
    b. ectopic pregnancy
    c. placenta membranacea
    d. placenta praevia
    e. placenta circumvallate

20. Fetal alcohol syndrome is associated with all of the following except:
    a. cardiac defects
    b. craniofacial anomalies
    c. maternal alcoholism
    d. mental retardation
    e. high birth weight

ANSWERS TO REVIEW QUESTIONS

---

1. b	2. a	3. c	4. e
5. a	6. c	7. d	8. c
9. d	10. b	11. e	12. c
13. a	14. b	15. a	16. a
17. e	18. c	19. d	20. e

CHAPTER

6

EARLY DEVELOPMENT OF EMBRYO

I. FIRST WEEK OF DEVELOPMENT

Cleavage starts about 24 hours after fertilization. The four-cell stage is reached 40 hours after fertilization; the 12-cell stage within 3 days; and the morula stage (16 to 32 cells) is seen on day 4. The cells of the morula segregate into trophoblasts (outer cell mass) and embryoblasts (inner cell mass). The cells (blastomere) at this stage are not determined cells. Their position in the morula determines their ultimate location. The outer larger cells with microvilli divide early to form the trophoblast. With the accumulation of fluid, the blastocyst is formed. The cells of the blastocyst are determined cells, and they differentiate into trophoblast and embryoblast. The embryoblast continues for a short time to add cells to the expanding trophoblast. During the first seven days, the cells undergo about eight divisions. The first week ends with onset of implantation.

II. SECOND WEEK OF DEVELOPMENT

A. Embryonic Disk

The embryoblast cells differentiate into two distinct populations. The tall cells adjacent to the trophoblast form the epiblasts or primitive ectoderm; the smaller flattened cells facing the blastocele form the hypoblasts, or the primitive endoderm (see Chapter 5, Figs. 5.1 and 5.2).

B. Formation of the Amniotic Cavity

Electron microscopic observations suggest that the amniotic cavity appears as a result of rearrangement of epiblast cells. A change in cell association within the epiblast separates cells into small amnioblasts and the ectoderm proper. The cavity starts as an upfolding of the margins of the epiblast. It is also speculated that small clefts appear between the trophoblast and the epiblast. These clefts coalesce to give rise to a single cavity. There is strong evidence that both methods, the upfolding and cavitation, contribute to the formation of this cavity. Amnioblasts were also observed to be added by delamination of adjacent trophoblast.

C. Formation of Amnion and Connecting Stalk

1. During development of the chorionic cavity, the extraembryonic mesoblast covers the amnioblast layer and separates the embryo (embryonic disk, yolk sac and amniotic cavity) from the trophoblast layer except at connecting stalk. A concentration of mesoblast in the connecting stalk indicates the caudal end of the embryonic disk (Chapter 5, Fig. 5.2). The amnioblast layer with the covering of a mesoblast layer is now called the amnion. The roof of the amniotic cavity is the amnion and the floor is the ectoderm (epiblast).
2. A diverticulum of yolk sac, the allantois, extends into the connecting stalk. The connecting stalk appears to be made up of two parts:
   i. The amnio-embryonic part connects the amniotic cavity to the chorionic mesoblasts.
   ii. The umbilical part anchors the caudal end of the embryonic disk to the chorion.

III. THIRD WEEK OF DEVELOPMENT

The structures that develop during the third week include: A) the trilaminar embryonic disk, B) the notochord, C) mesodermal somites, D) the neural tube, E) the embryonic folding, F) the intraembryonic coelom, G) the cytotrophoblastic shell, H) primitive blood vessels and I) the allantois. Formation of the cytotrophoblastic shell and allantois, and angiogenesis are described along with placentation in Chapter 5.

A. Trilaminar Embryonic Disk

The ectoderm of the midline area of the caudal half of the embryonic disk becomes thickened, and a faint groove, known as the primitive streak, appears in the midline (Fig. 6.1). The cephalic end of the primitive streak further swells to form a primitive node (Hensen's node). The ectodermal cells migrate toward the primitive streak and, after reaching it, invaginate and form a linear furrow, the primitive groove, in the primitive streak; and a primitive pit in the node. The migrating cells, now known as mesenchymal cells, begin to move laterally between the ectoderm and endoderm. This newly formed intermediate layer is called the intraembryonic mesoderm. The mesenchymal (mesodermal) cells extend laterally beyond the margin of the disk and finally fuse with the extraembryonic mesoderm.

At the cephalic end (prochordal plate) as well as at the caudal end of the embryonic disk, over a small area the ectoderm and endoderm remain fused. The mesodermal cells pass around these areas. The cephalic area later gives rise to the oropharyngeal membrane, and the caudal area gives rise to the cloacal membrane.

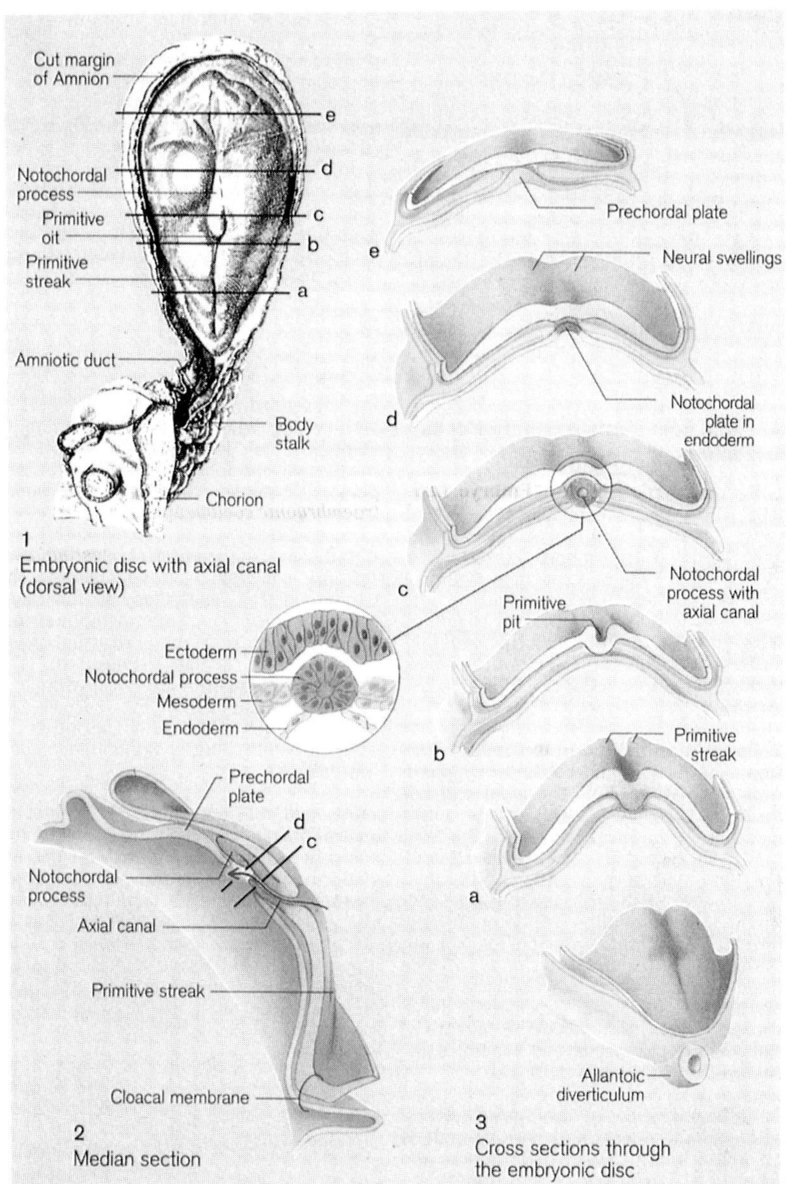

Figure 6.1

Stage 8: axial canal (see Color Plate 6.1). Reproduced with permission from Thieme Medical Publishers Inc., New York, 1995, *Color Atlas of Embryology,* Ulrich Drews, Chapter 2: Human Development.

The newly formed mesoderm soon becomes differentiated into three regions (Fig. 6.2):

1. The area immediately lateral to the notochordal process (developing notochord) becomes thickened and is known as the paraxial mesoderm.

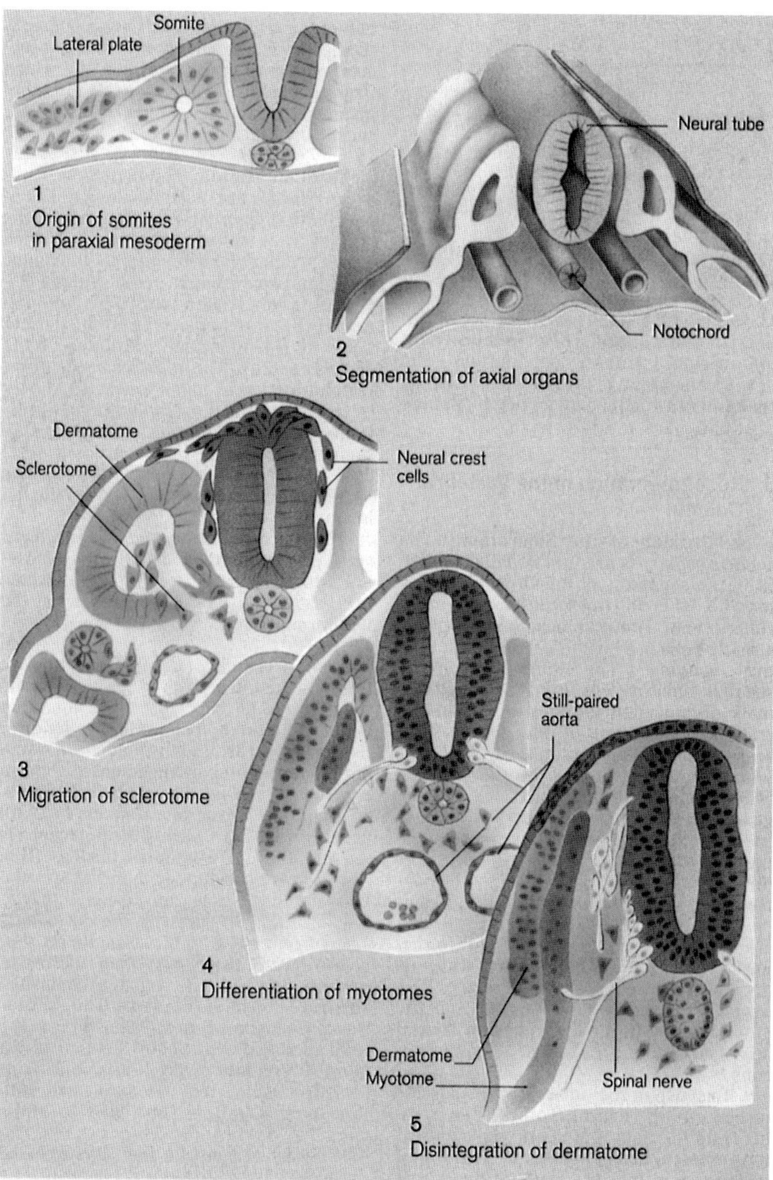

Figure 6.2

Differentiation of somites (see Color Plate 6.2). Reproduced with permission from Thieme Medical Publishers Inc., New York, 1995, *Color Atlas of Embryology*, Ulrich Drews, Chapter 3: General Embryology.

2. Lateral to the paraxial mesoderm is a zone of intermediate mesoderm.
3. The peripheral layer, which enters into the margins of the germ disk and becomes continuous with the extraembryonic mesoderm, is the lateral mesoderm.

B. Formation of the Notochord

A hollow tubular column of cells grows cephalically from the primitive node in the midline between the ectoderm and endoderm until it reaches the oropharyngeal membrane. This columnar growth of cells is known as the notochordal process. By day 18, the floor of the notochordal process fuses with the underlying endoderm (Fig. 6.1). The fused region of two layers disintegrates, which leads to a temporary communication between the amniotic cavity and the yolk sac through a small passage, the neurenteric canal, in the region of the primitive pit. The remaining notochordal process forms a grooved plate, the notochordal plate. The notochordal plate, starting at its cephalic end, infolds to form a solid column of cells, the notochord proper. Soon afterward, the continuity of the endoderm is restored at the site of breakdown. By the end of the fourth week, formation of the notochord is completed and the neurenteric canal is obliterated. The notochord is the structure around which the entire vertebral column forms, and it induces the formation of the nervous system from the overlying ectoderm.

C. Development of Somites

On about day 20, the paraxial mesoderm begins to break up into blocks of cells, the somites. The first pair of somites appears at a short distance from the cranial end of the notochord. New somites appear in a craniocaudal sequence, approximately three per day, totaling 42 to 44 at the end of the fifth week (Fig. 6.2).

D. Neural Tube Development

The ectoderm situated over the notochord thickens to form the neural plate (Figs. 6.1 and 6.2). On about day 18, the neural plate invaginates along its central axis to form the neural groove. The lateral edges of the neural plate, which are now elevated, form the neural folds. As the neural groove continues to deepen, the neural folds begin to move closer; finally, at the end of the third week, they fuse to form the neural tube. This fusion begins in the region of the future neck and extends in the cephalic and caudal directions. Some of the cells of the lateral edges of the neural folds are not incorporated into the tube and form neural crests.

E. Folding of the Embryo

The oropharyngeal membrane is cephalic to the developing neural tube. The newly formed intraembryonic coelom is situated dorsal to the cardiogenic area and is cephalic to the oropharyngeal membrane. The mesoderm continuous from the amnion to yolk sac at the cephalic end of the embryo is designated as the septum transversum.

At the caudal end of the embryo, the cloacal membrane lies caudal to the primitive streak (Fig. 6.3). The central part of the embryo, especially the central nervous system, grows more rapidly than the peripheral part. This results in longitudinal (cephalocaudal) and transverse (lateral) folding of the embryonic disk. The forebrain area grows beyond the oropharyngeal membrane. Consequently, the septum transversum, cardiogenic area and coelom related to this area, and oropharyngeal membrane turn under onto the surface. Some of the cephalic portion of the yolk sac becomes incorporated into the embryo as the foregut. Due to caudal

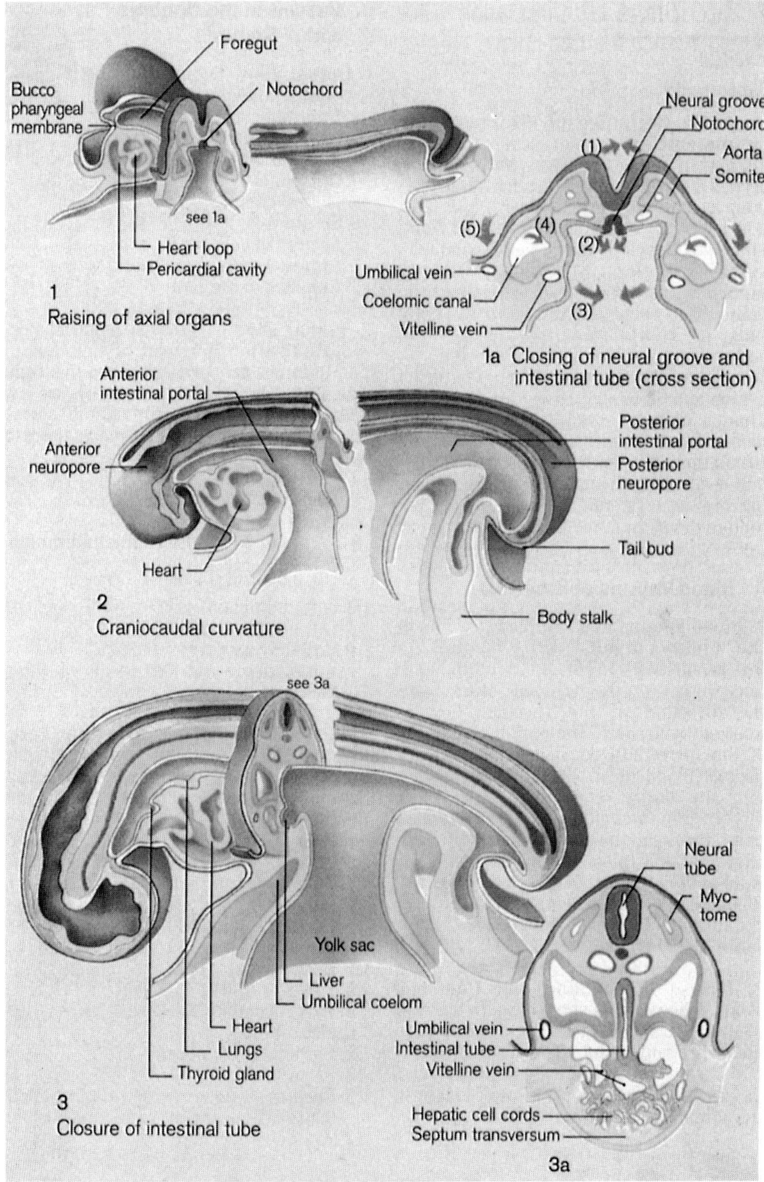

Folding from yolk sac and development of body form

Figure 6.3

Overview: folding (see Color Plate 6.3). Reproduced with permission from Thieme Medical Publishers Inc., New York, 1995, *Color Atlas of Embryology*, Ulrich Drews, Chapter 2: Human Development.

folding, some of the caudal portion of the yolk sac is incorporated as the hindgut. After caudal folding, the primitive streak becomes caudal to the cloacal membrane. Due to lateral folds, the lateral edges of the embryonic disk roll ventrally toward the midline, making the embryonic disk cylindrical and enclosing the embryo within the amniotic cavity.

F. Development of the Intraembryonic Coelom

Small, isolated spaces appear within the lateral plate of the mesoderm. These spaces coalesce to form a horseshoe-shaped cavity, the intraembryonic coelom. Formation of the intraembryonic coelom divides the lateral plate of mesoderm into two layers:

1. The somatic (parietal) layer is continuous with the extraembryonic mesoderm covering the amniotic cavity.

2. The splanchnic (visceral) layer is continuous with the extraembryonic mesoderm covering the yolk sac and the incorporated gut. The newly formed intraembryonic coelom opens to the extraembryonic coelom.

IV. THE AMNION AND AMNIOTIC CAVITY

A. Amnion

The amnion is a thin, transparent, nonvascular membrane that forms during the second week (see §IIc above). It is attached to the margins of the embryonic disk and forms the walls of the amniotic cavity. As a result of the foldings of the embryo, it becomes located at the ventral surface of the embryo, forming a sheath around the connecting stalk (Fig. 6.3).

B. Amniotic Cavity

After enclosing the embryo within it, the amniotic cavity expands further, obliterating the chorionic cavity (Chapter 5, Fig. 5.6). With this expansion, the amnion and chorion come together and fuse to form the amnio-chorionic membrane. The connecting stalk lengthens and becomes approximated with yolk sac, giving rise to the umbilical cord. As the umbilical cord lengthens, the blood vessels in it grow with it, providing a connection between the placental and embryonic blood vessels.

C. Amniotic Fluid

The amniotic cavity becomes filled with a pale, straw-colored, watery fluid, which increases from 30 ml at 10 weeks to 350 ml at 20 weeks, and to 1 liter at full term. Thereafter, it decreases by about 150 ml every week if delivery is delayed. In the early stages, it is secreted by amniogenic cells. It is later derived mostly from maternal blood by filtration. It is also filtered from fetal blood passing through the placenta. When the kidneys become functional, urine is added to the amniotic fluid.. A large volume moves through the placental membrane in both directions between the fetal and maternal circulations, so that it is changed every 3 hours. After rupture of the oropharyngeal membrane, it is swallowed by the fetus, where it is absorbed by the gastrointestinal tract into the bloodstream and passes by way of the placenta into the maternal blood. The additions are counterbalanced by the loss of fluid into the fetal gastrointestinal tract.

D. Fetal Abnormalities Resulting from Extreme Variations in Amniotic Fluid Volume

1. Polyhydramnios — a very marked increase in the volume of amniotic fluid. The fetus is unable to swallow the normal amount of fluid either due to malformation of the central nervous system (anencephaly and hydrocephalus) or due to congenital obstruction of the esophagus, duodenum or upper jejunum. In some instances, such maternal diseases as diabetes may cause this condition.
2. Oligohydramnios — a marked decrease in the volume of amniotic fluid. The fetus is unable to excrete urine into the amniotic cavity either due to renal agenesis or urethral obstruction. This condition may be associated with fetal deformities because the uterine wall exerts pressure on its developing fetus (see Chapter 14, Urinary System).

E. Functions of Amniotic Fluid

1. Provides a protective cushion against external injury and localized pressure from the uterine wall. This allows symmetrical growth and development.
2. Maintains the body temperature of the embryo (fetus).
3. Enables the fetus to move freely.
4. At the beginning of labor, the bag of fluid is pushed down the cervical canal to facilitate the early stages of labor.

F. Estimation of Fetal Maturity from Examination of Amniotic Fluid

1. Concentration of lecithin and sphingomyelin in the amniotic fluid reflects the maturation of fetal lungs.
2. Concentration of creatinine reflects the renal function of the fetus
3. Concentration of bilirubin reflects the metabolic activity of the liver.
4. The types of cells found in the amniotic fluid indicate the general maturity and karyotype of the fetus.

TABLE 6.1. Summary of Changes in General Body Form

Age (days)	Features
	Fertilization
1½–3	From 2 to about 16 cells
4	Free blastocyst
5–6	Attaching blastocyst
7–12	Implanted although previllous
7–8	Solid trophoblast
9	Trophoblastic lacunae
11–12	Lacunar vascular circle
13	Chorionic villi; primitive streak may appear
14	Chorionic villi
15	Primitive streak
16	Notochordal process
18	Primitive pit; notochordal and neurenteric canals
20	Somites first appear
22	Neural folds begin to fuse; two branchial arches; optic sulcus
24	Rostral neuropore closes; optic vesicle
26	Caudal neuropore closes; three branchial arches; upper limb bud appearing
28	Four limb buds; lens placode; optic vesicle
32	Lens pit and optic cup; endolymphatic duct distinct
33	Lens vesicle; nasal pit; antitragus beginning; and hand plate; trunk relatively wider; cerebral vesicles distinct
37	Nasal pit faces ventrally; retinal pigment visible in intact embryo; auricular hillocks beginning; foot plate
41	Head relatively larger; trunk straighter; nasofrontal groove distinct; auricular hillocks distinct; finger rays
44	Body more cuboidal; elbow region and toe rays appearing; eyelids beginning; tip of nose distinct; nipples appear; ossification may begin
48	Trunk elongating and straightening
50	Upper limbs longer and bent at elbows
52	Fingers longer; hands approach each other; feet likewise
54	Eyelids and external ear more developed
57	Head more rounded; limbs longer and more developed

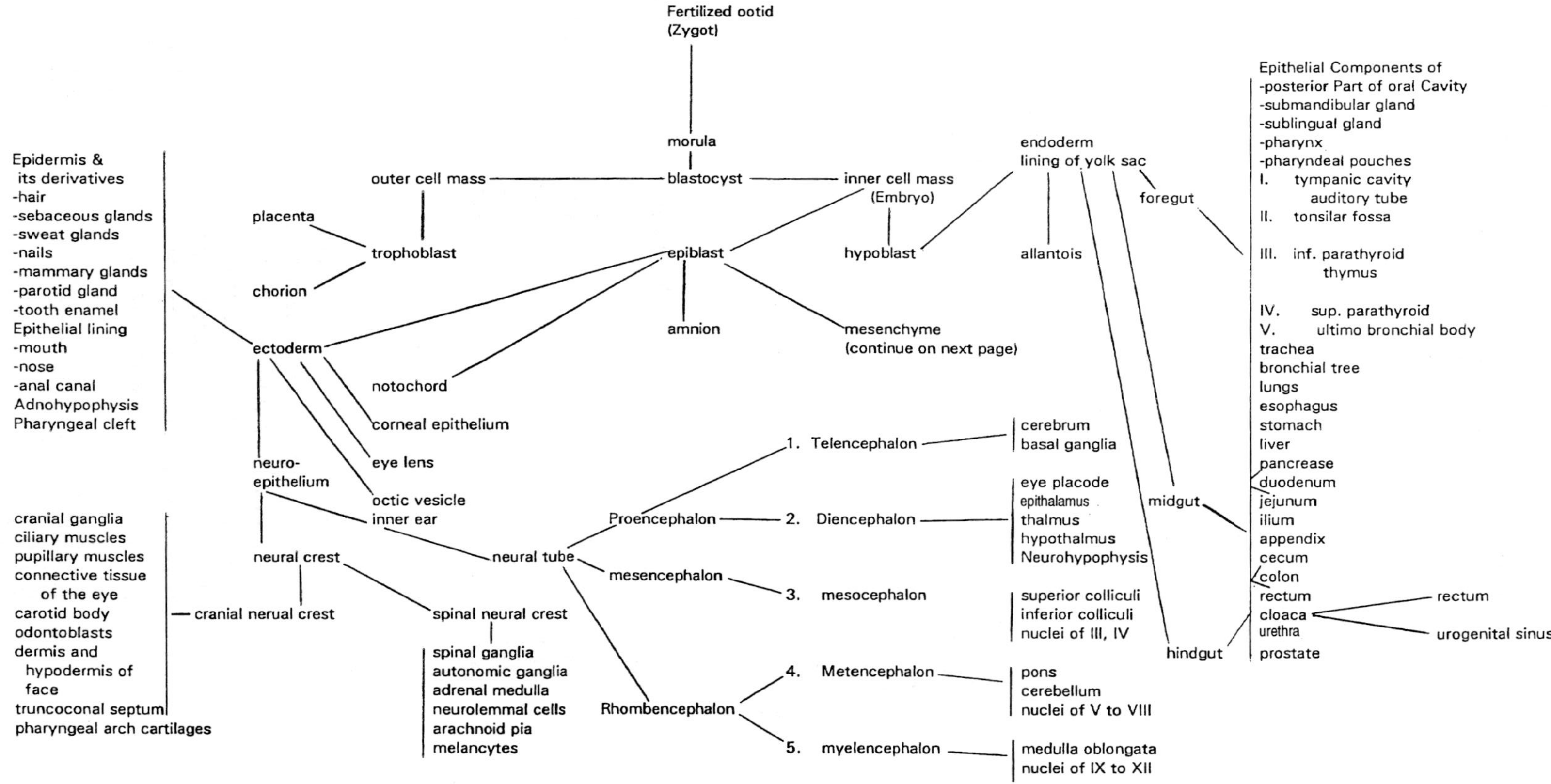

Fertilized ootid
(Zygot)
morula
outer cell mass
blastocyst
inner cell mass
(Embryo)
placenta
trophoblast
chorion
epiblast
hypoblast
amnion
mesenchyme
(continue on next page)
endoderm
lining of yolk sac
foregut
allantois
Epidermis &
its derivatives
-hair
-sebaceous glands
-sweat glands
-nails
-mammary glands
-parotid gland
-tooth enamel
Epithelial lining
-mouth
-nose
-anal canal
Adnohypophysis
Pharyngeal cleft
ectoderm
notochord
corneal epithelium
eye lens
neuro-
epithelium
octic vesicle
inner ear
neural crest
neural tube
cranial nerual crest
spinal neural crest
spinal ganglia
autonomic ganglia
adrenal medulla
neurolemmal cells
arachnoid pia
melancytes
cranial ganglia
ciliary muscles
pupillary muscles
connective tissue
of the eye
carotid body
odontoblasts
dermis and
hypodermis of
face
truncoconal septum
pharyngeal arch cartilages
Proencephalon
mesencephalon
Rhombencephalon
1. Telencephalon
2. Diencephalon
3. mesocephalon
4. Metencephalon
5. myelencephalon
cerebrum
basal ganglia
eye placode
epithalamus
thalmus
hypothalmus
Neurohypophysis
superior colliculi
inferior colliculi
nuclei of III, IV
pons
cerebellum
nuclei of V to VIII
medulla oblongata
nuclei of IX to XII
midgut
hindgut
Epithelial Components of
-posterior Part of oral Cavity
-submandibular gland
-sublingual gland
-pharynx
-pharyndeal pouches
I. tympanic cavity
auditory tube
II. tonsilar fossa
III. inf. parathyroid
thymus
IV. sup. parathyroid
V. ultimo bronchial body
trachea
bronchial tree
lungs
esophagus
stomach
liver
pancrease
duodenum
jejunum
ilium
appendix
cecum
colon
rectum
cloaca
urethra
prostate
rectum
urogenital sinus

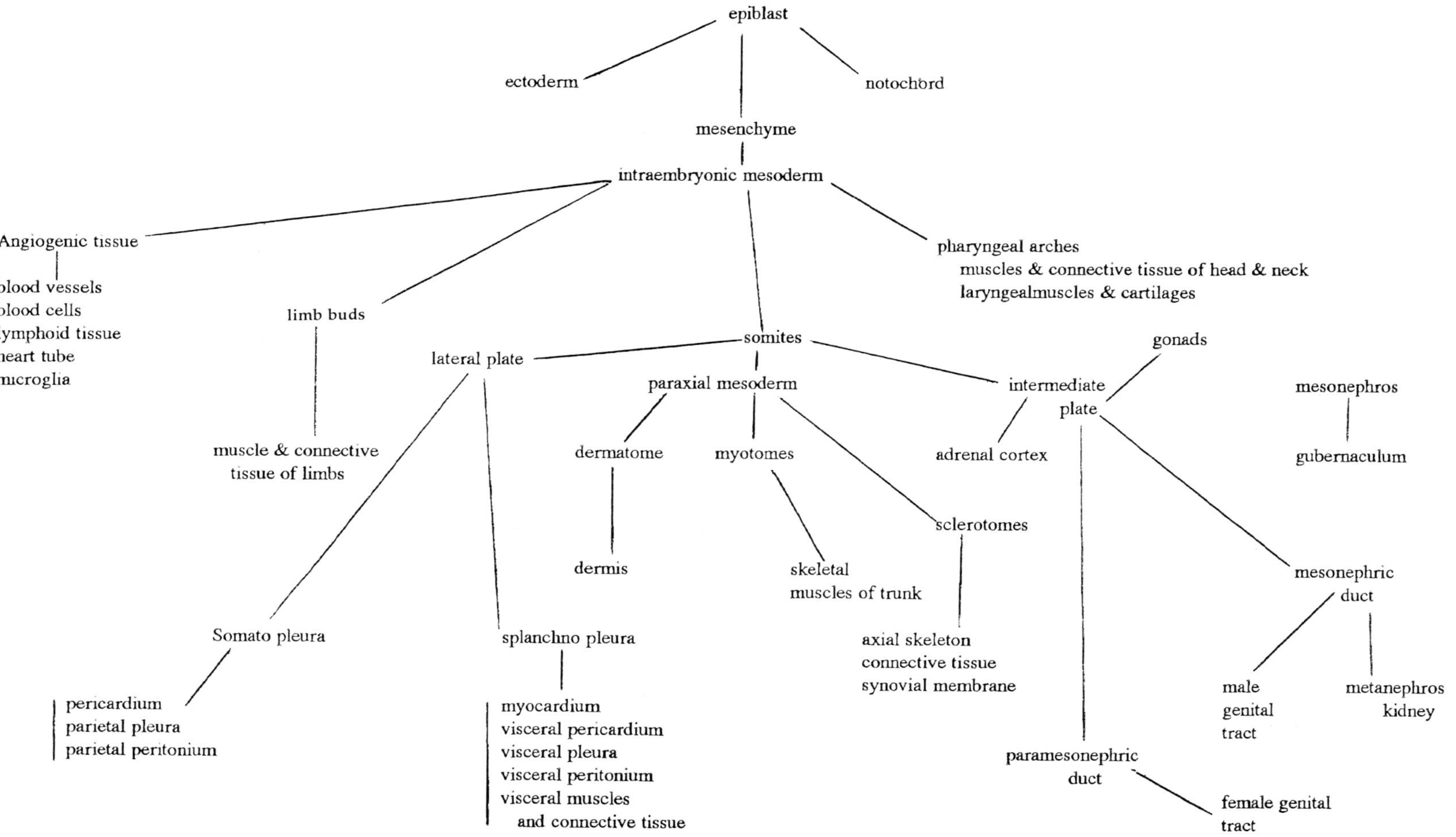
epiblast
ectoderm
notochord
mesenchyme
intraembryonic mesoderm
Angiogenic tissue
blood vessels
blood cells
lymphoid tissue
heart tube
microglia
limb buds
muscle & connective tissue of limbs
pharyngeal arches
muscles & connective tissue of head & neck
laryngealmuscles & cartilages
somites
lateral plate
paraxial mesoderm
intermediate plate
gonads
mesonephros
gubernaculum
dermatome
myotomes
adrenal cortex
sclerotomes
dermis
skeletal muscles of trunk
mesonephric duct
Somato pleura
splanchno pleura
axial skeleton
connective tissue
synovial membrane
pericardium
parietal pleura
parietal peritonium
myocardium
visceral pericardium
visceral pleura
visceral peritonium
visceral muscles
and connective tissue
male genital tract
metanephros kidney
paramesonephric duct
female genital tract

REVIEW QUESTIONS — Chapter 6

Select the most appropriate answer.

1. The migration of intraembryonic mesoderm occurs between the ectoderm and endoderm to give rise to:
   a. the notochordal process
   b. the oropharyngeal membrane
   c. the allantois
   d. the somites
   e. all of the above

2. The oropharyngeal membrane is made up of:
   a. embryonic ectoderm, mesoderm and endoderm
   b. endoderm fused with mesoderm
   c. ectoderm fused with mesoderm
   d. ectoderm fused with endoderm
   e. the cephalic end of notochordal process

3. The cloacal membrane:
   a. persists as hymen in the vaginal canal
   b. consists of embryonic ectoderm and endoderm
   c. consists of embryonic ectoderm, mesoderm and endoderm
   d. extends into the connecting stalk as allantois
   e. none of the above is correct

4. The primitive streak:
   a. extends from the primitive node to the parachordal plate
   b. is a primary inducer during organogenesis
   c. is a site of migration of epiblast cells to form the mesoderm
   d. persists as the parachordal plate
   e. all of the above are correct

5. The intraembryonic mesoderm differentiates into the:
   a. gut
   b. neural tube
   c. germ cells
   d. somites
   e. all of the above

6. The amniotic cavity appears on the 8th day as a slit-like space between the embryonic polar trophoblast and the:
   a. ectoderm
   b. extraembryonic mesoderm
   c. exocoelomic (Heuser's) membrane
   d. connecting stalk
   e. endoderm

7. The functions of amniotic fluid include all of the following except:
   a. provides a protective cushion
   b. allows free movement for symmetrical growth and development
   c. maintains the body temperature of the fetus
   d. provides nutrition in early embryonic stages
   e. exchanges fetal waste

8. Analysis of amniotic fluid could reveal all of the following except:
   a. hydramnios
   b. a concentration of a-fetoprotein reflecting CNS development
   c. a lecithin/sphingomyelin ratio reflecting lung maturity
   d. the severity of erythroblastosis fetalis
   e. the gender of the fetus

9. A mother, immediately after she discovers that she is pregnant, demands that amniocentesis be performed. Her physician should refuse to do that because:
   a. of his/her religious belief
   b. the amount of fluid is small
   c. the procedure may infect the fetus
   d. the home pregnancy test result may be a false positive
   e. any or all of the above

10. During the third week of development, the notochordal process starts as a column of cells migrating from the:
    a. primitive streak
    b. endoderm
    c. primitive node
    d. primitive groove
    e. paraxial mesoderm

11. During the 3rd week of development, a communication between the amniotic cavity and yolk sac is formed that is known as:
    a. the cloaca
    b. the neuroenteric canal
    c. the extraembryonic coelom
    d. the intraembryonic coelom
    e. the allantois

12. The functions of the yolk sac in the human embryo include all of the following except:
    a. it stores nutrients during pregnancy
    b. most of it is incorporated in the gut
    c. early hemopoietic activity occurs in it
    d. primordial germ cells are produced in it
    e. the allantois grows out of it
13. Much of the intraembryonic coelom (body cavity) is lined by the:
    a. paraxial mesoderm
    b. hypoblast
    c. intermediate mesoderm
    d. endoderm
    e. lateral plate mesoderm
14. The primary inducer during organogenesis is believed to be the:
    a. paraxial plate of somites
    b. intermediate mesoderm
    c. yolk sac proper
    d. notochord
    e. neural plate
15. Which of the following structures is *not* derived from ectoderm?
    a. the epidermis
    b. the neural tube
    c. the adenohypophysis
    d. the allantois
    e. the neurocrest
16. The paraxial plate of somites gives rise to all of the following except the:
    a. dermis in the skin over the trunk region
    b. muscles of the abdominal wall
    c. bodies of the cervical vertebrae
    d. lining of the abdominal wall
    e. vertebral arches in the lumbar region
17. All of the following are derived from mesoderm except the:
    a. blood vessels
    b. suprarenal medulla
    c. skeletal muscles
    d. endothelium
    e. blood cell
18. Which of the following is derived from endoderm?
    a. the kidney
    b. the gonads
    c. the hair
    d. the liver
    e. the muscles

19. Which of the following processes are necessary for the early development of the human embryo?
    a. cell proliferation
    b. cell migration
    c. cell fusion
    d. differential growth
    e. all of the above

20. Teratoma
    a. is caused by teratogens
    b. is a tumor of the placenta
    c. secretes excess HCG
    d. is a remnant of the primitive streak and contains tissue from all germ layers
    e. all of the above are correct

ANSWERS TO REVIEW QUESTIONS

1. d	2. d	3. b	4. c	5. d
6. a	7. d	8. d	9. b	10. c
11. b	12. a	13. e	14. d	15. d
16. d	17. b	18. d	19. e	20. d

CHAPTER

7

TERATOGENESIS

Teratogenesis is the development of a defect in the embryo caused by any substance that interferes with normal development. These substances are called teratogens, and they act directly or indirectly on the developing embryo and may cause one or more abnormalities. The type and extent of the defect depend on: i) the specific kind of teratogen, ii) the amount or dose, iii) the duration of exposure, iv) the mode of action, v) the developmental stage of the embryo, vi) genetic predisposition and vii) maternal health. The mother's early history, such as childhood chronic illness and their treatment, may also alter the course of pregnancy and embryonic development. In humans, maternal metabolic diseases and abnormal anatomy of pelvic organs are known to harm the fetus. The permeability of the placenta to diverse compounds may also be a factor.

In general, the etiology of congenital anomalies can be divided into I) environmental, II) genetic and III) unknown or miscellaneous factors. This categorization is based on epidemiological classification. Due to new molecular biology, the interpretation of the etiology of congenital anomalies should not be made rigid. This may be the reason why more than two-thirds are classified as being of unknown etiology. However, there are certain indications that bring about concern for possible fetal congenital anomalies. Mother's history (e.g., advanced maternal age, previous miscarriages, stillbirths and babies born with congenital anomalies) and known genetic disorders in a family can indicate the possibility of certain anomalies in a growing fetus. Other indicators, such as maternal exposure to teratogens, abnormal fetal movements, abnormal heart rate, and abnormalities in the placenta and amniotic fluid, should also be considered. Most malformed embryos and fetuses (80–90%) are spontaneously aborted. This greatly reduces the number of malformed babies born alive. Although gross structural defects are easily detected at birth, cellular and molecular abnormalities, inborn errors of metabolism, mental retardation and physiological disturbances usually escape detection and manifest themselves later in infants and children.

I. ENVIRONMENTAL FACTORS

Some of the known teratogens in humans are: A) ionizing radiation, B) hyperthermia, C) mercury, D) polychlorinated biphenyls (PCBs), E) infections (cytomegalovirus, herpes virus, rubella virus, varicella zoster virus, toxoplasmosis, syphilis and AIDS), F) drugs, e.g., alcohol, anticonvulsants, cocaine, heroin, antithyroid drugs, folic acid antagonists, alkylating agents, streptomycin, tetracycline, thalidomide, diethylstilbestrol (DES), androgens, aminoglycosides and vitamin A derivatives, and G) maternal illness (diabetes mellitus, thyroid disease, lupus, myasthenia gravis, virilizing tumors, phenylketonuria, cretinism and fever). Uterine compression due to multiple gestation, uterine anomalies and oligohydramnios can also cause fetal malformation. Many other agents, like anesthetic gases, dioxin (agent orange), lead, lithium, penicillin, spermicides and diazepam (valium), have also been indicated as possible teratogens. Fetal growth may also be influenced by smoking, and by deficiencies of zinc, vitamin A or folic acid.

II. GENETIC FACTORS

A. Single-Gene or Mendelian Defects

Genes are units of heredity, located within the structure of DNA, that can be passed from generation to generation in predictable ratios. These are contributed by both parents, preserved intact and passed to children. Defects are caused by mutant genes. The causative gene may be located either on an autosome or the X chromosome. If these genes are expressed when present on only one chromosome of a pair, they are called dominant. If these genes are expressed when present on both chromosomes of the pair, they are referred to as recessive.

1. Autosomal Dominant Trait

This is usually characterized by some obvious anomaly in the individual who carries the abnormal allele. This abnormality may be present at birth or may develop later. This trait will usually appear in every generation in 50% of the offspring. The sex ratio of affected individuals is equal. The gene for this trait can be transmitted to children by either parent. It may involve structural protein abnormalities or may affect the control of the gene. The explanation of this trait, especially when there is variability in expression, is not as simple as implied above. One individual in a family may be severely affected and someone else mildly; some may even be apparently unaffected (phenotypically normal). This phenotypically normal individual may carry the abnormal allele or gene.

There are many variations in the expression of abnormal genes in terms of severity, time of onset and type of tissue affected. There may also be multiple, apparently unrelated, clinical manifestations of an abnormal gene. Furthermore, the mechanism causing these abnormalities is not fully understood. New mutations in genetic information occur on a regular basis. Examples of autosomal dominant trait are achondroplasia, sickle cell anemia, neurofibromatosis I, hereditary (Huntington's) chorea and Marfan's syndrome.

2. Autosomal Recessive Trait

These disorders are fully expressed clinically when two copies of abnormal genes, usually on two chromosomes of the same pair, are present. Occasionally, the absence of a normal gene on one chromosome with an abnormal gene on another may also be observed in this trait. The carriers of this trait usually appear normal because the clinical symptoms in these individuals are not seen unless they are exposed to stressed conditions. Individuals having both copies of abnormal genes generally manifest the abnormalities. Autosomal recessive traits may be manifested in some siblings, while other members of the family with the abnormal gene may become carriers. On average, one of four children of two carriers will be affected and two of four will be carriers. The parents of an affected person may have been related before marriage. The gender ratio in affected individuals is equal. However, in some diseases (e.g., polydactyly, imperforate anus, congenital heart disease, polycystic kidneys, intestinal anomalies), females may have corresponding different manifestations like vaginal atresia, and males will have urethral stenosis. In a specific family, the affected children may show similar manifestation and age of onset. Expression could also be variable within a family. Marriage between close relatives (consanguineous marriage) plays an important role in producing a homozygous individual. It increases the likelihood of both parents being carriers for relatively a rare autosomal recessive gene. The chance of manifestation of an abnormal gene is increased. Some examples of an autosomal recessive trait are cystic fibrosis, gangliosidosis, GM_2 (Tay–Sachs), camptomelic dysplasia and Meckel's syndrome (posterior encephalocele, polycystic kidney and liver, cleft palate and ambiguous genitalia).

3. X-Linked Dominant Trait

X-linked traits are those in which the abnormal gene is located on an X chromosome. The male has only one X chromosome; therefore, the abnormal gene located on it usually manifests clinically. As the male cannot contribute an X chromosome to its male offspring, male-to-male transmission of this trait does not occur. A woman transmits this trait to both males and females with equal frequency, and an affected male transmits to all his daughters. For these reasons, in families with an X-linked dominant trait there are twice as many affected females as males. Many conditions caused by this trait are lethal in the male but show a patchy involvement in the female. The explanation for this is that an abnormal gene in these individuals may be located on an inactive X chromosome and not be expressed fully. Examples of these are focal dermal hypoplasia, orofacial digital syndrome type I and incontinentia pigmenti. These are lethal for males, but females show some involvement of bone, skin and eyes. Females with this trait have an increased spontaneous abortion rate. This may represent the miscarriage of affected male fetuses. Examples of other X-linked dominant traits are vitamin D-resistant rickets and Alport's disease (renal failure, progressive neurosensory hearing loss and occasional ocular abnormalities).

4. X-Linked Recessive Trait

In this trait, the affected females usually do not manifest clinical disease but are carriers. The abnormal genes on an X chromosome are transmitted from fathers to their daughters and will be transmitted to 50% of affected daughters' sons. Examples of this trait are otopalatodigital dysplasia, hydrocephalus caused by type I aqueductal stenosis, FG syndrome (macrocephaly, prominent forehead, hypotonia, agenesis of corpus callosum and anal anomalies), hemophilia and (Duchenne's) muscular dystrophy.

It is suggested that single-gene mutations can affect the master genes that control the activity of many other subordinate genes.

B. Chromosomal Disorders

When a whole chromosome or part of a chromosome is missing, the karyotype of an individual is said to be abnormal. The presence of an extra chromosome or an extra part of a chromosome is also considered abnormal. In these disorders, there are always congenital anomalies involving many organs. In some instances, an individual may have a normal amount of genetic material but it may be abnormally arranged, such as by inversion or translocation. Chromosomal abnormalities may be caused by nondisjunction and anaphase lag during cell division (meiosis), which lead to formation of abnormal gametes. Loss, deletion, of part of any chromosome causes a variety of abnormalities. Chromosomal disorders are discussed in Chapters 1 (Cell Division) and 3 (Fertilization and Cleavage).

III. MISCELLANEOUS FACTORS

Approximately 15% of congenital anomalies are identified as environmental, 15% as genetic, while the etiology is unknown in 70% of congenital anomalies. These may represent the interaction of many different factors, some of which are called polygenic because they result from the additive effect of a number of abnormal genes. For instance, isolated clubfoot is considered to be caused by many environmental and genetic factors. Other anomalies considered multifactorial include isolated congenital heart disease, cleft lip with or without cleft palate, isolated cleft palate, dislocated hip, pyloric stenosis and isolated neural tube defects. These abnormalities when associated with others (multiple anomalies) are not considered to be multifactorial. The specific diagnosis is important because isolated anomalies respond differently to treatment than multiple anomalies. For instance, isolated clubfoot responds well to physical therapy treatment, but it worsens with use and physical therapy in diastrophic dysplasia (autosomal recessive trait) in which clubfoot is included. Some congenital defects appear unpredictably and only occur by chance. As there is no known history for these in a family, there is no known risk of their appearing in other family members. Embryological development passes through many critical threshold points. A precise event must occur at a precise time; otherwise, a structural anomaly will occur. The nature of many of the factors that must come together to carry the development over the threshold point is either not fully understood or is unknown. More analysis of normal development is needed to understand these anomalies. Approximately 2% of newborns show major congenital anomalies, and 15% show minor congenital anomalies at birth. The problem is severe and quite complex, but there are some helpful methods for evaluating the embryonic development.

A. Growth Factors

A number of growth factors have been isolated, purified and experimentally observed to have major effects on growing tissues. They appear to be specific for particular tissues. These factors may act as inducers for differentiation and growth promoters or be essential for cell division. They may act on the environment surrounding the cell or act intercellularly. Many factors have been observed in different concentrations at different times during normal development of a particular tissue or organ. These factors may act alone or in combination with other factors to promote tissue growth. A majority of growth factors are secreted by the embryonic tissue itself. Early differentiated cells produce growth factors to influence differentiation of late-growing cells. If the early cells fail to produce growth factor, they will either be abnormal or fail to grow.

B. Cytoplasmic Factors

Many proteins in the cytosol of the cell influence the functions of cell organelles. Inhibition or disturbance in the synthesis of these molecules may alter the functions of the cell organelles and the cell itself. Especially during embryonic development, this may lead to congenital anomalies. Cell organelles are involved in specific metabolic pathways. Proper functioning of organelles is necessary for normal metabolic pathways. Any alteration in these pathways leads to inborn errors of metabolism and may result in many congenital anomalies. Even single-enzyme defects can lead to major clinical manifestations. The organelles which are now known to be involved are 1) mitochondria, 2) lysosomes and 3) peroxisomes.

1. Mitochondrial Disorders

It was shown recently that mitochondria are involved in some of the inborn errors of metabolism. In general, mitochondria are involved in energy metabolism, propionic acid metabolism, the urea cycle and fatty acid oxidation (formation of ketone bodies). Any disorders in mitochondrial transport, the citric acid cycle, and utilization of substrate and respiratory chain or any other pathway may lead to major congenital abnormalities. Disturbance in oxidative phosphorylation leads to a) nuclear DNA mutation that manifests as infantile mitochondrial myopathies and cardiomyopathies; b) point mutation of mitochondrial DNA that manifests as optic neuropathy, myoclonic epilepsy, mitochondrial myopathy, encephalomyopathy and lactic acidosis; and c) mitochondrial DNA deletion or duplication that manifests as primary myopathy, myoclonic epilepsy, and marrow and pancreas syndrome. Severe respiratory chain defects are incompatible with intrauterine life. It is now believed that the mother (oocyte) contributes the entire complement of mitochondria to the developing embryo. Any mutation in mother's mitochondrial DNA will be passed to the offspring. As most of the mitochondrial disorders cause myopathies, it may also be involved in growth retardation.

The urea cycle is responsible for detoxification of ammonia. Mitochondrial and cytosolic enzymes are involved in this. Deficiency of the initial two enzymes (carbamyl phosphate synthetase I and ornithine carbamyl transferase OTC) in this cycle results in severe disorders. These deficiencies in neonates appear as hyperammonemia and coma. The deficiency involving OTC is an X-linked recessive trait. It is less severe in female children. Prenatal diagnosis of this deficiency can be done by intragenic DNA probes on fetal liver or biopsy of the chorionic villi. Maternal carrier status should be established before performing these tests.

Other disorders involving mitochondria are associated with pyruvate dehydrogenase (an enzyme required for complete oxidation of glucose and fatty acid and formation of acetyl coenzyme A), glutaryl-CoA dehydrogenase and β-hydroxyisobutaryl CoA decyclase deficiencies.

2. Lysosomal Disorders

Lysosomal hydrolases are capable of breaking down large molecules like glycoproteins, glycogen glycolipids and glycosaminoglycans (mucopolysaccharides). Deficiency of a single hydrolase will result in an inability to degrade macromolecules. Furthermore, the pathway from gene transcription to the final lysosomal hydrolase is a multistep process, and each step is susceptible to interference. The situation is further complicated by the addition of sidechains. The enzymes undergo cleavage before becoming active, and activator proteins must also be present in the lysosomes. Mutation in any of these will cause deficiency of hydrolase. Any fault in lysosomal membrane transport will lead to accumulation of hydrolases in this organelle and may lead to storage diseases.

The lesions produced by lysosomal storage disease may be very small, observable only by electron microscopy, or may manifest as growth retardation ascites, dysmorphisms or hydrops fetalis. The neuropathic lysosomal disorders exclusively affect the nervous system of the fetus, especially in the developing nervous system. The accumulation of gangliosides (glycosphingolipids) in the cell membrane of neurons leads to neuronal loss. Accumulations of many other substances have been observed in spleen, liver, bone marrow and amniotic fluid cells.

3. Peroxisomal Disorders

Peroxisomes (microbodies) are spherical membrane-bound organelles of various diameter (0.5 to 1.2 μm). The peroxisomes are involved in both the anabolic and catabolic pathways. Their matrices contain oxidases that oxidize a variety of substrates, leading to production of hydrogen peroxide. Although H_2O_2 is necessary in many cellular reactions, it is capable of killing microbes and is tolerated in only low concentrations by cells. After generation of H_2O_2, the catalase present in peroxisomes converts H_2O_2 to water and oxygen. Aside from other enzymes, these contain D and L oxidases, urate oxidases and hydroxyacid oxidases. Unlike mitochondria, where β-oxidation produces ATP, the oxidation in peroxisomes leads to formation of H_2O_2. The anabolic functions of peroxisomes include cholesterol, bile acid and plasmalogen synthesis. In addition to lipid metabolism, they are also involved in deamination of amino acids.

As the sperm does not contribute any membrane-bound organelles to conceptus, genetic defects of peroximal formation in the oocyte could be maternally transmitted. The most well known disorder associated with peroximal dysfunction is the cerebrohepatorenal (Zellweger's) syndrome. Peroxisomes are abundant in hepatocytes, renal tubular cells and macrophages. At birth, this syndrome presents with severe hypotonia, seizures, high forehead, malar hypoplasia, micrognathia and shallow supraorbital ridges. In liver there is a variable degree of fibrosis. Lipid lamellar bodies are seen in adrenals, liver, kidneys and Leydig cells. Dysplasia of the brain and renal tubular cysts are also present. In other generalized less severe peroximal disorders, some of these features are present, while some are missing. It is now thought that the primary defect is in the stability of the peroxisomal membrane, which may be caused by mutation in an integral protein of the membrane. Many other diseases, like rhizomelic chondrodysplasia punctata, X-linked adrenoleukodystrophy, hyperoxaluria, acyl-CoA oxidase deficiency and thiolase deficiency, are associated with peroximal dysfunction.

IV. INFECTIONS DURING EMBRYOGENESIS

The true extent of the effects of infections on the embryo and fetus is not known. Infections during very early development (first month) may cause spontaneous abortion, which may not be noticed. The effects of infections on implantation are also not known. It has been suggested that chronic infections of the endometrium may prevent implantation. It is now acknowledged that infections by rubella, mycoplasmas and toxoplasma are especially damaging in the first trimester of pregnancy and may lead to unrecognizable miscarriages.

A. Bacterial Infections

1. LISTERIA MONOCYTOGENES. This is a β-hemolytic organism. Infection in the mother may show influenza-like symptoms, but it may have a devastating effect on a fetus or newborn infant. Sepsis and pneumonia are predominant features. This infection has also been shown to be associated with prematurity, obstetric complications and low birth weight. It can cause microabscesses, septicemia and widespread granulomatous lesions in the fetus and newborn. Prompt treatment of an infected expectant mother may result in the birth of a normal child.

2. MYCOBACTERIUM TUBERCULOSIS. Congenital tuberculosis is rare. Maternal infection spreads to the fetus through umbilical vessels, and fetal liver becomes the principal site of infection. Infection of the lungs is less severe, but bones, lymph nodes, kidneys, spleen, skin and gastrointestinal tract are all involved.

3. NEISSERIA GONORRHEA. A two to five times increase in premature birth due to gonococcal infections has been reported. Ophthalmia neonatorum is the

most commonly known in neonates born to mothers with this infection. Most cases are treated by silver nitrate drops.

4. BORRELIA BURGDORFERI (LYME DISEASE). Rash, syndactyly, cortical blindness and intrauterine death have been reported in untreated maternal Lyme disease.

5. SYPHILIS. Young, promiscuous, intravenous drug-using mothers are high-risk individuals. Fetal infection can lead to abortion, neonatal death and premature delivery. Tissue damage results from an inflammatory response (after 16 weeks of gestation). Multiple organs are involved in congenital syphilis that may manifest as enlargement of liver and spleen, hydrops fetalis, hemolytic anemia, nephrosis, lymphadenopathy, osteochondritis, meningitis and intestinal obstruction. Newborns also show a rash due to cutaneous hematopoiesis. Other CNS involvement has also been reported.

6. MYCOPLASMAS. Mycoplasmas do not have rigid walls like other bacteria and are seen in variable shapes. Many mycoplasmas are common human flora, especially in the female genital tract. Ascent of these from the vaginal canal may infect the placenta and the fetus. Genital mycoplasmas are found associated with prolonged rupture of the fetal membrane, low birth weight, intrauterine growth retardation and preterm delivery. Congenital fetal malformation has not been associated with infection.

B. Viral Infections

The effects of viral infection during pregnancy on the embryo and fetus depend on many factors, including type, virulence and dose of infectious virus, the gestational age of the conceptus and the mode of transmission of the agent. Immunological competence and genetic factors are also important.

1. Rubella Virus (Togaviridae)

Fetal rubella infection could occur following maternal infection at any stage of gestation. The outcome of fetal infection varies with the age of the fetus. If the mothers were infected in the first trimester, about 80% of children born show congenital heart disease and deafness. The risk decreases to about 50% when fetal infection was in weeks 13 to 16, and further decreases to about 25% by the end of the second trimester. The classical clinical manifestations with this infection are: 1) congenital heart disease, 2) sensory neural deafness and 3) cataracts. Intrauterine growth retardation is also associated with most cases. In addition to these, almost any organ system may be involved. There is an extensive list of malformations that have been reported to be caused by this infection. It has been reported that this virus primarily damages the endothelial cells, which could act as infectious emboli. Maternal viremia leads to placental infection, which in turn leads to fetal viremia. Spontaneous abortion in about 5% of the infected cases was reported.

2. Human Immunodeficiency Virus (HIV)

Human immunodeficiency virus is a retrovirus. Retroviruses have the ability to convert single-stranded viral RNA to DNA with virus-coded reverse transcriptase.

This complementary DNA strand is integrated into the host cell DNA as proviral DNA, which can interfere with normal cellular functions. The integrated retrovirus then multiplies with host cell DNA during cell division. Infection of an embryonal cell with retroviruses has the potential to cause congenital malformations. Although many retroviruses have been associated with a variety of diseases, HIV-1 is important in embryogenesis. Infection with HIV-1 affects the fetus and leads to abortion, stillbirth, or congenital anomalies. The course of congenital HIV infection is different from that of adult infection. Some of the reported clinical features of congenital HIV infection include generalized lymphadenopathy, enlargement of spleen and liver, low birth weight, digital clubbing, opportunistic infections, pyrogenic infections, neurological abnormalities and failure to thrive. Many other features have been reported in HIV-infected children born to drug-abusing mothers.

3. Herpes Simplex Virus

Herpes virus particles can potentially reach the embryo or fetus by an ascending infection through the cervix, or spread through blood. Herpes simplex transplacental transmission may occur by either viremia of the mother or intraaxonal transport of virus from infected dorsal root ganglia to the endometrium and placenta. Congenital infection with this virus may result in microcephaly, severe mental retardation, intracranial calcification, microphthalmia, retinal dysplasia, patent ductus arteriosus and short digits.

4. Cytomegalovirus

Gestational infection with this virus is associated with a wide variety of placental changes. Transmission of virus may occur during any stage of pregnancy. Infection during organogenesis leads to profound congenital abnormalities. The clinical features include microcephaly with encephalitis, intracranial calcification, sensory neural deafness, hepatitis, hepatosplenomegaly and petechiae.

5. Varicella Zoster Virus (Chicken Pox)

Maternal infection early in pregnancy affects the fetus. Spontaneous abortion and stillbirths have been reported in mothers with severe varicella pneumonia. Congenital malformations resulting from this include scar-like skin lesions, bare (denuded) skin, atrophy and scarring of limbs, microcephaly, chorioretinitis, neurological herpes zoster and renal anomalies.

6. Epstein–Barr Virus

Epstein–Barr virus can infect human B-lymphocytes and cord blood lymphocytes. Intrauterine infection might occur by a transfer of infected B-lymphocytes across the placenta. Congenital malformations have been reported in neonates born to mothers who developed mononucleosis during pregnancy. The congenital anomalies reported include hypotonia, cataracts, thrombocytopenia, monocytosis, proteinuria, persistent atypical lymphocytosis and CNS deficits.

7. Influenza Virus

No clear-cut evidence exists for congenital malformations caused by influenza virus. Development of viremia is rare in infected mothers. Furthermore, transplacental acquired maternal antiinfluenza IgG in the fetus protects or modifies gestational infection by a related strain.

C. Mycoses

Candida species are human commensal flora, found especially in the female genital tract. Although candidal vaginitis is common during pregnancy, systemic maternal disease is rare. Involvement of the umbilical cord and the fetal side of placenta has been reported. Fetal infection has been associated with middle ear and gastrointestinal tract infections, premature delivery and prolonged rupture of membranes and subsequent fetal death. Many other fungal infections have been reported, but congenital malformations have not been conclusively documented.

D. Parasitic Infection

In underdeveloped and tropical countries, numerous parasites contribute to prenatal diseases. In the Western world, toxoplasmosis and trichomoniasis are recognized as significant diseases in pregnancy.

1. Toxoplasma Gondii

This is the most extensively studied protozoan that causes disease during pregnancy. Felines are primary hosts; humans and other mammals are its intermediate hosts. After ingestion, its oocytes release spores that infect intestinal epithelial cells. In its unicellular form, it can spread to any tissue. This unicellular form can initiate intracellular or extracellular disease or may become latent or form cysts in the cell. It also can be transmitted by eating poorly processed meat of another intermediate host (cow or pig). Transfusion of contaminated blood products is another possible method of transmission.

The organisms disseminate into many tissues after the mother develops parasitosis. More commonly, a parasitosis is found in the placenta, where villous edema and fibrosis have been seen. Following placental infection, parasites can disseminate in the fetal blood and other fetal tissues. Congenital manifestations present as a triad: 1) hydrocephalus, 2) chorioretinitis and 3) intracranial calcifications. Infection has also been reported in the heart, kidneys, lungs, adrenal cortex, pancreas, gonads, skeletal muscles and brain. Central nervous system disease is manifested as necrosis of cortex and basal ganglia. Vascular congestion and vasculitis are commonly found in necrotic areas. Cortical calcification may follow necrosis. The meninges and spinal cord may be involved. The initial site of infection in the eyes is the retina. Following infection, the inflammatory response may disrupt the retinal layer and infection can pass to other layers. Fetal infection during the first or second trimester will result in severe consequences. At birth, the baby may have fever, cutaneous rash, intrauterine growth retardation, jaundice, anemia, visual defects, enlargement of liver and spleen, lymphadenopathy, microcephaly and seizures.

Some manifestations of the disease, like mental retardation, may be delayed and observed later in early childhood. Antenatal therapy with antibiotics appears to reduce the severity of symptoms.

2. Trichomonas

Trichomonas vaginalis is a common pathogen found in the female genital tract. Only low birth weight has been associated with this infection.

3. Malaria

Maternal disease is more severe with *Plasmodium falciparum*. This parasite primarily infects placental structures. Focal necrosis, loss of microvilli and thickening of basement membrane have been associated with this infection. Placental dysfunction may lead to low birth weight, spontaneous abortion and stillbirth. Intrauterine acquired infection of the fetus manifests at birth as anemia, jaundice, liver and spleen enlargement.

4. Others

In different parts of the world many other parasites have been associated with fetal involvement. Schistosomiasis in Africa, Asia and South America, and Trypanosomiasis in Africa and South America have been associated with congenital malformations.

CHAPTER

8

SKELETOMUSCULAR SYSTEM

The skeletomuscular tissues arise from the mesenchyme, which contains mesodermal and neural crest cells. In this chapter, we discuss the development of two major subdivisions of the skeletal system (the axial skeleton of the trunk and appendicular skeleton of the limbs) and development of limbs.

As mesodermal tissue becomes organized into somites, the individual vertebrae are derived from the sclerotomes of two adjacent somites. Ribs are derived from the lateral mesenchymal condensations. The skeletal muscles of the trunk differentiate from the myotomes of the somites.

The upper limb buds appear on day 24 as small bulges at the lateral body wall; the lower limb buds appear on day 26. Each limb bud is made up of a mesenchymal core covered by ectoderm, the apical ectodermal ridge that induces differentiation and growth of the limbs. The upper limbs rotate 90° laterally, and the lower limbs rotate 90° medially. The visceral musculature differentiates from the splanchnic pleura; cardiac muscle develops from the mesenchyme of the cardiogenic plate.

The skeletomuscular components are derived from the mesenchyme. Most bones first appear as mesenchymal condensations, which either develop into membranes and undergo intramembranous ossification and form the flat bones, or undergo chondrification to give rise to hyaline cartilage models that in turn are replaced by bones under endochondral ossification. The development of axial skeleton and appendicular skeleton and related musculature is discussed here. Formation of the skull is discussed along with craniofacial development (Chapter 12).

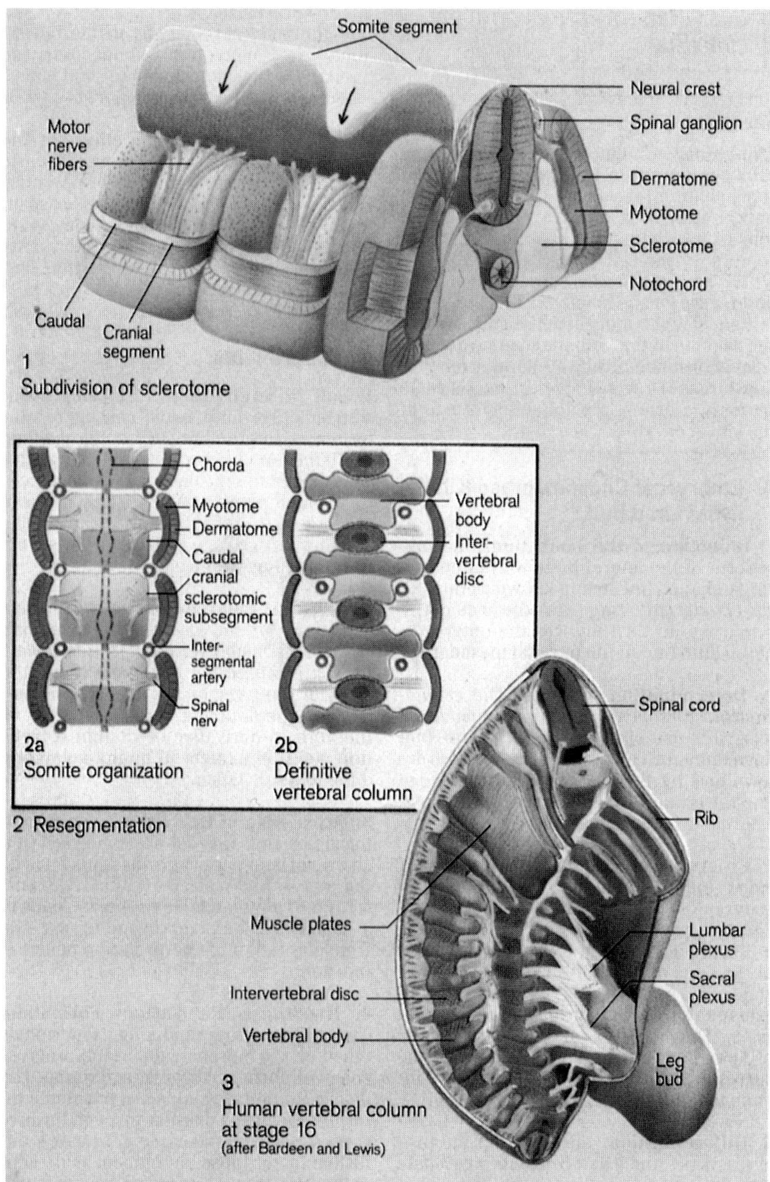

Figure 8.1

Resegmentation of axial organs (see Color Plate 8.1). Reproduced with permission from Thieme Medical Publishers Inc., New York, 1995, *Color Atlas of Embryology*, Ulrich Drews, Chapter 3: General Embryology.

I. DIFFERENTIATION OF SOMITES

Each somite shows three distinct regions: 1) a paraxial plate on each side of the notochord, 2) an intermediate plate, which will give rise to most of the urogenital organs, and 3) a lateral plate, which will form the lining of the intraembryonic coelom (Chapter 6, Fig. 6.2). In the paraxial plate of each somite, a central cavity develops that is now surrounded by pseudostratified epithelium. Soon the cavity of somites becomes occupied by loose core cells. These core cells and the medioventral wall cells form a distinct mass of loosely connected cells, the sclerotomes. The cells of dorsolateral wall give rise to dermomyotomes. The sclerotomes migrate toward the notochord and develop a neural tube. The dermomyotomes become arranged into an external (lateral) dermotome layer and internal (medial) myotome layers. The sclerotomes develop into the vertebral column and ribs. The myotomes give rise to the body wall and back muscles; the dermotomes give rise to the dermis of the skin (Fig. 8.1).

II. THE SCLEROTOMES

FORMATION OF THE VERTEBRAL COLUMN. During the fourth week of gestation, cells from sclerotomes migrate around the notochord and neural tube. Chondroitin sulfate-like molecules secreted by the notochord appear to induce formation of vertebral body cartilages, whereas the formation of neural arches is induced by substances secreted by the neural tube. Growth factors such as transforming growth factor beta (TGF_{β}), fibroblast growth factor (FGF) and the TGF_{β}-like molecule activin-β are shown to induce mesodermal formation. Some of these are suggested to be gene activators.

A. Vertebral Bodies

After its migration around the notochord, each sclerotome segment appears as an upper (cephalic) loose zone and a lower (caudal) dense zone of cells. The segmental nerve and artery traverse the loose cephalic zone of the sclerotome. This helps to keep the loose and dense zones separated. With further development, the cephalic loose zone of each sclerotome fuses with the caudal dense zone of the upper segment. The cephalic zone of the number one segment fuses with the caudal zone of the fourth occipital sclerotome to form the condylar region of the occipital bone; the cephalic zone of the second segmental sclerotome fuses with the dense zone of the first segment to give rise to the centrum of the first cervical vertebra. Thus, the centra of all vertebrae are formed by the fusion of the cephalic and caudal zones from two different segmental sclerotomes. This guarantees the periodic pattern of the vertebrae and associated nerves and blood vessels (Fig 8.1).

B. Neural Arches

Zonation of sclerotome segments also occurs in the dorsally migrating cells, but the vertical shift is not well defined in these segments. The caudal dense zone develops into neural arches. It is also suggested that neural processes from the dense zone of each sclerotome segment extend dorsally on each side of the neural tube to form neural arches. In any case, the left and right neural arches later come together dorsal to the neural tube and unite to complete the neural arch. Segmentation into neural arches depends on the presence of developing spinal ganglia.

C. Intervertebral Disks

Fibrous intervertebral disks develop between the vertebral centrum largely from the dense zone of the sclerotomes. The part of the notochord that is incorporated within the centrum becomes the mucoid streak, and later is not recognizable. The notochord between the centrum forms a cellular aggregation as a core of each disk. Later, the notochord expands and blends with the surrounding tissue; the nucleus pulposus of each disk is formed from the combination of both tissues (Fig. 8.2). The peripheral tissue transforms into fibrocartilaginous annulus fibrosus. Notochordal cells then gradually are replaced by sclerotomal cells and completely disappear at about 10 years of age.

Note. Prenatal movements are essential for the proper development of the vertebral column and other skeletomuscular components.

D. The Ribs

The costal processes, which are the lateral mesenchymal condensations of sclerotomal cells, develop in continuation with the dense zone of the sclerotome, and hence become associated with the developing neural arches and intervertebral disks. These costal segments probably develop at all the levels but remain short except in the thoracic region. In the thoracic region, the distal part of the costal processes grows ventrally to form ribs, which come into contact with sternal condensation. Chondrification of ribs starts at about 45 days, and at the end of the embryonic period there is a well-defined cartilaginous rib cage. Each rib at this time also shows a center of ossification at its angles. The ventral (distal) tips of the costal processes remain cartilaginous and give rise to costal cartilages. Outside the thorax, the costal processes give rise to the lateral components of the vertebrae.

E. The Sternum

Two mesenchymal condensations appear near the median plane of the ventral body wall of the thoracic region. At 7 weeks, chondrification occurs in them, and bilateral cartilaginous sternal bars form. The sternal bars become continuous with seven cranial costal cartilages of ribs and with the clavicle condensation. The two bars approach each other and fuse with each other starting cranially. The manubrium forms first, followed by the body and ziphoid processes. Fusion is complete at the end of the embryonic period.

III. THE DERMOTOMES

The dermotomes give rise to the dermis of the neck, the back and trunk. The dermis in other areas is derived from somatopleura of the lateral plate.

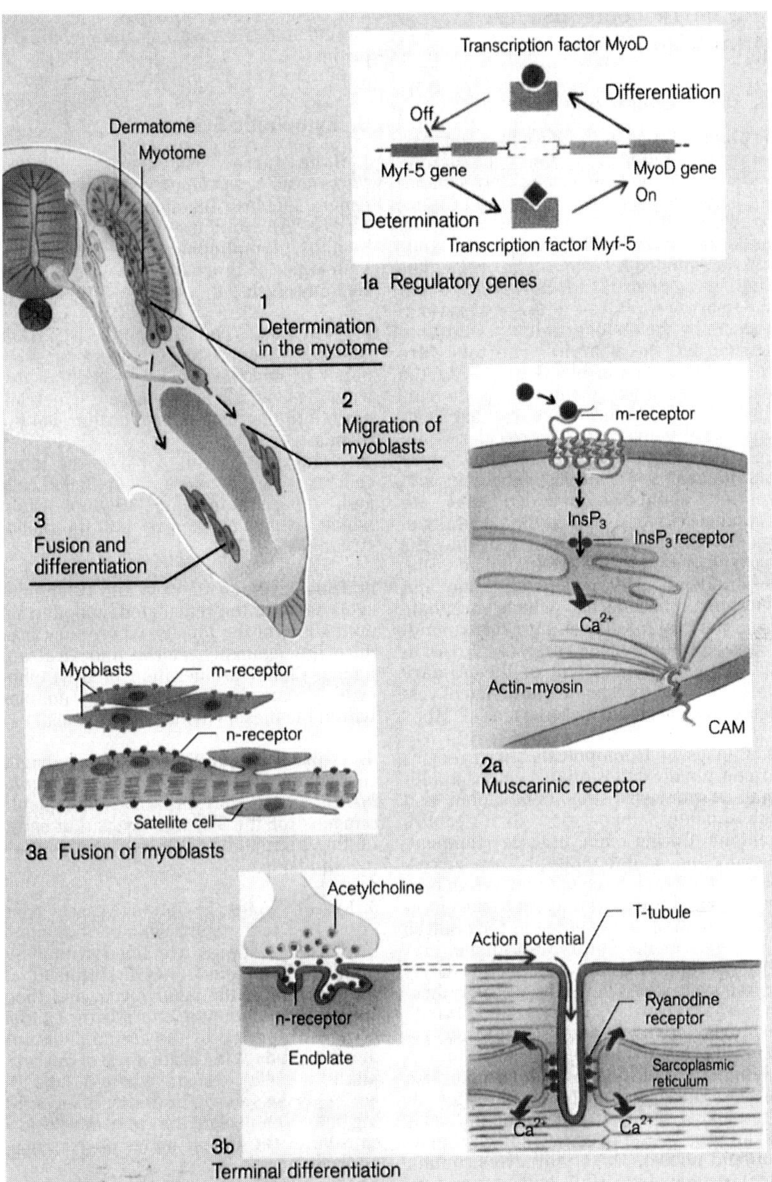

Figure 8.2

Skeletal musculature (see Color Plate 8.2). Reproduced with permission from Thieme Medical Publishers Inc., New York, 1995, *Color Atlas of Embryology*, Ulrich Drews, Chapter 3: General Embryology.

IV. THE MYOTOMES (Fig. 8.3)

In the human, the cells from myotomes aggregate into two distinct masses: the epimere dorsally and the hypomere ventrally. At this stage the cells in these masses resemble fibroblasts. A gene on chromosome 11 transfers fibroblasts to myoblasts.

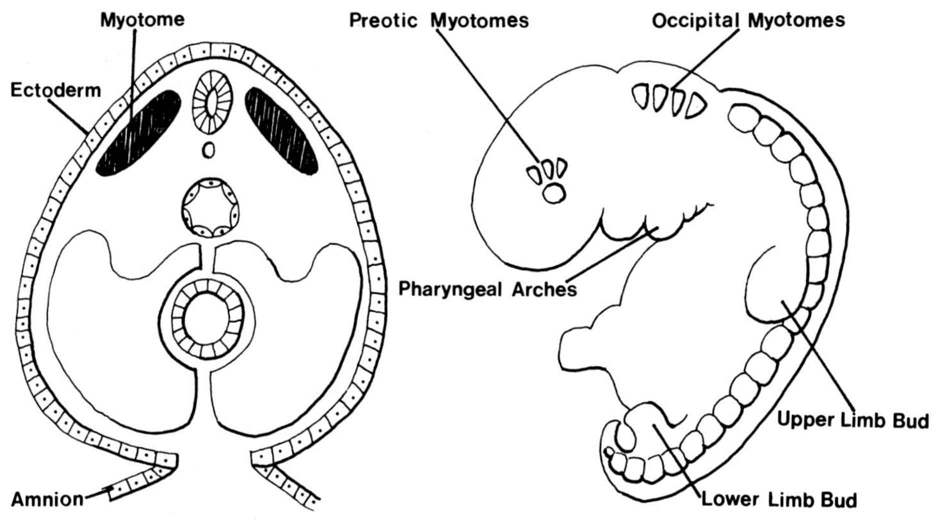

Figure 8.3
The myotomes.

A. The Epimeres

The dorsally located epimeres give rise to the deep muscles of the back, including the erector spinae and transverse spinalis muscle (Fig. 8.4).

B. The Hypomeres

The hypomere mass, growing laterally and ventrally, splits into three sheets. In the cervical region it gives rise to scalene muscles. In the thoracic region it gives rise to intercostal muscles. In the abdominal region it gives rise to oblique muscles. In the lumbar region it gives rise to quadratus lumborum.

C. The Median Aggregation

The myoblasts aggregate in the middle of the ventral body wall. Some authors suggest that these cells originate from the somatopleura of the lateral plate of the somites, while others believe that they are part of the hypomeres. This median mass differentiates into:

1. The infrahyoid muscles and geniohyoid in the cervical region.
2. The sternalis (which disappears) in the thoracic region.
3. The rectus abdominis in the abdominal region.
4. The pyramidalis (which is seen in 20% of cadavers) in the pelvic region.

It is suggested that the muscle of the body wall also receives some myoblasts from the somatopleura of the lateral plates. Other skeletal muscles will be discussed with the development of limbs and development of the face, as they come from

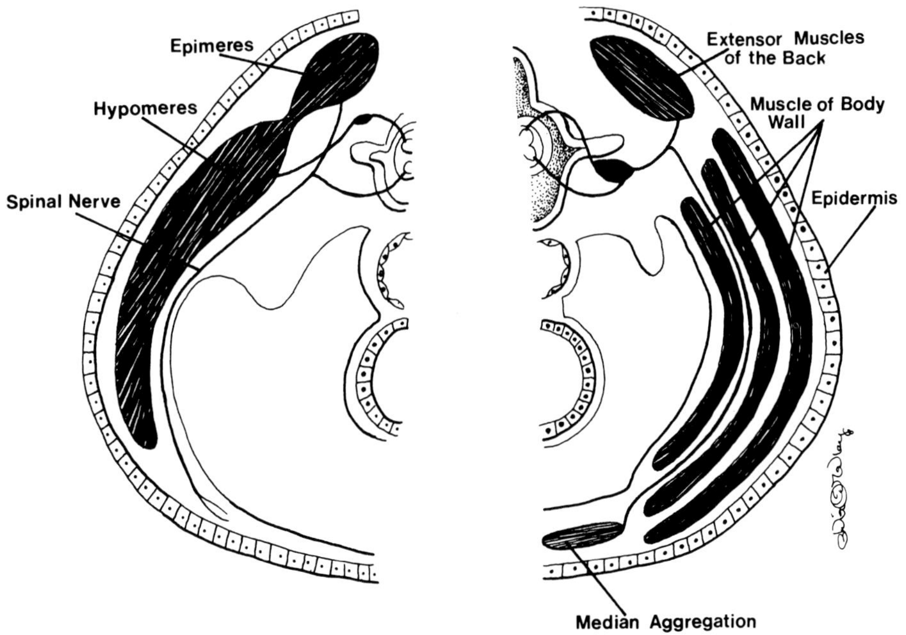

Figure 8.4
Migration and division of a myotome to form the muscles of the body wall.

different sources. Despite their differences in origin, further differentiation of the myoblast is similar. Mononucleated myoblasts in skeletal muscle divide quickly and then fuse to make multinucleated cells. Some nuclei move to the periphery of the syncytium, and striations appear. At 18 weeks, most of the muscle fibers are multinucleated as in adults. Some of the myoblasts remain mononucleated satellite cells. Although early differentiation of muscle does not need innervation, muscle maintenance and function require a nerve supply. Fetal movements are essential for proper development of the muscular system. It is generally agreed that no new muscle fibers are added in the skeletal muscles after birth. Postnatal growth is caused by an increase in size of the preexisting muscle fibers.

V. THE LIMBS

A. The Limb Buds (Fig. 8.5)

The mesodermal cells of a limb bud are derived from the somatopleura (body wall) and paraxial plate (the somite). Limb development is influenced by a series of inductive interactions. In response to a signal from the somites, the cells in the somatopleura proliferate in the apical region of the future limbs. The ectoderm covering the apical region is induced by the underlying mesoderm to become thick (stratified). This ectodermal thickening is known as the apical ectodermal ridge and is an essential inductor of further limb development. It is shown experimentally

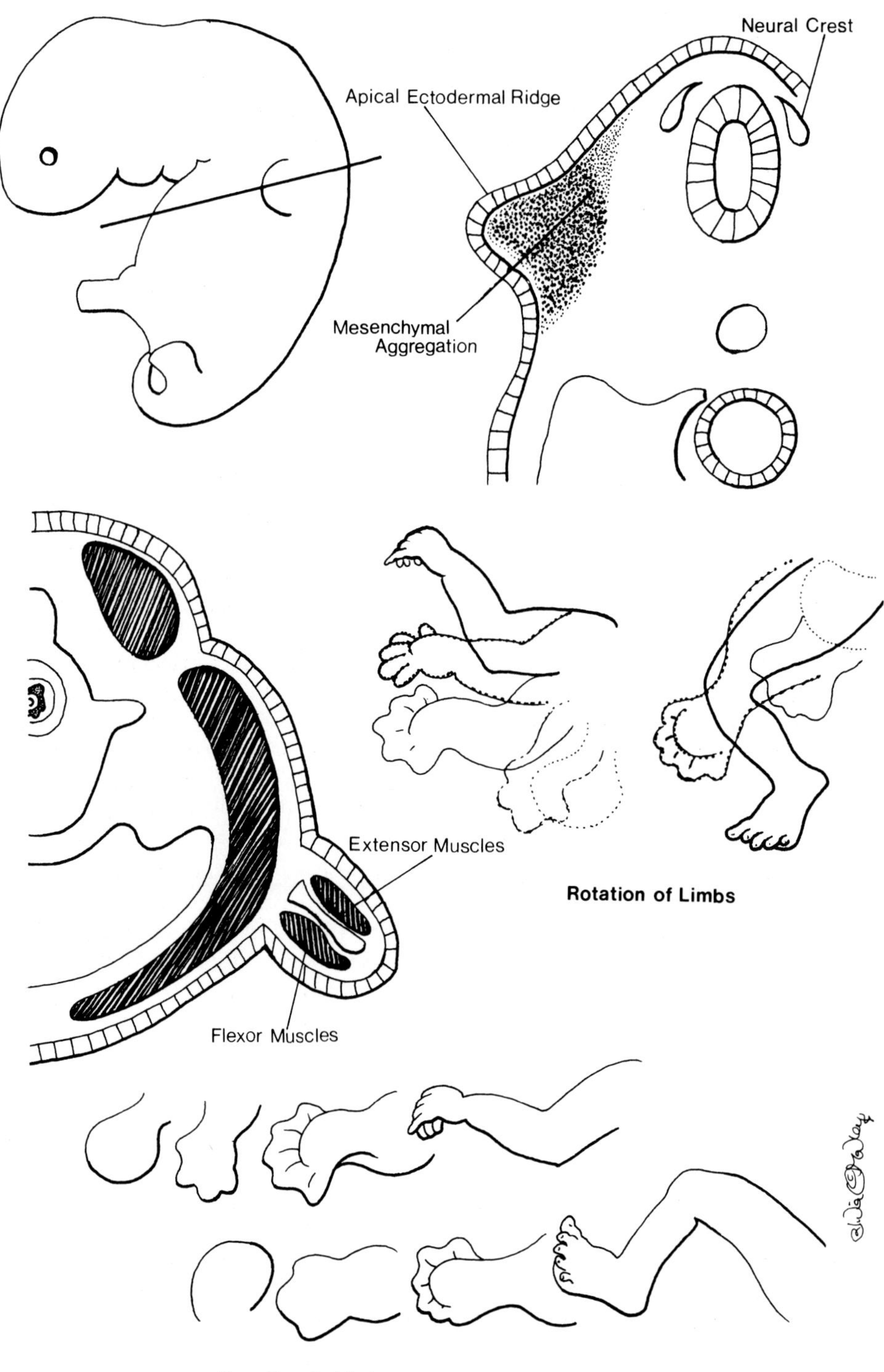

Figure 8.5
Different stages in the development of limbs.

that removal of the ridge results in limbless animals. The mesodermal cores covered by apical ectodermal ridges form the limb buds. The upper limb buds appear on the 24th day, 2 days earlier than the lower limb bud. The amount of time the mesodermal cells spend under the influence of the apical ridge in part determines which segment of limb will be formed from it. For instance, the late-formed tip of the elongating limb bud gives rise to the distal segment of the limb, whereas the early formed mesoderm at the base of the limb bud gives rise to the proximal segment of the limb. The apical ridge also affects normal programmed cell death and participates in the maintenance of the vasculature of the limbs. In the human embryo, skeletal development is followed by ingrowth of blood vessels, which is followed by ingrowth of nerves, which is immediately followed by muscular differentiation.

B. Elongation of Limbs and Skeletal Elements

Elongation of the limbs takes place from the fifth to eighth weeks. The skeletal elements first appear as mesenchymal condensation, which undergoes chondrification and then ossification. At about the 29th day, the upper limb bud appears flattened and paddle-shaped; at 32 days, the arm and shoulder region can be distinguished. The distal ends of limb buds show hand and foot plates. At 37 days, the forearm, carpal region and digital plates are formed in the upper limb; the thigh, leg and foot are distinct in the lower limb. At 38 days, the digital rays are seen. At 44 days, the digital rays deepen through programmed cell death, and the toe rays appear. At this time, the limbs are divided into upper (arm, thigh) and lower (forearm, leg) parts by horizontal flexion. There is one mesenchymal condensation in each upper part and two condensations in each lower part for long bones. Elbows and knees are obvious. By 52 days, due to further programmed cell death, the fingers and toes are separated. At 56 days, all regions of the upper and lower limbs are well defined. At zones between the different parts of the limbs, the mesoderm does not undergo chondrification and develops into joint elements. During fetal development, the upper limbs rotate laterally 90° on their longitudinal axes, so that the elbows point backward; the lower limbs rotate medially 90° on their longitudinal axes, so that the knees point forward. At the time of birth, most of the primary centers of ossification for limbs are present.

C. Vasculature

Initially, an axial artery develops in each limb bud along the central axes of the limb. In the upper limb, it gains connection with the seventh thoracic intersegmental artery; in the lower limbs, it joins the fifth lumbar intersegmental artery. In the upper limb, the axial artery forms the subclavian, axillary and brachial arteries. In the lower limb, most of the axial artery degenerates. A new branch from the fifth lumbar arises as external iliac and supplies most of the leg. The remnants of the axial artery are the inferior gluteal and sciatic arteries.

D. Innervation

The limbs are innervated by the branches of the ventral primary rami of the spinal nerves, C_5 to T_1 (upper limbs) and L_4 to S_3 (lower limbs) (Fig. 8.6).

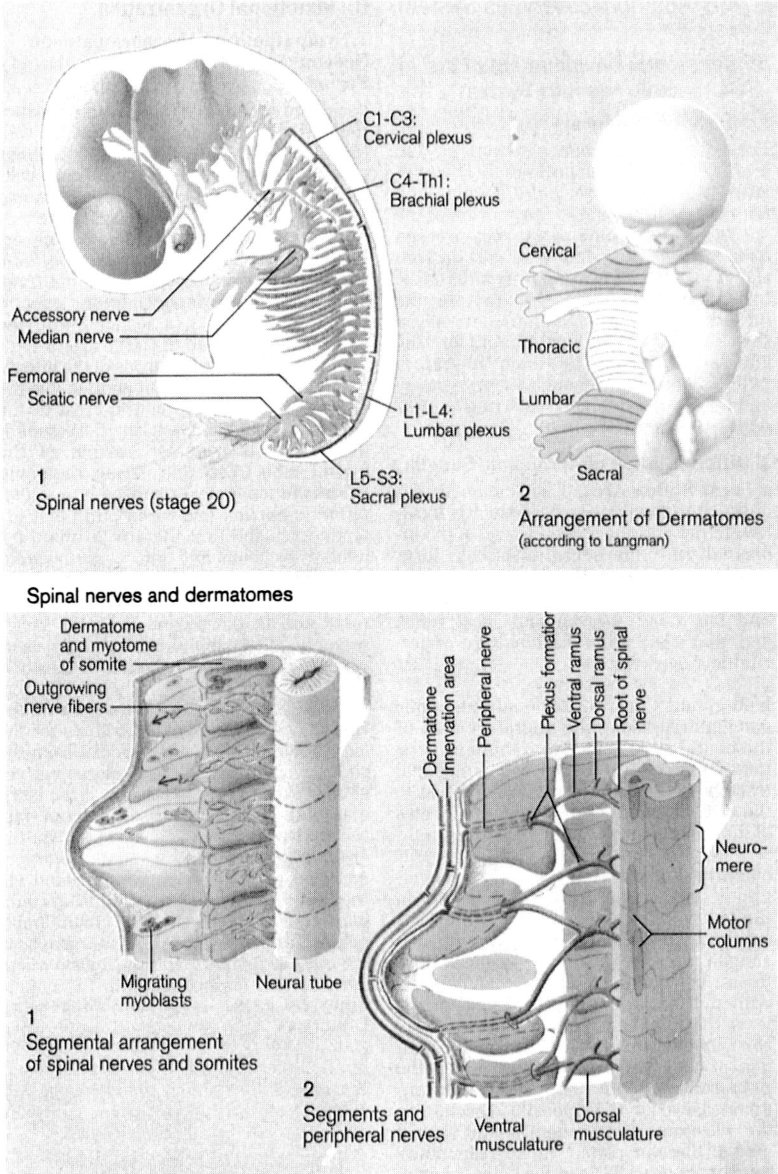

Figure 8.6

Dermatomes (see Color Plate 8.6). Reproduced with permission from Thieme Medical Publishers Inc., New York, 1995, *Color Atlas of Embryology,* Ulrich Drews, Chapter 4: Nervous System.

E. Musculature

The mesenchymal cells from somites migrate into the limb buds. The differentiating myoblasts aggregate into the dorsal and ventral mass in relation to developing bone models in the limb bud. The dorsal mass gives rise to the extensors and suppinators of the upper limbs, and the extensors and abductors of the lower limbs. These become innervated by posterior divisions of the primary ventral rami, forming brachial and lumbosacral plexus. The ventral mass gives rise to the flexors and pronators of the upper limb and the flexors and adductors of lower limbs (Fig. 8.2).

VI. VISCERAL MUSCULATURE

Smooth muscle differentiates from the splanchnic pleura of the lateral plate, which surrounds the primitive gut and derivatives. Myoblasts do not fuse to form syncytium but become arranged muscle sheets. The smooth muscle of blood vessels develops with endothelium from local mesoderm.

VII. CARDIAC MUSCLE

The cardiac muscle develops from the cardiogenic plate. The differentiation of cardiac muscle as the myocardial mantle starts at the third week. Myosin has been detected in early (precardiac) muscle fibers of chick and mouse. The early fibers divide by mitosis, develop myofibrils and remain mononucleated. Soon after, cells join with each other by intercalated disks. Conducting fibers can be detected at 4½ weeks, and at 8 weeks the conducting system is well developed.

VIII. CONGENITAL MALFORMATIONS

A. Vertebral Anomalies

Vertebral anomalies vary. Many are caused by derangement of vertebral models. Some that were classified as congenital are now considered to be postnatal in origin. The most frequent congenital anomalies associated with vertebrae are (Fig. 8.7):

1. Spina Bifida

Failure of the neural tube to close disrupts closure of the neural arches. The arches remain underdeveloped and fail to fuse along the dorsal midline. In mild cases, only the dura and arachnoid may protrude from the vertebral canal in the affected region, causing a meningocele. In other cases, neural tissue along with meninges may protrude, resulting in meningomyelocele. Spina bifida is usually localized in the cranial or lumbosacral region. The portion of spinal cord and spinal nerve involved in the meningomyelocele fail to develop properly, resulting in malfunction of the organs and muscles innervated by these nerves. In a very mild condition, where only one vertebra is involved and the contents of vertebral column do not protrude, development of the neural tube and spinal nerves is normal. This condition is then known as spina bifida occulta. Numerous studies have shown that, the higher the dose of folic acid given a mother, the lower the percentage of babies born with spina bifida and neural tube defects.

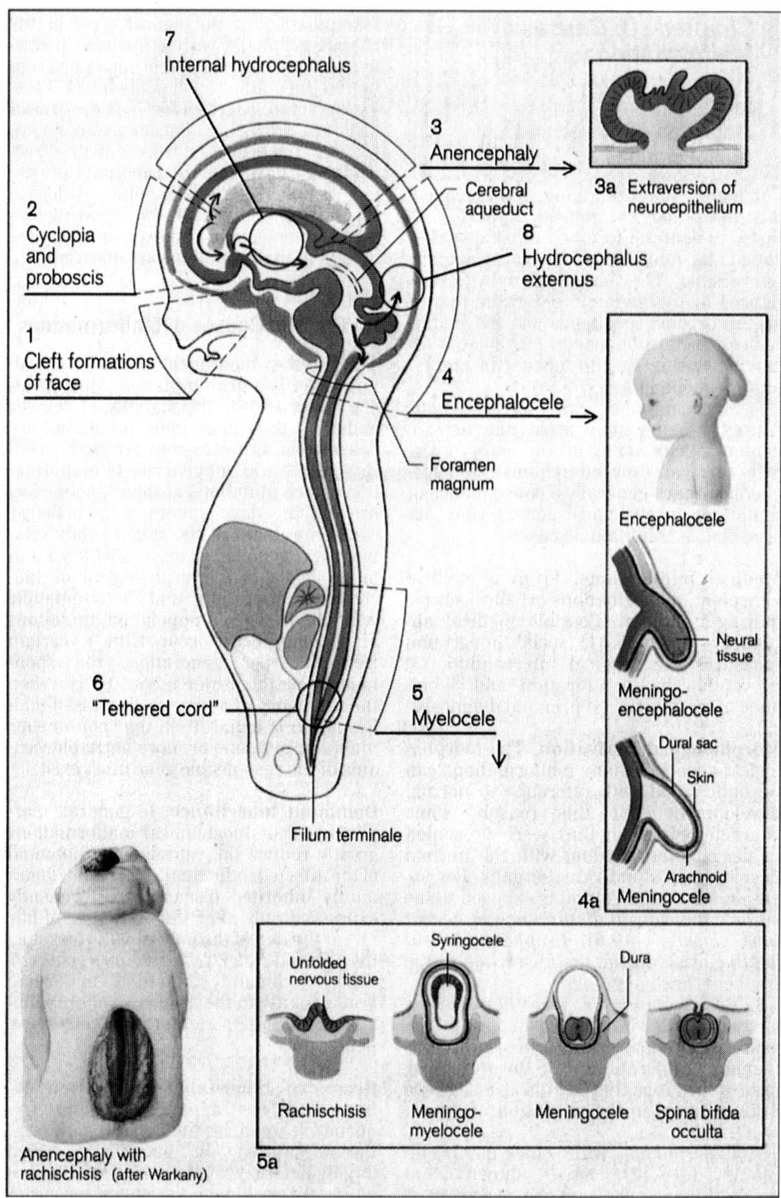

Figure 8.7

Malformations (see Color Plate 8.7). Reproduced with permission from Thieme Medical Publishers Inc., New York, 1995, *Color Atlas of Embryology,* Ulrich Drews, Chapter 9: Head.

2. Scoliosis

Lateral curvature (deviation) of the vertebral column may be caused by hemivertebrae or wedge vertebrae. A hemivertebra appears as a wedge between two other vertebrae. It is caused by a failure in the formation of one of the two centers of ossification in the body of the vertebra. It may also be an absence of chondrification of that portion. Lateral deviation of the notochord may cause abnormal induction. In any case, half of the body of the affected vertebra does not develop. Rarely, a wedge vertebra is located dorsally, resulting in dorsal deviation of the vertebral column, causing kyphosis. Both centers of ossification fail to develop, leading to absence of a vertebral body. Kyphosis is often associated with abnormalities of ribs.

3. Block Vertebra

This is the result of failure of separation between two or more vertebral bodies. It usually occurs in the lumbar region. It may arise from complete chondrification of intervertebral zones.

4. Congenital Brevicollis (Klippel–Feil Syndrome)

This involves shortness of the neck caused by nonsegmentation of two or more cervical vertebrae. It is associated with a low hairline and restricted head and neck movement. The head appears to rest on the shoulders. About 65% of the cases are female. Abnormal segmentation of the mesenchyme in the dense and loose zones in the cervical region results in short and fused cervical vertebrae. Notochord defects have also been implicated. Both environmental and genetic factors play a role in this condition.

B. Rib Anomalies

1. Accessory Rib

An extra rib usually found in the cervical or lumbar region may be unilateral, bilateral, complete or incomplete. In most cases, it is asymptomatic, but a cervical rib may cause neurovascular compression by pressing on the lower trunk of the brachial plexus and the subclavian artery. Symptoms generally occur later in life.

2. Short-Rib Polydactyly Syndrome

The heterogeneous short-rib polydactyly syndrome is usually lethal in the prenatal period. It is associated with chondroplasia, is autosomal recessive and is characterized by shortening ribs, which may profoundly restrict thoracic size and result in hypoplasia of the lungs. This leads to the death of live-born infants. Aside from short ribs and polydactyly, short extremities, cystic renal dysplasia, heart defects, urogenital tract anomalies, pancreatic fibrosis and GI tract anomalies have been reported. The basic cause appears to be impaired chondrocyte proliferation, which leads to irregular and poor columnization and results in abnormal bone formation.

3. Pigeon Chest

This results from congenital overgrowth of ribs, which was caused by enlargement of the right ventricle in congenital heart disease.

4. Funnel Chest

In this condition, there is depression of the lower end of the sternum and the lower costal cartilages, and the small central tendon of the diaphragm exerts an excessive pull upon the lower end of the sternum.

C. Limb Anomalies

The main causes involved in malformation of limbs are: 1) arrested growth of components, 2) failure of differentiation of components into primordia, 3) undergrowth of components, 4) overgrowth of primordia, 5) duplication of components, 6) localized defects caused by other developing structures and 7) generalized skeletal abnormalities that could involve limbs. These may manifest as missing an entire limb (amelia) or a part of a limb (meromelia, hemimelia) or a very short limb having all its components (micromelia). In a duplication defect, a supernumerary limb component is present. The term phocomelia is used when only hands or feet are present, and these seem to be directly attached to the trunk. The major factors causing anomalies of the limbs are: 1) a genetic factor (dominant or recessive trait), 2) chromosomal abnormalities, 3) environmental teratogens and 4) the intrauterine environment.

1. Genetic Factors

These defects often run in families.

a. Polydactyly
This is the presence of extra digits, which usually are incompletely formed. It is caused by formation of extra digital rays. After interdigital cellular death occurs, an extra segment forms. This shows an autosomal dominant, autosomal recessive or X-linked association. Polydactyly in many instances is part of a more general condition like Biedl–Bardet syndrome, where mental deficiency is combined with obesity. It may also be associated with intraatrial septal defect or may be present with short limbs and small stature.

b. Syndactyly
Fusion of the digits is caused by the persistence of tissue in the digital rays. It is inherited as a dominant or recessive trait. It could show as a skin web between digits or as nonseparation of bones in the digits. It occurs more often in the foot than the hand.

c. Cleft hand or foot (lobster claw)
One or more of the central digits are missing. The remaining digits remain partially or completely fused. There is extensive programmed cell death in the affected region.

d. Clubfoot (taliped equinovarus)
Although hereditary factors have been suggested, this condition may result from abnormal positioning of the foot in utero. Uterine pressure due to oligohydramnios may also cause this condition. The foot may be plantar flexed, inverted and adducted. It is more common in males.

e. Congenital dislocation of the hip
This may result from several factors:

i. generalized joint laxity, which may be inherited as a dominant trait,

ii. joint laxity caused by hormonal factors,

iii. underdevelopment of acetabulum,

iv. breech position of fetus may in turn affect the development of the acetabulum (more frequent in females and characterized as flat acetabulum).

f. Achondroplasia
This is the most common cause of dwarfism. The condition is autosomal dominant. The infants have a large head, a flat nasal bridge, a small chest, a protruding abdomen and short limbs. The shafts of the long bones are abnormal. There is minimal proliferation of chondrocytes. Although the trunk is of normal size, the extremities remain very short due to the early fusion of epiphyseal plates.

g. Arthrogryposis
This term is used for multiple stiffness of the joints (ankylosis) of the limbs. It causes articular rigidity. Its origin is not well known.

h. Arachnodactyly
This is manifested in tall stature, long slender limbs with "spider" fingers and joint laxity.

i. Marfan's syndrome
Arachnodactyly combined with abnormality of the aorta. This is an autosomal disorder of connective tissue.

2. Chromosomal Factors

a. Fetal-neonatal scoliosis is seen in triploidy, trisomy 18, trisomy 4, trisomy 9 (mosaic) and trisomy 20.

b. Fetal clubfoot is seen in trisomy 18, trisomy 13, triploidy, trisomy 4, trisomy 9, trisomy 9 (mosaic), partial trisomy 10, trisomy 20, deletion 4p, deletion 9p, deletion 13q and deletion 18q.

c. Fetal polydactyly is reported in trisomy 13 and trisomy 18.

3. Environmental Teratogens

These include drugs ingested by the mother during pregnancy, metabolic poisons and other environmental teratogens. The agents that influence cell metabolism and cell proliferation, if taken during the period of limb development, may cause limb defects. Some of these agents/causes are:

a. Acetazolamide — a carbonic anhydrase inhibitor.

b. Cadmium — inhibits the zinc metabolism.

c. 5-fluoro-2-deoxyuridine — thymidylate synthetase inhibitor.

d. Triethylene melamine — alkylating agent.

e. Dimethadone — an anticonvulsant.

f. Aspirin.

g. Thalidomide — a spasmolytic and anticonvulsant.

h. Hyperthermia.

Most of these are known to produce limb defects in experimental animals, and many have been reported as teratogens in humans.

4. Intrauterine Environment

Limb deformities can be caused by:

a. Oligohydramnios.

b. Local vascular disruptions.

c. An amniotic band.

d. Uterine wall compression (bicornate uterus).

D. Congenital Anomalies of Muscles

Many of the muscular anomalies may be caused by deficiencies in other systems, such as vascular insufficiency, fetal disorders in the central nervous system, lower motor neurons and the peripheral nervous system. These may also be associated with chromosomal anomalies like trisomy 8 (mosaic), trisomy 9 (mosaic), partial trisomy 10q, trisomy 12p, trisomy 13, trisomy 18, trisomy 20p, 4p, 9p, and 18q, Prader–Willi syndrome and sex chromosome aneuploidies. Localized muscular deficiencies may be seen in abdominal wall defects, amniotic band sequence defects and intrauterine infection (fetal varicella). Mechanical limitation of intrauterine movements as in oligohydramnios or by neuromuscular disease also influences skeletal muscular development. Other limitations in the movement of limbs as in arthrogryposis, abnormal flaps of skin grown across joints (pterygia) and floppy limbs (increased joint mobility, poor resistance to passive movements, poor withdrawal response) involve muscular deficiencies. Connective tissue disorder and metabolic disorder also influence muscular development. Only a few of these are discussed here.

1. Localized Muscular Anomalies

Muscle agenesis, partial or complete absence of a muscle or muscle group, is commonly seen in abdominal wall defects such as prune belly syndrome, exstrophy and omphaloceles. Other muscles frequently missing are the pectoralis major and palmaris longus. Deficiency in the diaphragm musculature may cause herniation or respiratory distress because of a small diaphragm. Unilateral shortening of the sternocleidomastoid muscle, which may result from tearing of the muscle during difficult delivery or an intrauterine tumor in this muscle, causes the head to tilt on the affected side.

2. Congenital Myopathies

There is a long list of disorders that may be recognized at birth or later in life. These are associated with extreme hypotonia, contractures and respiratory insufficiency. They are categorized by their abnormal histological findings. For instance, myotubular myopathy shows a large number of immature muscle fibers that appear as myotubes with centrally placed nuclei surrounded by a clear zone. Nemaline rod myopathy shows rod-like structures located in a subsarcolemmal position or throughout the myofibers.

3. Congenital Muscular Dystrophy

This is a group of disorders seen during the neonatal period. It is reported that fetuses show muscular pathology at weeks 14 to 21 of gestation. However, there are no specific chemical or pathological markers on the amniocytes, and at present there is no feasible prenatal detection of the disease and no precise criteria to confirm the diagnosis in a fetus.

In these conditions, muscle fibers tend to be small and rounded with poor fiber-type differentiation. Due to degeneration and regeneration of the muscle fibers, one type of this disorder is known as pseudohypertrophic (Duchenne's) muscular dystrophy. The amount of perimysial, endomysial and fat tissue is progressively increased. Some infants develop severe contractures and die from respiratory complications.

Duchenne's muscular dystrophy is characterized by weakness in the shoulder and pelvic girdles that progressively worsens and causes difficulty in walking. It is an X-linked recessive trait. In this condition, a protein, dystrophin, is significantly reduced. This protein has been identified as the product of the Duchenne's muscular dystrophy gene. Many muscular dystrophies have only muscular involvement; others have multisystem involvement. Many are transmitted as an autosomal recessive trait. Intrauterine infection has also been implicated.

E. Osteogenesis Imperfecta

This is a group of congenital malformations characterized by abnormal collagen (especially type I). A common feature in all forms is thin, fragile bones. Osteogenesis imperfecta appears to be caused by defective collagen production by osteoblasts. Abnormal genes caused by point mutation or deletion lead to formation of abnormal α_1 (1) and α_2 (1) procollagen chains that may be subjected to excessive modification. This could result in enhanced degradation of procollagen chains and may delay collagen secretion. Consequently, type I collagen in bones is deficient. Although the epiphyseal plates in growing bones are normal, the bone trabeculae are abnormally thin and small.

Ultrasonographs and radiographs after 16 weeks of gestation show multiple fractures. Callus formation at the fracture site is often evident. A characteristic radiological feature is underossification, usually best seen in the skull, where there may be a total absence of ossification, and many small fragments of bone are present. The ribs are flat, thin and notched. The vertebrae are undermineralized and flattened. Long bones are usually broad, short, and wavy due to intrauterine fractures. This results in short and bowed limbs. Not all of these features are present in every case. And the degree of severity is also quite variable. Therefore, osteogenesis imperfecta is classified in four major types:

Type I: Bone fragility and blue sclerae.

Type II: Very short, broad long bones and notched ribs. It is perinatally lethal.

Type III: Multiple fractures at birth with progressive deformities and imperfect dentinogenesis.

Type IV: Bone fragility with white sclera.

The incidence of all forms is from 1:25,000 births to 1:62,000 live births. Obstetric complications are common, and a newborn's bones may fracture further during delivery. The placenta and cord may be easily fragmented.

TABLE 8.1. Summary of Development of Skeletal System

Age (days)	Embryonic structure
20	Appearance of somites
26	Upper limb buds
28	Lower limb buds
34	Appearance of myotomes in limbs
42	Appearance of cartilaginous models and digital rays; subdivision of limbs
48	Appearance of primary centers of ossification
56	Separation of digits

REVIEW QUESTIONS — Chapter 8

1. Fetal movements are essential for the proper development of the:
   a. nervous system
   b. notochord
   c. intraembryonic coelom
   d. muscular system
   e. all of the above

2. Two mesenchymal condensations appear at the median plane of the ventral body wall and give rise to the:
   a. sternum
   b. infrahyoid muscles
   c. rectus abdominis
   d. pyramidalis
   e. all of the above

3. Chondroitin sulfate-like molecules secreted by the notochord appear to induce formation of:
   a. vertebral body cartilages
   b. neural arches
   c. ribs
   d. limb buds
   e. all of the above

4. During early development of the limbs, the apical ectodermal ridge interacts with the underlying mesoderm so that:
   a. the apical ectodermal ridge promotes outgrowth of the developing limb
   b. the overall shape of the limb is determined by the mesoderm
   c. the apical ectodermal ridge prevents differentiation of the distal mesodermal cells
   d. a maintenance factor preserves the integrity of the apical ectodermal ridge
   e. all of the above statements are correct

5. The nerve supply to a skeletal muscle may indicate:
   a. the germ layer from which it arose
   b. its maturation
   c. the level of its origin and migration
   d. the time of its differentiation
   e. all of the above

6. The vertebral bodies are:
   a. formed by the migration of sclerotome cells that undergo intramembranous ossification
   b. derived from the sclerotomes of two adjacent somites
   c. the remnants of the notochord
   d. derived from the somatic pleura and neural crest cells
   e. all of the above

7. The cells from the sclerotomes of the somites migrate to give rise to the:
   a. long bones of the limbs
   b. extensor muscles of the back
   c. dense connective tissue under the skin
   d. muscles in the limbs
   e. ribs

8. The core of the intervertebral disk is formed by:
   a. aggregation of cells from the notochord
   b. loose segments of sclerotomes
   c. dense segments of sclerotomes
   d. neural crest cells
   e. dorsal migration of sclerotomes

9. The smooth muscles of embryonic blood vessels are derived from the:
   a. paraxial mesoderm
   b. blood islands
   c. lateral plate mesoderm
   d. local mesoderm
   e. cardiogenic plate

10. Most of the notochord is incorporated in the:
    a. dermomyotomes
    b. vertebral body
    c. neural tube
    d. neural crest
    e. neural arches

11. The dermotomes give rise to the dermis of the:
    a. trunk
    b. upper limbs
    c. lower limbs
    d. head region
    e. all of the above

12. The epimeres of myotomes of regular somites give rise to the:
    a. muscles for eye movements
    b. muscles of the tongue
    c. muscles of facial expression
    d. deep muscles of the back
    e. all of the above

13. Achondroplasia is characterized by all of the following except:
    a. short limbs
    b. large head
    c. flat nasal bridge
    d. protruding abdomen
    e. ambiguous external genitalia

14. Osteogenic imperfecta is a group of congenital malformations characterized by abnormal type I collagen secretion. Ultrasonographs and radiographs after 16 weeks of gestation show:
    a. short limbs
    b. intrauterine hip dislocation
    c. multiple fractures
    d. short skull
    e. all of the above

ANSWERS TO REVIEW QUESTIONS

---

1. d
2. e
3. a
4. e
5. c
6. b
7. e
8. a
9. d
10. b
11. a
12. d
13. e
14. c

CHAPTER

9

CARDIOVASCULAR SYSTEM

In the mesoderm of the yolk sac walls, the blood islands give rise to vitelline blood vessels. Simultaneously, blood islands in the embryonic disk give rise to the cardiogenic tube and dorsal aortae that are connected by aortic arches. Originally, bilateral endocardial tubes fuse in the midline to form the heart tube. Specific regions of the heart tube can be identified. Starting from the inflow tract that receives blood from the yolk sac, the embryonic disk and the placenta are the sinus venosus, the atrium, the ventricle and the outflow tract. Differentiation of the heart tube commences at inflow and progresses to outflow. The tube undergoes remolding and septation. The remolding separates atria from ventricles. Septation separates the right side from left side of the heart.

Within the aortic arches, the first and second pairs mostly disappear. The third pair, with the remnant of the first two pairs, forms the arteries that supply the head region. The fourth pair develops asymmetrically, the left forming part of the aorta, and the right forming the brachiocephalic artery. The fifth pair disappears, and the sixth pair becomes associated with the pulmonary arteries.

The three pairs of veins that bring the blood to the sinus venosus undergo complex modification during which they give rise to the main veins in the adult. Major malformations of the cardiovascular system are discussed in this chapter.

I. BLOOD VESSEL FORMATION

At the middle of the third week, mesenchymal cells in the body of the embryo, in the wall of the yolk sac, in the chorion and in the connecting stalk differentiate into isolated clusters (blood islands), the angioblasts. The peripherally located cells flatten and give rise to the endothelial cells. These cells join together to form angiogenetic cords. The angiogenetic cords grow rapidly and join to develop vascular plexuses that spread throughout the embryo. New blood vessels also arise by budding from existing blood vessels. The large channels develop by enlargement and fusion of small channels. The vessels from which the blood is diverted disappear. The final vascular pattern depends on genetic and hemodynamic influences.

Three separate circulations soon become interconnected (Fig. 9.1):

1. The vitelline circulation develops in the wall of the yolk sac.
2. The chorionic circulation gives rise to umbilical vessels delivering blood to and from the placenta.
3. The intraembryonic circulation, common cardinal veins, transports blood throughout the body of the developing embryo.

At first, intraembryonic angiogenetic cords are located on the lateral sides, but they rapidly spread in the cephalic direction forming a horseshoe-shaped plexus. The underlying endoderm induces it to become the cardiogenic plate (Fig. 9.1). The intraembryonic coelomic cavity located over the central portion of this plexus develops into the pericardial cavity. In addition, another plexus develops parallel and closer to the midline of the embryo. This gives rise to a pair of longitudinal vessels, the dorsal aortae.

II. DEVELOPMENT OF THE ENDOCARDIAL TUBE

With the rapid growth of the central nervous system, the oropharyngeal membrane and cardiogenic tube including the pericardial cavity rotate 180° along a transverse axis. As a result, the central portion of the cardiogenic plate and the pericardial cavity become located ventrally and caudally to the oropharyngeal plate.

The cephalic end of the paired dorsal aortae that are attached to the cardiogenic plate becomes bent around the cephalic end of the embryo, forming the aortic arches (Fig. 9.1). Because of the lateral folds of the embryo, the two lateral limbs of the cardiogenic plate approach each other and fuse to form a single endocardial heart tube, which remains attached to the pericardial cavity by dorsal mesocardium (Fig. 9.2). The single endocardial heart tube is connected caudally to the umbilical, vitelline and common cardinal veins, and cephalically to the dorsal aortae by means of aortic arches.

The mesoderm surrounding the endocardial tube differentiates into the myocardial mantle, which is separated from the endothelium by the cardiac jelly secreted by myocardium. Later, the jelly is invaded by the mesenchymal cells, and the cardiac tube then consists of endocardium, myocardium and epicardium (Fig. 9.3).

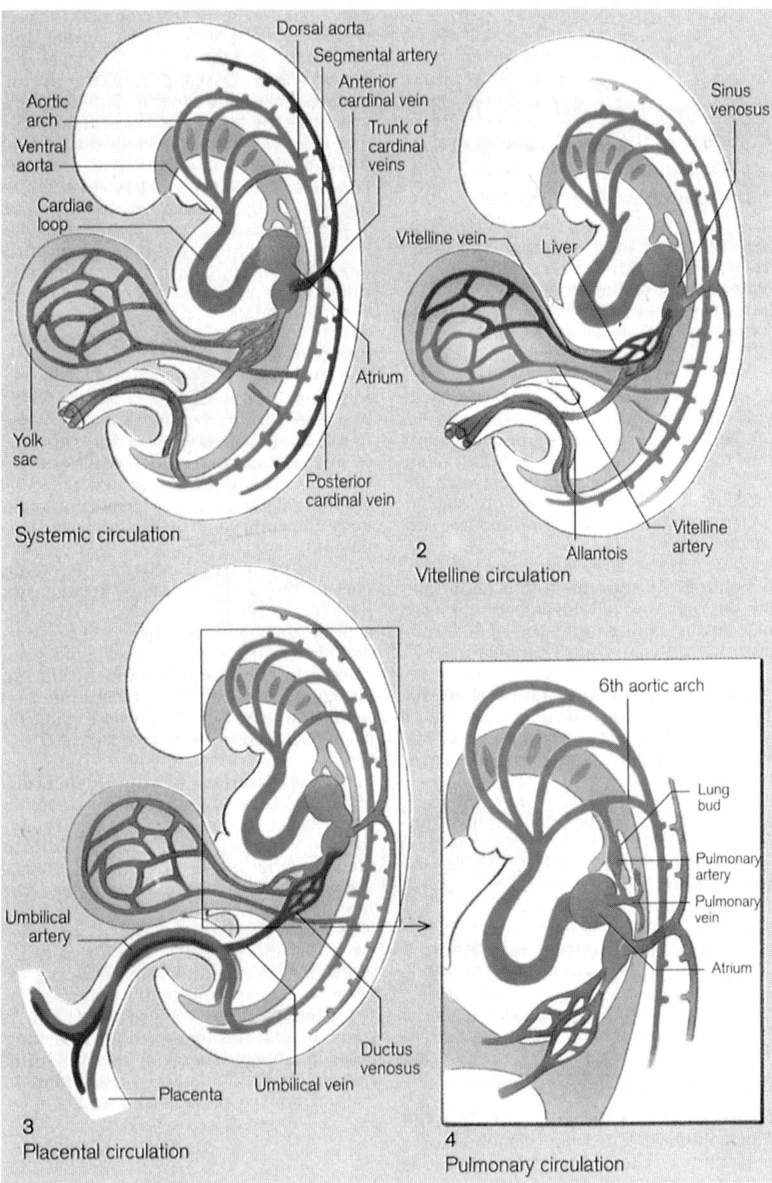

Figure 9.1

Circulatory systems in the embryo (see Color Plate 9.1). Reproduced with permission from Thieme Medical Publishers Inc., New York, 1995, *Color Atlas of Embryology*, Ulrich Drews, Chapter 6: Heart and Vessels.

Four layers become apparent: 1) the endothelium, 2) the myocardium, 3) the cardiac jelly, a thick acellular matrix, and 4) the epicardium. The myocardium arises from the cardiogenic plate. Cardiac muscle cells divide and remain mononucleated. Special cell junctions develop into intercalated disks.

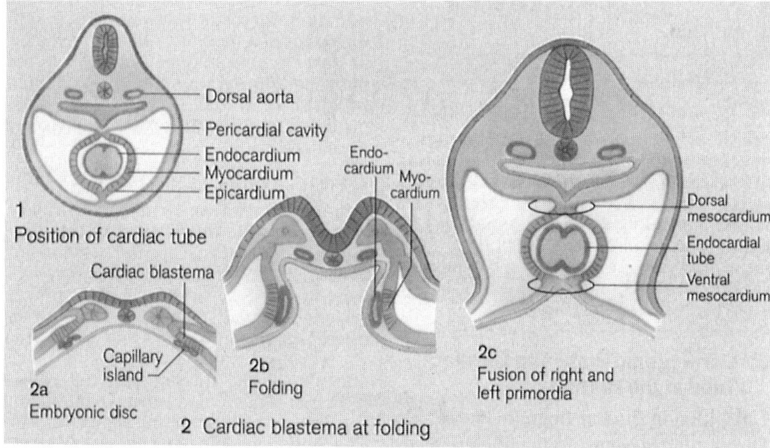

Origin of cardiac tube in the visceral mesoderm

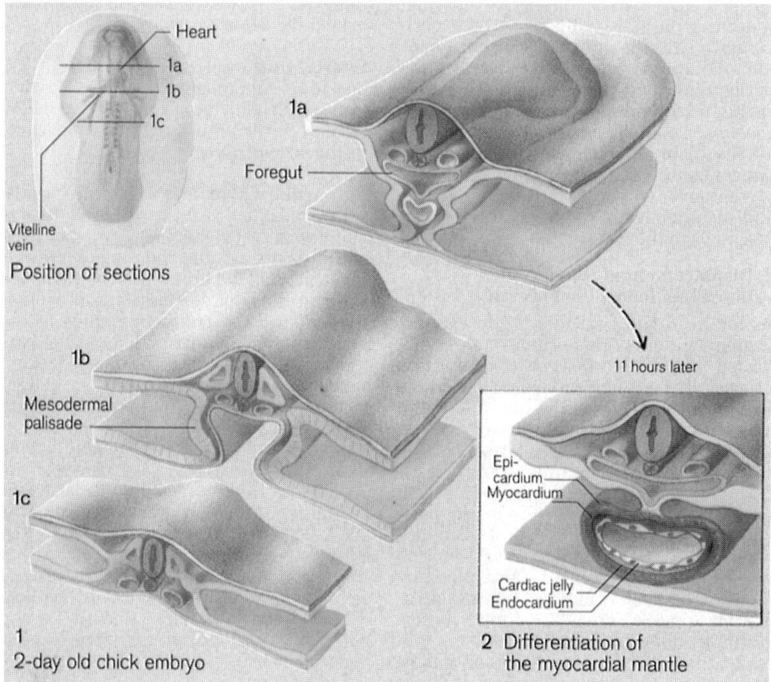

Development of the chick heart during embryonic folding

Figure 9.2

Formation of the heart tube (see Color Plate 9.2). Reproduced with permission from Thieme Medical Publishers Inc., New York, 1995, *Color Atlas of Embryology*, Ulrich Drews, Chapter 6: Heart and Vessels.

III. DEVELOPMENT OF THE HEART

The heart (endocardial) tube undergoes a process of remodeling, folding and septation to form the four-chambered heart. The dorsal mesocardium degenerates,

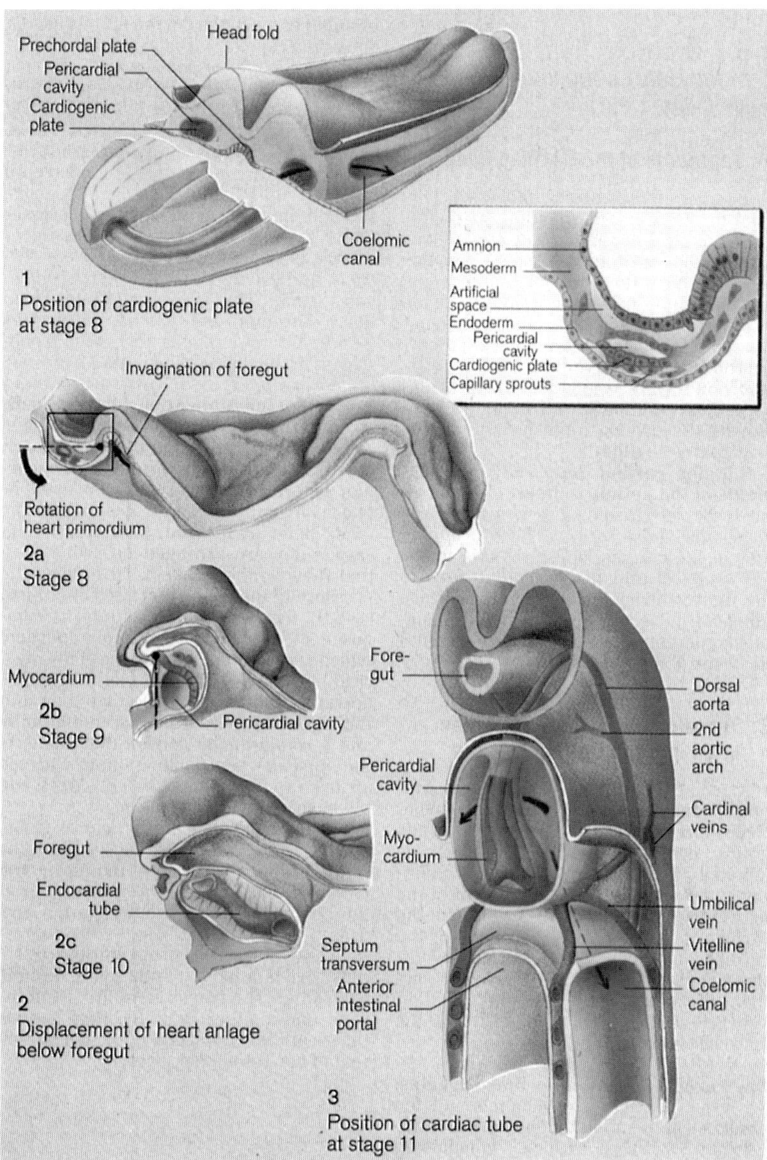

Cardiogenic plate and cardiac tube in the human

Figure 9.3

Cardiac tube in the human (see Color Plate 9.3). Reproduced with permission from Thieme Medical Publishers Inc., New York, 1995, *Color Atlas of Embryology,* Ulrich Drews, Chapter 6: Heart and Vessels.

leaving the heart tube attached within the pericardial cavity only at its cephalic (or arterial) and caudal (venous) ends. The heart tube elongates and becomes subdivided by alternate dilatations and constrictions into four distinct regions (Fig. 9.4):

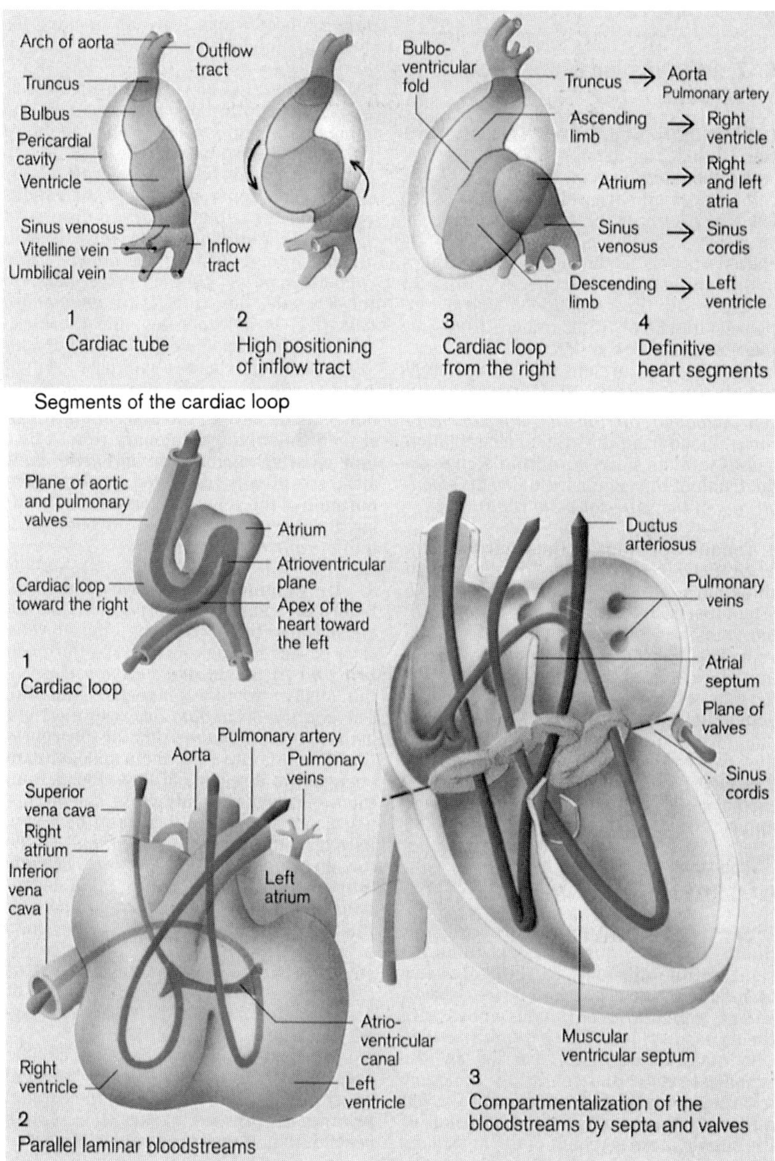

Figure 9.4

Overview: transformation of the heart loop (see Color Plate 9.4). Reproduced with permission from Thieme Medical Publishers Inc., New York, 1995, *Color Atlas of Embryology,* Ulrich Drews, Chapter 9: Head.

1. The bulbus cordis (B) defined by outflow.
2. The ventricle (V) separated from the bulbus cordis by interventricular sulcus.
3. The atrium (A).
4. The sinus venosus (SV), defined by the inflow of umbilical veins.

The distal part of the bulbus cordis, which is continuous with aortic arches, is called the truncus arteriosus. The sinus venosus receives the umbilical, vitelline and common cardinal veins. The differentiation starts at this inflow end.

The bulboventricular portion of the heart tube grows much more rapidly than other regions, but the growth of the pericardial cavity is slow. The heart tube bends over itself, forming a loop. This bending leads to displacement of the bulbus cordis inferiorly, ventrally and to the right; the ventricle goes to the left; the atrium moves posteriorly and superiorly. Other factors contributing to bending of the heart tube are active migration of the myocytes and a change in cellular shape and differential growth. The heart beats by the third week.

As the heart tube continues to grow and bend, the bulbus and ventricle come to lie side by side and ventral to the atrium. The atrium, which is now dorsally placed, becomes pressed against the bulbus and bulges laterally on each side of the bulbus as right and left primitive atria (Fig. 9.4). The bulbus cordis can now be divided into three portions:

1. The distal portion
   The truncus arteriosus will form the roots and the proximal portion of the aorta and pulmonary artery.
2. The midportion
   The conus cordis will form the outflow tract for both ventricles.
3. The proximal portion
   Develops into the trabeculate part of the right ventricle, and its outflow joins the sixth aortic arch.

The ventricular portion of the heart tube will develop into the left ventricle, and its outflow joins the fourth aortic arch.

At the end of loop formation, the venous and arterial ends are brought together as in adults. The sinuatrial orifice shifts to the right and is guarded by the right and left venous valves, which fuse superiorly as the septum spurium. At this stage (week 4), each sinus horn receives umbilical, common cardinal and vitelline veins.

IV. PARTITIONING OF THE ATRIUM

A. Division of the Atrioventricular Canal

Communication between the atria and ventricles, the atrioventricular canal, shifts to the right and is divided into the right and left canals by endocardial cushions that develop as thickening of subendocardial tissue in the dorsal and ventral walls of the heart. The endocardial cushions grow toward each other and fuse (Fig. 9.5).

B. Septation of the Primitive Atrium

This is the beginning of the separation of the pulmonary and systemic circulations. During this process, a continuous communication between the left and right

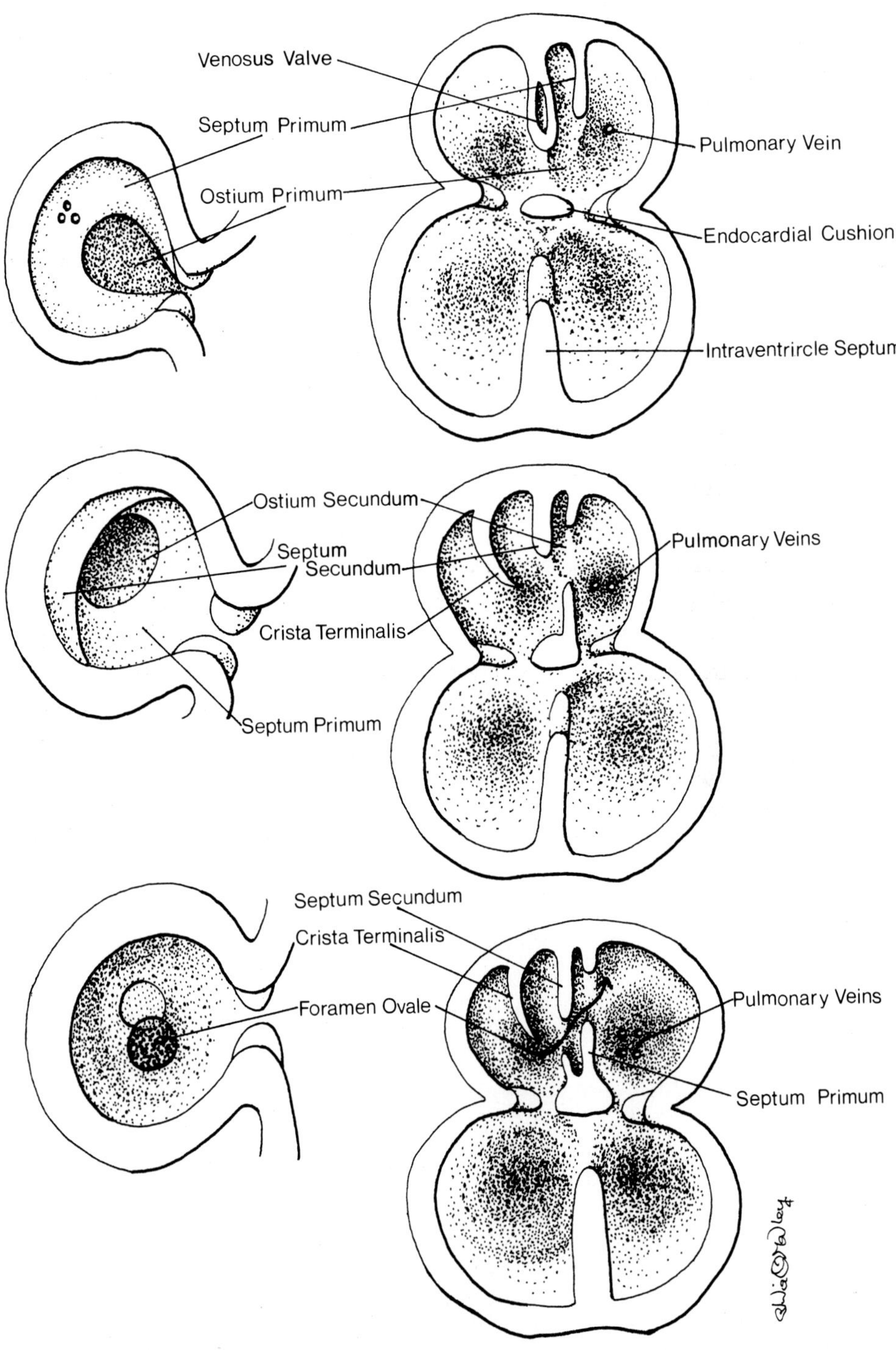

Figure 9.5
Stages in the partitioning of the atrium.

sides is maintained that permits blood to flow directly from the right atrium to the left atrium, bypassing the nonfunctioning lungs. In the rapidly expanding primitive atrium, a thin sickle-shaped membrane, the septum primum, appears on its dorsocranial wall (Fig. 9.5).

The septum primum gradually grows toward the endocardial cushions, leaving a large opening, the foramen primum, between its lower free edge and the endocardial cushions. The septum primum continues to grow toward the endocardial cushions and finally fuses with them. Before the foramen primum is obliterated, the upper part of the septum primum becomes perforated. The perforations coalesce to form the foramen secundum. After establishment of the foramen secundum, another thick membrane, the septum secundum, appears on the ventrocranial wall at the right of the septum primum. The septum secundum gradually extends downward to cover the foramen secundum. The septum secundum does not extend all the way to fuse with the endocardial cushions, but leaves an oval passage, the foramen ovale, through which blood from the right atrium can flow toward the left atrium. Flow of blood in the opposite direction is prevented by the flap valve formed by the remaining septum primum.

C. Remodeling of the Right Atrium

The right sinus horn is incorporated into the posterior wall of the right atrium. It also pulls part of the left horn to the right atrium. The primitive trabeculated part remains as an auricle; the new part, the sinus venerum, forms the smooth part of the atrium. The left venous valve (on sinoatrial opening) fuses with the septum secundum; the right valve forms the valve to the inferior vena cava and coronary sinus. The most superior part of the right venous valve gives rise to the cresta terminalis, which separates trabeculated and smooth parts. The cresta terminalis carries conducting fibers from the sinoatrial (SA) node to the atrioventricular (AV) node.

D. Remodeling of the Left Atrium

A single pulmonary vein develops as an outgrowth of the posterior wall of the left atrium. It divides into four major branches that gain connection with the blood vessels supplying the developing lungs. During further development, the main stem and proximal parts of its branches become incorporated into the posterior wall of the left atrium, forming the smooth part of the left atrium (Fig. 9.5). This smooth part is then displaced ventrally and to the left. The shift of the atrioventricular canal to the right aligns the atria to the respective ventricles.

V. SEPARATION OF VENTRICLES

Originally, the bulbus and ventricles are in craniocaudal sequence to each other. As the heart tube bends, the bulbus and ventricle come side-by-side. The communication between the two chambers, the interventricular foramen, becomes partially closed by the bulboventricular fold, which is formed by the bend in the heart

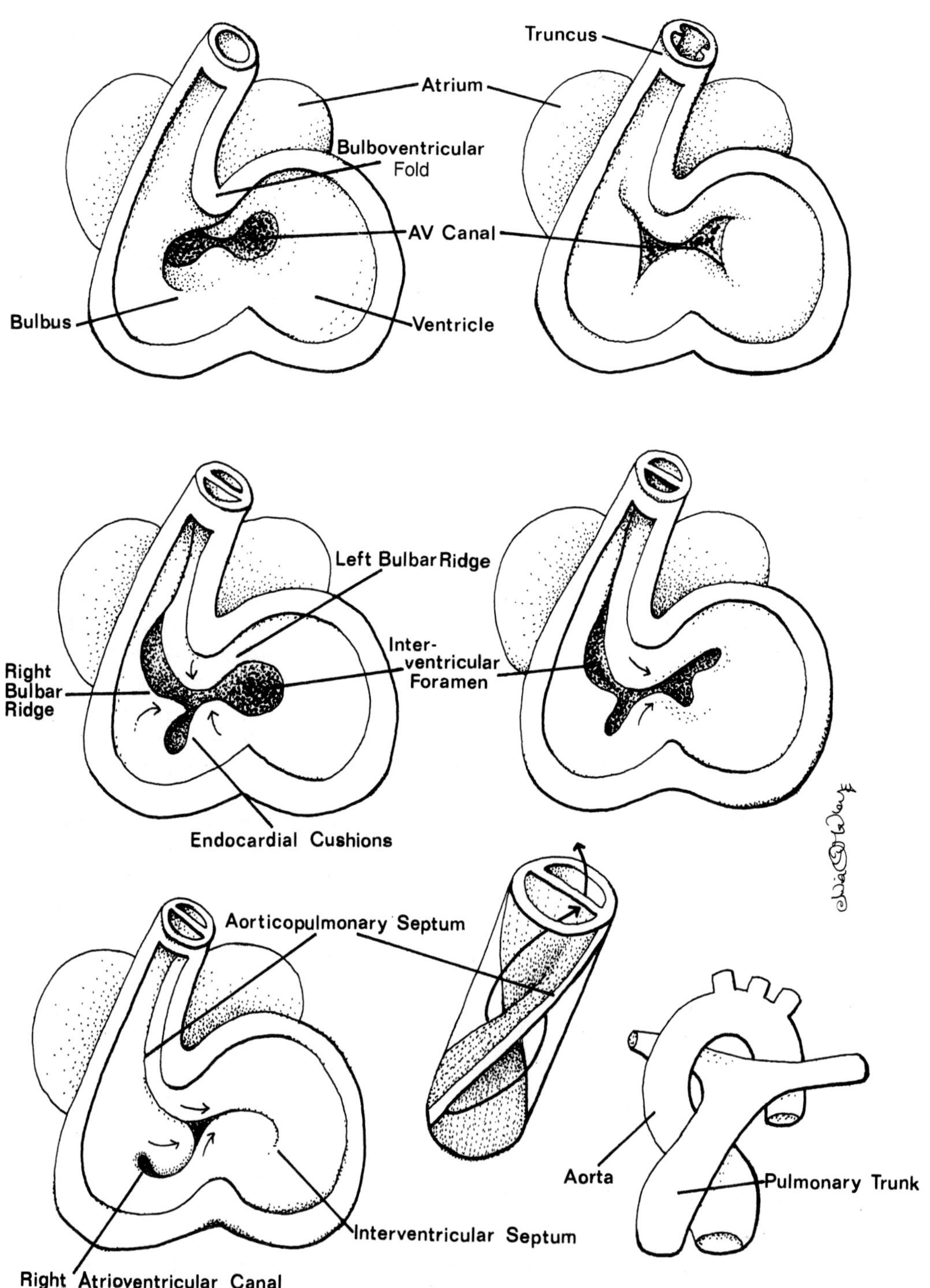

Figure 9.6
Development of the truncal ridges and the closure of the interventricular foramen.

tube (Fig. 9.6). The ventricles formed in this way start to dilate, and their medial walls at the lower part become opposed to each other and finally fuse to form the interventricular septum. As the heart grows, reduction of the bulboventricular fold widens the distal portion of the bulbus, which becomes incorporated into the outflow tract of both the right and left ventricles. The differential expansion of the right wall of the right ventricle and the left wall of the left ventricle further aligns the ventricles to the outflow. The interventricular septum, with further growth, becomes the muscular part of the interventricular septum in the adult. Separation of the two ventricles and completion of the interventricular septum result from fusion of the subendocardial tissue from the following areas:

A. the right bulbar ridge
B. the left bulbar ridge
C. the dorsal endocardial cushion

VI. DIVISION OF THE TRUNCUS

Two spiral mesodermal (truncal) ridges grow from the inner walls of the truncus arteriosus on the opposite wall. Similar ridges form in the bulbus cordis and become continuous with the truncal ridges (Fig. 9.6). The ridges are thick and fill much of the lumen. Early spiral blood flow determines the helical arrangement. The mesenchyme from the ridges fills the area between the flow of blood. The neural crest cells migrate and contribute to this septum. Opposing ridges twist around each other and eventually fuse to form a spiral aorticopulmonary septum. The septation starts at the inferior end of the truncus and proceeds superiorly and inferiorly.

The aorticopulmonary septum divides the conus cordis and truncus arteriosus into two channels, the conus arteriosus for the right ventricle and the aortic vestibule for the left ventricle. The truncus arteriosus forms the outflow for the ventricles. The blood from the right ventricle now flows into the pulmonary trunk, and blood from the left ventricle flows into the aorta. The tissue from the proximal portions of the ridges and the tissue from the dorsal endocardial cushion fuse to form the membranous part of the interventricular septum.

A large percentage of cardiac defects derive from errors in this complex process. The pacemaker and conducting system develop early to coordinate the beating of the heart. In the heart tube, the ventricle appears to serve as a pacemaker. Later, the SA node, which develops in the right sinus venosum, takes over. Cells in the anterior cardinal cushion give rise to the AV node.

VII. DERIVATIVES OF THE AORTIC ARCHES

Simultaneous with heart tube development, paired dorsal aortae form in the dorsal mesenchyme of the embryo on each side of the notochord and connect the heart tube through the arch arteries before folding of the embryo. Folding causes them to form dorsoventral loops that become continuous with the aortic sac (Fig. 9.7). First one, then several of these arteries are formed. Ventrally, they arise from

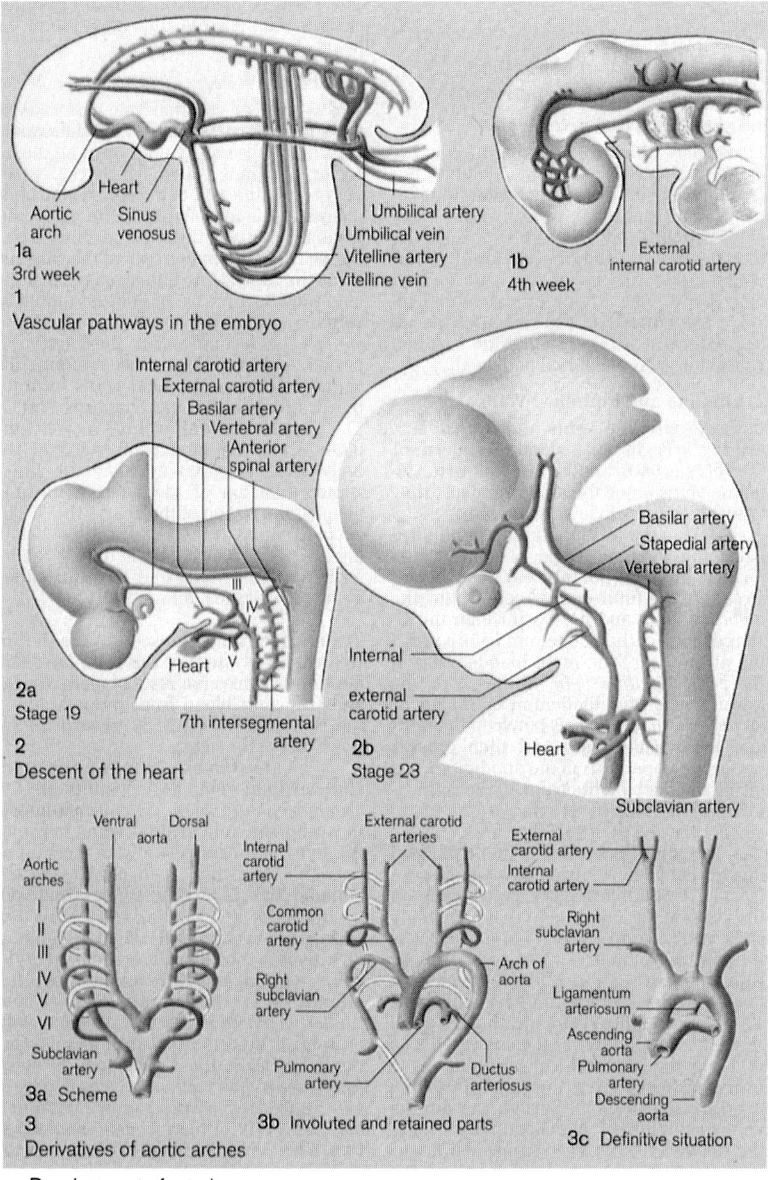

Figure 9.7

Arteries (see Color Plate 9.7). Reproduced with permission from Thieme Medical Publishers Inc., New York, 1995, *Color Atlas of Embryology*, Ulrich Drews, Chapter 6: Heart and Vessels.

the aortic sac, an expansion at the cranial end of the truncus arteriosus. The neural crest is believed to be important in the normal development of aortic arch arteries.

Although in this manner six pairs of aortic arteries develop, they are not all present at the same time. By the time the sixth pair is formed, the first, second and fifth have already disappeared:

A. The first pair largely disappears; a small portion remains as the maxillary arteries.

B. The second pair similarly disappears; a small portion persists as the hyoid and stapedial arteries.

C. The proximal parts of the third pair form the common carotid arteries, and the distal portions join with the dorsal aortae to form the internal carotid arteries. The external carotid arteries sprout from the common carotid arteries.

D. Derivatives of the fourth arch
The left fourth arch forms part of the arch of the aorta. The proximal part of the arch of the aorta is formed by the aortic sac, the middle part by the left fourth arch, and the distal part develops from the left dorsal aorta, which continues with the single midline aorta formed by fusion of both dorsal aortae. The right fourth aortic arch forms the proximal part of the right subclavian artery. The distal part of the right subclavian artery is formed from the right dorsal aorta and the seventh intersegmental artery.

E. The fifth pair completely disappears.

F. The fate of the sixth arch
The proximal part of the right sixth aortic arch gives rise to the proximal part of the right pulmonary artery. The distal part of this artery degenerates. The proximal part of the left sixth aortic arch persists as the proximal part of the left pulmonary artery. The distal part of this artery forms the ductus arteriosus, which serves to reduce blood flow to the lungs. Its patency is maintained prenatally by the influence of placenta-derived prostaglandins. The distal portions of the pulmonary arteries develop from the buds of the sixth aortic arches, which grow toward the developing lungs. The dorsal aortae fuse from the fourth thoracic segment to the fourth lumbar segment, forming the descending aorta. The vitelline arteries eventually lose their connection to the yolk sac and become branches of the aorta.

VIII. VENOUS SYSTEM

In the embryo three major sets of veins return the blood to the heart tube:

A. The vitelline vein drains the yolk sac.

B. The umbilical vein brings oxygenated blood from the placenta.

C. The common cardinal veins drain the head and neck and body wall of the embryo (Fig. 9.8).

The vitelline and umbilical veins course through the septum transversum, where they are interrupted by the growing hepatic cords. In the septum transversum, portions of these veins transform into hepatic sinusoids. A single oblique channel is selected to align with the left umbilical vein and become dominant as the ductus venosus. On its superior end, the ductus venosus joins with the terminal portion of the right vitelline vein (terminal portion of the inferior vena cava), which enters

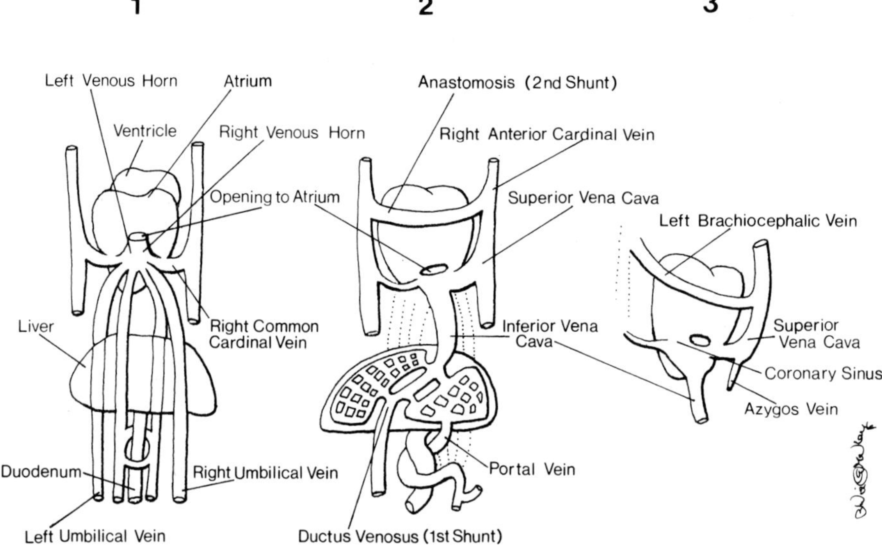

Figure 9.8
Stages of development of the venous system.

the right sinus horn. The oxygenated blood from the left umbilical vein courses through the ductus venosus and terminal part of the right vitelline vein to enter the right sinus horn. This shift of blood return to the right sinus horn (right atrium) initiates the major reshaping of the systemic venous system as discussed below.

A. The Fate of the Vitelline Veins

1. The portion of the right vitelline vein between the septum transversum and the right sinus horn becomes the terminal portion of the inferior vena cava (as mentioned above).
2. The portion of the left vitelline vein above the septum transversum disappears.
3. Both the right and left vitelline veins form anastomoses around the developing duodenum and give rise to the portal vein (draining into liver sinusoids), superior mesenteric vein and splenic vein. The inferior mesenteric vein then joins the portal vein (Fig. 9.8).

B. The Fate of the Umbilical Veins

1. The portion of both umbilical veins above the septum transversum disappears.
2. The portion within the septum transversum contributes to the hepatic sinusoids.
3. The left umbilical vein, which joins the ductus venosus, brings the oxygenated blood from the placenta and persists during intrauterine life.
4. The right umbilical vein below the septum transversum disappears (Fig. 9.8).

C. The Cardinal System

The cardinal system is intraembryonic and gives rise to most of the systemic veins. Each sinus horn receives the common cardinal vein, which is formed by the precardinal (anterior cardinal) vein, which drains blood from the head and neck region, and the postcardinal (posterior cardinal) vein, which drains the lower body.

1. The cranial portion of the precardinal veins gives rise to the cerebral veins, intracranial dural sinuses and internal jugular veins.
2. The cervical portions of the precardinal veins form anastomoses between them. Due to the right shift of venous return, the left sinus horn becomes very small; a single oblique channel in the neck dominates and shunts the blood from the left precardinal vein to the right precardinal vein. This way, all the blood from the head and neck region returns to the right sinus horn through the right common cardinal vein. The left common cardinal vein and terminal portion of the left precardinal vein (below the anastomosing channel) disappear. The portion of the precardinal veins above the anastomoses is now the left and right brachiocephalic veins, which drain into the superior vena cava.
3. The superior vena cava is thus formed by the terminal portion of the right precardinal vein and the right common cardinal vein.
4. Three sets of new veins associated with the cardinal system appear (Fig. 9.9):
    a. The subcardinal veins sprout from the postcardinal veins associated with the mesonephros and appear medially. The two veins form anastomoses and drain the kidneys, gonads and suprarenals. The right subcardinal vein becomes dominant and develops into the renal segment (between liver and kidney) of the inferior vena cava. This joins the terminal portion of the inferior vena cava (the right vitelline vein segment).
    b. The supracardinal veins appear lateral to the sympathetic trunk and join the postcardinal veins. In the thoracic region, they give rise to azygos and hemiazygos veins. In the pelvic region, both supracardinal veins join to form the pelvic segment (below the kidneys) of the inferior vena cava.
    c. The sacrocardinal veins join the supracardinal veins and postcardinal veins, and contribute to the lower part of the inferior vena cava and continue as the common iliac veins.

Thus, the inferior vena cava develops from: 1) the right vitelline vein, 2) the right subcardinal veins, 3) the supracardinal veins, 4) the sacrocardinal vein and possibly a very small portion of the postcardinal veins. As new cardinal veins become established, most of the postcardinal veins disappear.

D. The Fate of the Sinus Venosus

Because of the shift, the blood returns to the right, the right horn becomes larger and the sinoatrial opening consequently moves to the right and opens into the right primitive atrium. The left horn loses its importance and becomes smaller. With disappearance of the terminal portions of left vitelline, the left umbilical and left

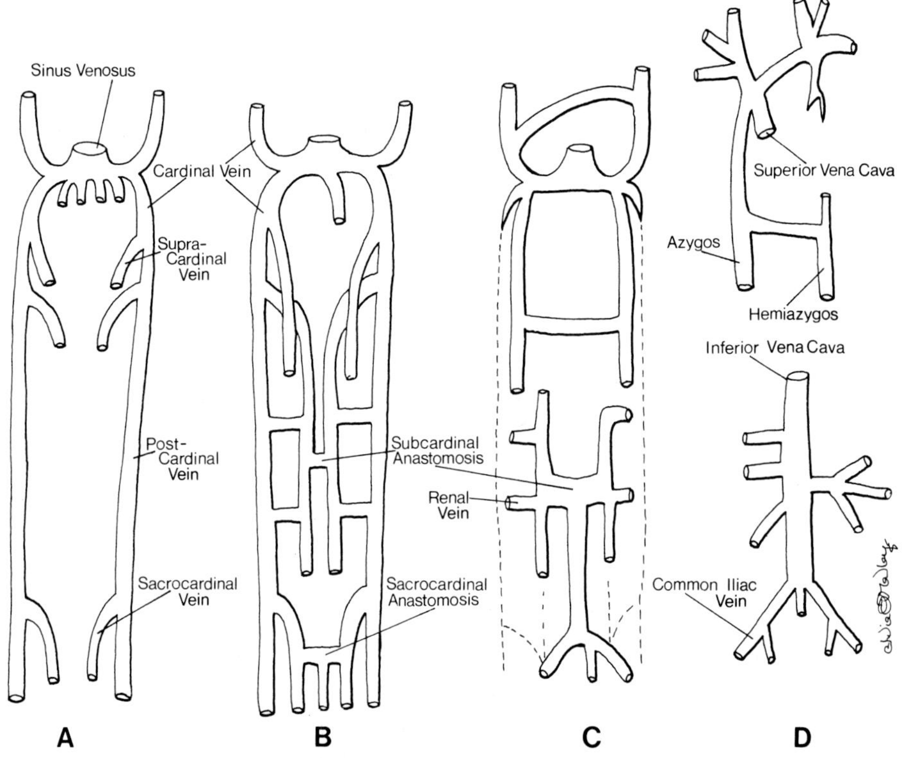

Figure 9.9

Development of the venous system. The transformation of the cardinal, supracardinal, subcardinal and sacrocardinal systems.

common cardinal veins, no blood returns to the left horn. Its terminal portion eventually persists as coronary sinus and the oblique vein of Marshall.

IX. FETAL CIRCULATION (Fig. 9.10)

The umbilical vein carries oxygenated blood from the placenta, shunting through the ductus venosus to the inferior vena cava. The inferior vena cava empties into the right atrium. Some deoxygenated blood from the lower part of the body drains into the inferior vena cava, and there is some mixing of oxygenated and deoxygenated blood in the right atrium. Deoxygenated blood in the superior vena cava also enters the right atrium, but the situation of the valve of the inferior vena cava allows very minimal mixing. Most of the blood from the superior vena cava passes through the right atrium to the right ventricle. The remaining blood in the right atrium passes through the foramen ovale to the left atrium and then to the left ventricle. Blood from the left ventricle is pumped out through the aorta and distributed to the fetal body. The blood from the right ventricle is shunted through the pulmonary trunk and ductus arteriosus to the aorta, bypassing the nonfunctioning lungs. Some blood reaches the lungs and is brought back to the left ventricle by the pulmonary vein. From the aorta, blood is returned to the placenta for oxygenation via umbilical arteries.

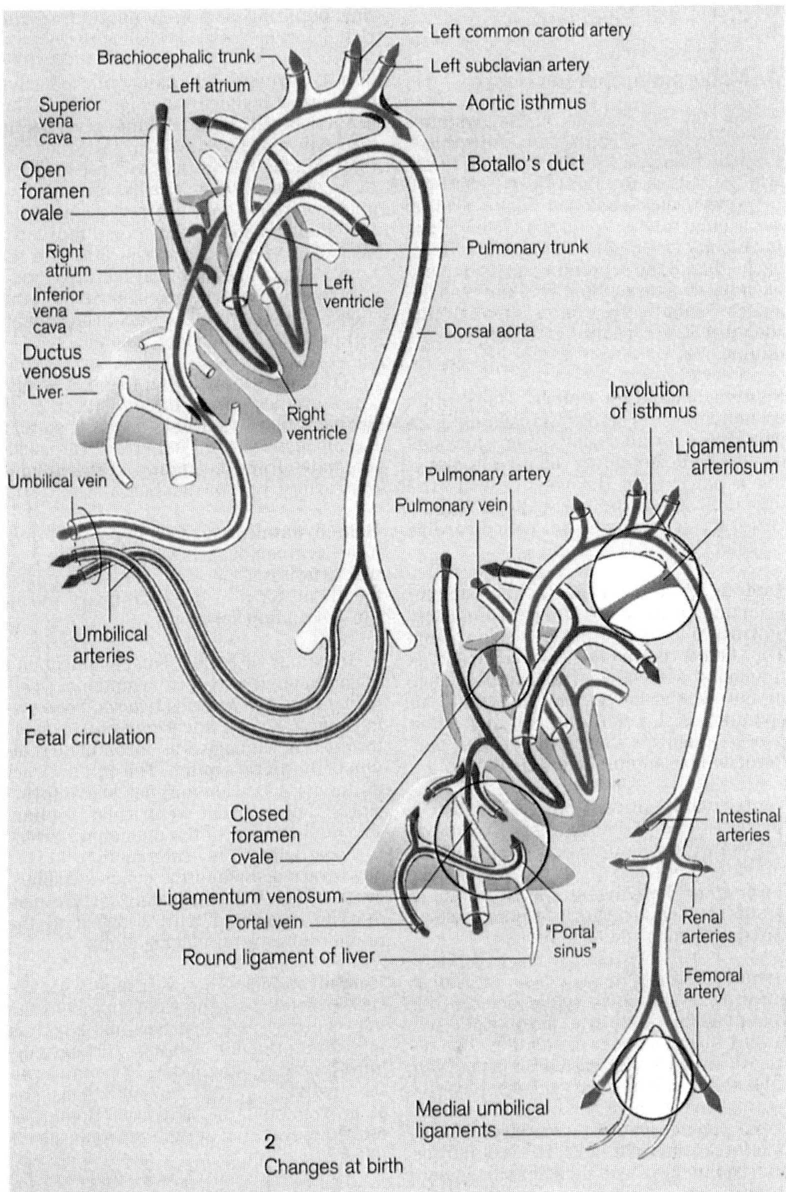

Figure 9.10

Changes in the circulation at birth (see Color Plate 9.10). Reproduced with permission from Thieme Medical Publishers Inc., New York, 1995, *Color Atlas of Embryology,* Ulrich Drews, Chapter 6: Heart and Vessels.

X. CIRCULATORY CHANGES AFTER BIRTH

The circulatory changes after birth are caused mainly by a decrease in pulmonary resistance, increased pulmonary blood flow, increased volume of the left atrium, and ligation of the umbilical vessels. There is also an active contraction of tunica media of the umbilical vessels.

1. The umbilical arteries (Fig. 9.9) constrict and later become the medial umbilical ligament. The proximal portions of these persist as the internal iliac or hypogastric arteries.
2. The umbilical vein and ductus venosus no longer receive blood and become the ligamentum teres hepatis and ligamentum venosus.
3. With expansion of the pulmonary capillary bed, the aortic pressure exceeds the pulmonary pressure. Closure of the ductus arteriosus is effected by contraction of the smooth muscle stimulated by increased oxygen saturation. The ductus arteriosus becomes constricted and later forms the ligamentum arteriosus.
4. With the corresponding increase in volume of returning blood in the left atrium, the pressure in the left atrium is increased, which causes the flap of the foramen ovale to close (Fig. 9.10).

XI. LYMPHATIC SYSTEM

At the end of the fifth week, the lymph vessels begin to develop in a manner similar to that of the blood vessels. Six primary lymph sacs develop: two jugular lymph sacs, two iliac lymph sacs, one retroperitoneal lymph sac, and one cisterna chyli.

The lymphatic vessels grow out of these sacs along the main blood vessels. The right and left thoracic ducts connect the jugular sacs with the cisterna chyli. The thoracic duct is formed from the upper part of the cisterna chyli. Surrounding mesenchymal cells invade the lymphatic sac (except the cisterna chyli) and break up the main channels into sinuses and give rise to lymph nodes. Small lymphocytes (T-cells) from the thymus gland circulate to seed the nodes and other lymphatic organs.

XII. CONGENITAL MALFORMATIONS OF THE CARDIOVASCULAR SYSTEM

The previous discussion shows that development of the cardiovascular system involves highly coordinated processes of cellular proliferation, adhesion, cell migration, cell differentiation and cell death under complex genetic, hemodynamic and cytochemical influences. Understanding the pathogenesis requires study on the organ, tissue, cellular, subcellular and molecular levels. The most vulnerable period with regard to major cardiac malformation is between days 17 and 56, when major events of cardiogenesis occur, but as morphogenesis continues this system is under risk throughout gestation.

Malformation of the cardiovascular system affects about 1% of live births, but is found in 18% of spontaneously aborted and stillborn fetuses. Many inherited and sporadic syndromes, and teratogenic and infectious agents, have been associated with cardiac anomalies. Some cardiac malformations can arise through more than one possible pathogenic pathway. So many points in normal development are unclear, that the embryological basis for these malformations is still speculative.

The most frequent types of congenital cardiac anomalies are ventricular septal defects (45%), coarctation of the aorta (17%), atrial septal defects (15%), atrioventricular septal defects (endocardial cushion defects 10%), tetralogy (6%), transposition of the great vessels (3%), persistent ductus arteriosus (2%), and persistent truncus arteriosus (1%). Only 10% of congenital cardiac malformation cases are identified as caused by chromosomal abnormalities, single-gene defects and teratogens; the other 90% are attributed to multiple factors. Thalidomide and rubella are examples of teratogens that, with a slight genetic predisposition, are capable of causing a high frequency of maldevelopment.

The effect of malformation depends on the nature and size of the defect and on its effect on the pulmonary and systemic circulations. The nature of malformation depends on the timing of the insult relative to development and the target of the insult. The presence of an abnormal communication between the chambers of the heart (a shunt) is an important feature. Blood is shunted from the chamber with high pressure to the chamber with low pressure. If the shunt is from left to right, there is no cyanosis. It causes an overload of work for the left ventricle. The pulmonary circulation is increased, which leads to an increase in pulmonary pressure, in turn causing thickening of the wall and narrowing of the lumen of the pulmonary arterioles. This causes difficulty in breathing. The child has a raised sternal angle, flared nostrils and excessive sweating.

A right-to-left shunt causes intermixing of oxygenated and deoxygenated blood, leading to reduced oxygen saturation in the blood supplied to the body, causing cyanosis. Furthermore, the pulmonary circulation is reduced. Cyanosis frequently leads to clubbing of the fingers and toes. Low oxygen saturation may lead to polycythemia, which causes increased viscosity of the blood and increased peripheral resistance. All this leads to increased workload for the right ventricle. Absence of the spleen (asplenia) is frequently associated with cyanotic heart disease. Many cardiac malformations do not develop a shunt and are acyanotic.

For simplification, the major anomalies are divided into three categories:

A. Malformations with a left-to-right shunt (acyanotic).

B. Malformations with a right-to-left shunt (cyanotic).

C. Malformations without a shunt (acyanotic).

A. Malformation with Left-to-Right Shunt (Acyanotic)

1. Interventricular Septal Defect (VSD)

This is the most common congenital cardiac anomaly. It can occur as an isolated defect but is often associated with other cardiac anomalies. Most of those involve the membranous part of the interventricular septum. The size of the defect and

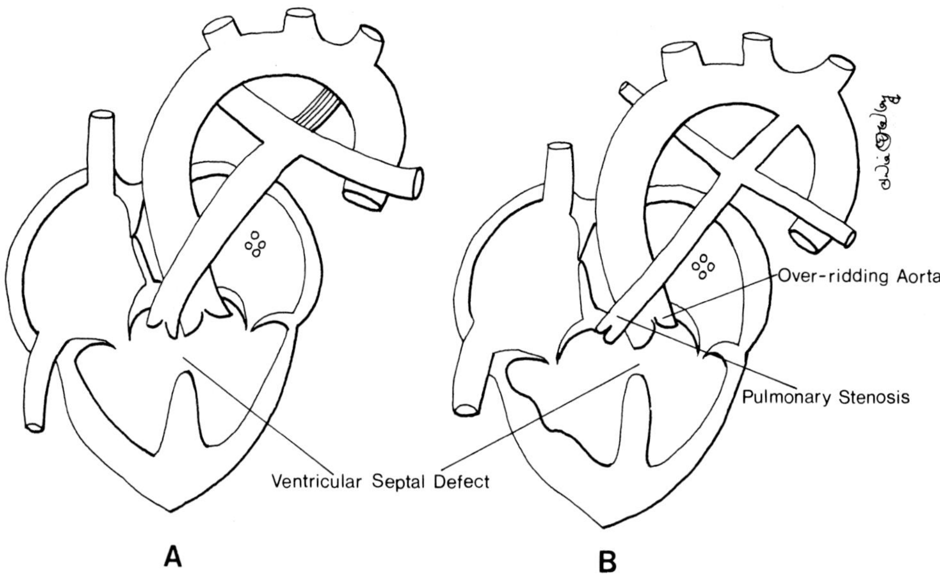

Figure 9.11
(A) Failure of the membranous part of the ventricular septum to develop. (B) Unequal division of the truncus arteriosus (tetralogy of Fallot).

pulmonary blood flow resistance are important. It is suggested that increased programmed cell death in the conotruncal region delays fusion of the truncal and conal ridges, leaving the defect (Fig. 9.11). A defect in the muscular part of the interventrical septum does not involve the membranous part, and the defect is completely bound by the myocardium. Although muscular defects are single, VSDs may be multiple.

2. Interatrial Septal Defect (ASD)

A normal right-to-left shunt through the foramen ovale is essential during prenatal life, but it is functionally closed after birth. Postnatally, single or multiple defects in the interatrial septum cause left-to-right shunt. These defects are often associated with other cardiac malformations. If isolated, these cause serious hemodynamic consequences in the fetus. These defects are classified by their location within the septum.

a. Ostium secundum defect (Fig. 9.12A)
This is the most common of the interatrial septal defects (93%). This high opening in the septum represents either incomplete closure of the ostium owing to inadequate size of the septum secundum, or excessive resorption of septum primum forming a large ostium. The defect is occasionally at the junction of the superior vena cava and the right atrium. In this case, the superior vena cava overrides the defect, and the two pulmonary veins empty into the right atrium. The mechanism for this is not clear.

b. Common atrium (Fig. 9.12B)
This is the virtual absence of an atrial septum. It is a combination of an atrioven-

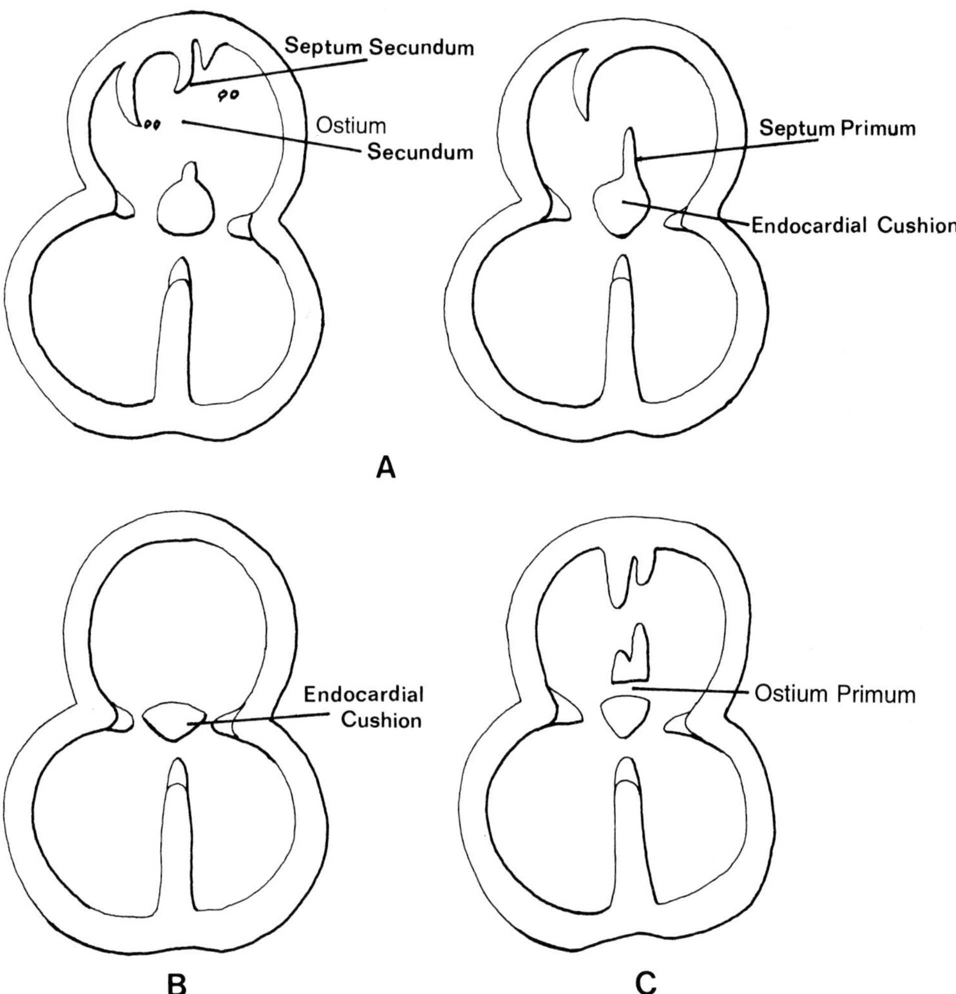

Figure 9.12
Atrial septal defects: (A) foramen secundum defect; (B) common atrium; (C) foramen primum defect.

tricular septal defect and a large ostium secundum defect. A remnant of the septum primum as a strand tissue is often found.

c. The ostium primum defect (Fig. 9.12C)
This low opening may combine with a common atrioventricular canal and is the most frequent cardiac anomaly associated with trisomy 21. This may represent a defect in atrioventricular septation.

3. Persistent Ductus Arteriosus

The ductus arteriosus forms by the sixth aortic arch and is normally retained as a connection between the pulmonary trunk and aorta throughout prenatal life. Approximately 35% of blood passes from the pulmonary trunk through the ductus arteriosus to the aorta, bypassing the lungs. Histologically, the ductus has more medial smooth muscle than the pulmonary trunk and aorta. Close to term, increasing accumulation of glycosaminoglycans in its media start to break down the elastic lamina, and medial smooth cells proliferate to further thicken the media.

Postnatal closure occurs within a few hours of birth by contraction of the medial smooth muscle. It begins at the pulmonary end and proceeds to the aortic end. Later (2 to 3 weeks), increasing fibrosis closes the ductus definitively. The most common anomaly of ductus is delayed closure. In these cases, persistent intact internal elastic lamina and subendothelial elastic laminae keep the ductus open. A change in hemodynamics after birth establishes a left-to-right shunt. The magnitude of blood through it depends on its diameter, the pressure difference between the pulmonary trunk and the aorta, and the systemic and pulmonary vascular resistances. Closure of the ductus may occur spontaneously or with prostaglandin antagonist (indomethacin, ibuprofen or aspirin) treatment. In premature infants, prostaglandin E_1 or E_2 is administered to maintain patency when there is a ductal-dependent cardiac defect. Persistent ductus arteriosus may occur as an isolated anomaly or may be associated with other cardiac malformations. It is more common in females than in males. It is the most common cardiac anomaly associated with maternal rubella infection during early pregnancy.

An absence of ductus arteriosus is seen commonly in tetralogy of Fallot. Prenatal closure of ductus has serious consequences. The ductal shunt is obstructed, and all right-ventricular blood is forced through the restrictive fetal pulmonary circulation. Excessive volume load on the right ventricle may lead to hydrops fetalis (generalized fetal edema) and consequent intrauterine death. The ductus arteriosus is very sensitive to in utero administration of indomethacin, ibuprofen or aspirin, which may cause prenatal closure. On the other hand, increased frequency and severity of ductal laceration, hemorrhage and edema have been reported in newborns treated with prostaglandins.

4. Hypoplastic Left Heart Syndrome (HLHS)

Hypoplastic left heart syndrome is the most lethal of the cardiac anomalies. Most affected newborns die within 7 days of birth. It is always associated with either aortic stenosis or aortic atresia, which could result from malformation of the aortic valves. There is marked right-side hypertrophy and dilation, and there is enlargement of the tricuspid annulus, pulmonary valve annulus and pulmonary trunk. Hypoplasia of the ascending aorta and aortic arch is also present. Systemic and coronary circulation are compromised (Fig. 9.13).

B. Malformation with Right-to-Left Shunt (Cyanotic)

1. Tetralogy (Fig. 9.11B)

In 1888 Dr. Fallot described four main features of this anomaly:

a. pulmonary stenosis,

b. interventricular septal defect,

c. overriding of the aorta,

d. right-ventricular hypertrophy.

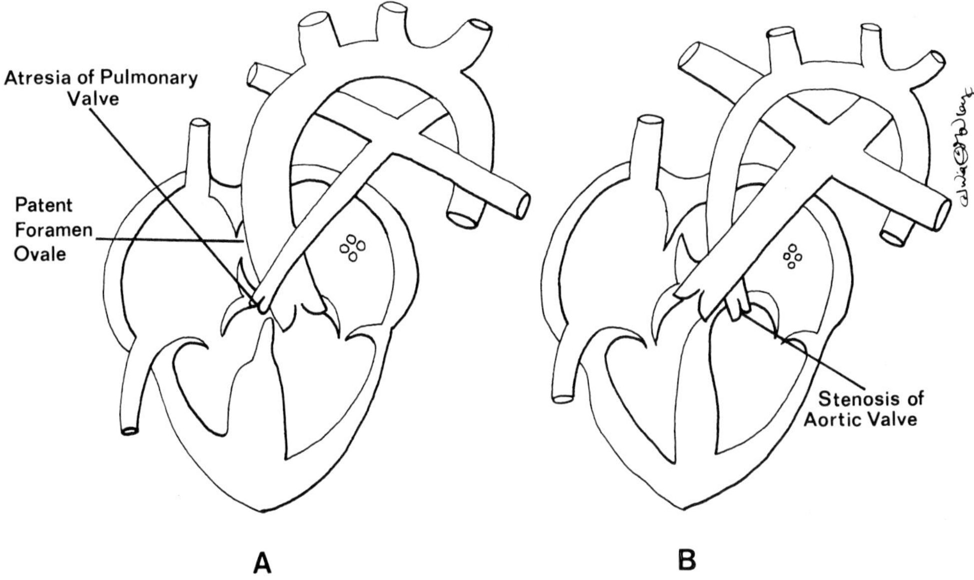

Figure 9.13
Abnormalities of the cardiac valves: (A) pulmonary stenosis; (B) aortic atresia.

This is the most frequent cause of congenital cyanosis. Pulmonary stenosis is considered to be the primary lesion. It is suggested that hypoplasia of the infundibular portion of the ventricular septum leads to malalignment of muscular ventricular septum. Another theory stresses malseptation of the truncus, which divides the blood stream from the right ventricle, leading to migration of the aorta to the right (morphologically seen as unequal division of the truncus arteriosus). The pulmonary stenosis caused thus leads to overload and hypertrophy of the right ventricle. Deoxygenated blood from the right ventricle passes through the interventricular septal defect and mixes with oxygenated blood in the left ventricle, which causes cyanosis. The overriding aorta receives blood from both ventricles, reducing pulmonary circulation, and further mixing of blood adds to the severity of the cyanosis. Preliminary treatment for this is to improve pulmonary circulation by anastomosing the left subclavian to the pulmonary trunk. Other corrections, such as relieving stenosis and closure of septal defects, are performed later.

Overriding aorta without pulmonary stenosis, raising of both the pulmonary trunk and aorta from the right ventricle (double outlet right ventricle) and overriding pulmonary trunk can also occur and lead to congenital cyanosis.

2. Complete Transposition of the Great Arteries

In this anomaly, the aorta arises from the morphologic right ventricle, and the pulmonary trunk arises from the morphologic left ventricle. The atria are in con-

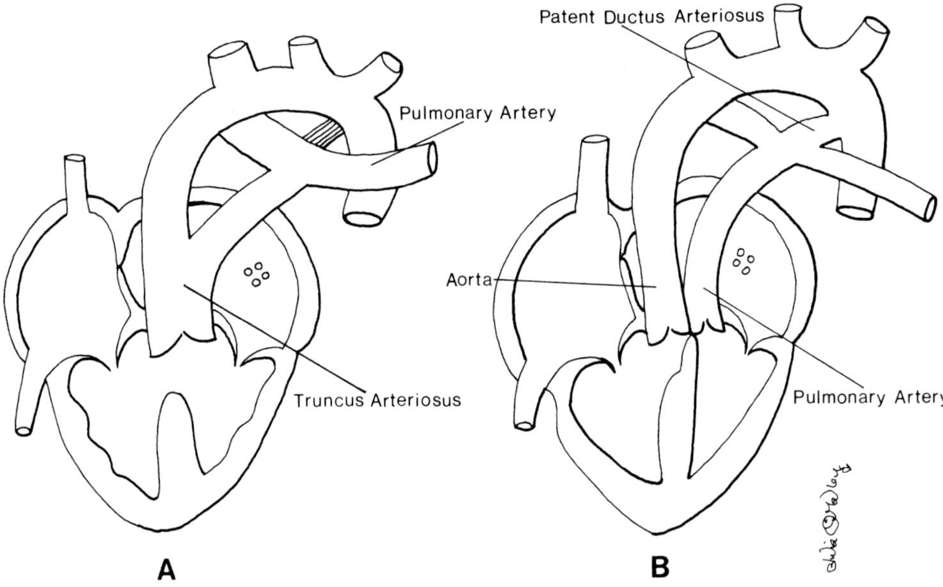

Figure 9.14
Abnormalities of the truncus and conus: (A) persistence of truncus arteriosus; (B) complete transposition of the great vessels.

cordance with the ventricles. The interventricular septal defect, which is present in about half of cases, is an important component of this condition. It allows mixing of the blood. Persistent ductus arteriosus, which is also common, somewhat compromises the fetus. Coarctation of the aorta is also associated with this condition (Fig. 9.14B). The oxygenation in transposition is deficient, and both ventricles are under an excessive workload. This condition results from failure of the truncal septum to spiral. The mechanism of spiraling failure is obscure. Another condition, congenitally corrected transposition (anatomically corrected malposition), is less common. In this condition, the morphologic right atrium opens into the left ventricle, which flows into the pulmonary trunk, and the morphologic left atrium opens into the right ventricle, which supports the aorta. The ventricles are reversed. The two discordances produce an effectively normal hemodynamic. The interventricular septal defect is always present, which causes a left-to-right shunt (acyanotic).

3. Persistent Truncus Arteriosus (Common Truncus Arteriosus)

This is characterized as a single trunk leading to both ventricles. It also has a single atrial root with a single semilunar valve that is often dysplastic and incompetent. It is classified according to the origin of pulmonary arteries, i.e., whether they arise as a common trunk or arise separately as left and right pulmonary

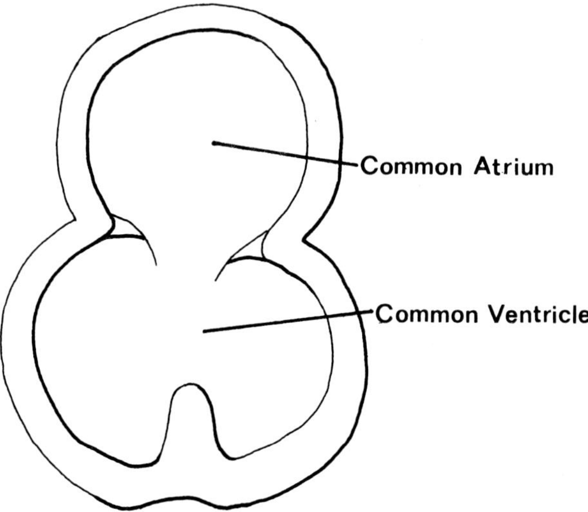

Figure 9.15
Endocardial cushion defect.

arteries, or they are absent and the pulmonary circulation is supported by collaterals. Aortic arch anomalies are common in this condition. An interventricular septal defect is always present, which leads to mixing of blood, resulting in cyanosis. Malposition of ventricles that causes abnormal septation and an outflow relationship has been suggested. Failure of neural crest cell migration to the truncoconal region is also suggested (Fig. 9.14A).

4. Atrioventricular Septal Defects (Endocardial Cushion Defects)

The atrioventricular septum is the portion of the cardiac septum that is above the tricuspid annulus and below the mitral annulus. Hence, this lesion involves the interatrial septum, the membranous interventricular septum, and the mitral and tricuspid valve leaflets. A complete atrioventricular septal defect has a single valve that covers both ventricles. A large defect permits interatrial and atrioventricular communications. A partial defect has separate right and left atrioventricular valve annuli and is accompanied by an ostium primum interatrial septal defect and an interventricular septal defect (Fig. 9.15).

The atrioventricular septal defect has been attributed to maldevelopment of the endocardial cushions and bulboventricular folds. The most significant association of this is with trisomy 21 (Down's syndrome). About 40% of children born with trisomy 21 have cardiovascular anomalies, and half of those have atrioventricular septal defects. The condition is also highly associated with asplenia cases.

Note. Absence of spleen (asplenia) is frequently found with cyanotic heart disease.

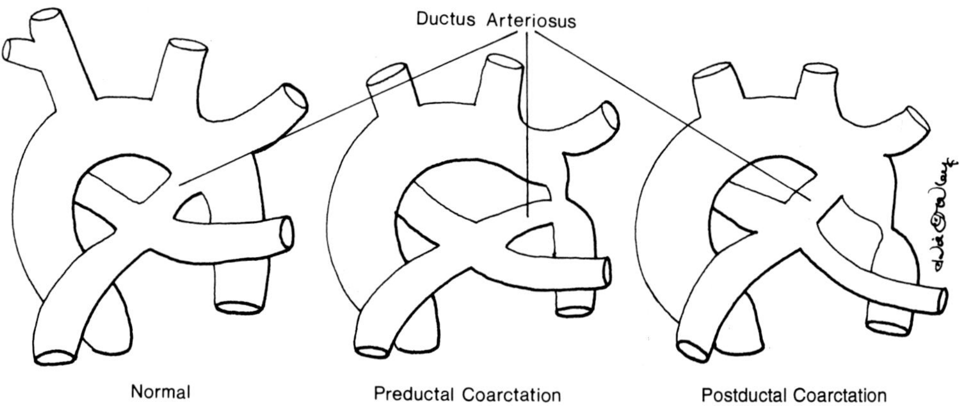

Figure 9.16
Coarctation of the aorta.

C. Anomalies of the Aortic Arches (No Shunt, Acyanotic)

Aortic arch malformations result from either persistent primitive derivatives that normally regress or the regression of structures that normally persist.

1. Coarctation of the Aorta

Due to abnormal constriction, narrowing of an aortic arch segment occurs close to the point of entry of the ductus arteriosus. The two principal lesions are described herein (Fig. 9.16).

a. Preductal coarctation (infantile tubular hypoplasia of the aortic arch) is narrowing of the aorta above the entry of the ductus arteriosus. It is often associated with other serious congenital heart disease. Prenatally, the blood from the right ventricle passes through the ductus arteriosus and supplies the aorta below the lesion, and blood circulation is not seriously disturbed. Postnatally, with closure of the ductus, the blood flow to the body is impeded, having fatal consequences. It is thought to be an abnormal development of the left fourth aortic arch.

b. Postductal coarctation is narrowing of the aortic segment below the entry of the ductus arteriosus. Blood flow through the aorta to the body is markedly reduced. To compensate for this, collateral circulation mainly through the intercostal arteries is established during fetal life. Enlargement of the intercostal arteries causes notching ribs. In some cases, the subclavian and internal thoracic arteries also take part in collateral circulation. Histologically, the constricted portion of the aorta is composed of ductal tissue without elastic media. This suggests that a large expansion of ductal tissue occurs in the wall of the aorta and, by fibrosis, may exert a pull on the aorta and interfere with its growth. The process is independent of closure of the ductus arteriosus. An abnormal infolding (kinking) of the posterior wall of the aorta to produce stenosis (narrowing) has also been suggested. Although it is more common in males, its frequency is high in Turner's syndrome (45,XO).

2. *Interrupted Aortic Arch*

This is an absence or narrowing of the aorta distal to the left subclavian artery. Narrowing could occur in other segments of the aortic arch. As an interventricular septal defect is usually present, it may result from a left-to-right shunt. Fifteen percent of infants born with interrupted aortic arch have DiGeorge's syndrome (hypoplasia or absence of the thymus and parathyroid glands).

3. *Double Aortic Arch*

This is characterized by bifurcation of the aortic trunk that forms a vascular ring around the esophagus and trachea. It may produce compression of the trachea and esophagus, causing difficulties in breathing and swallowing. It may result from failure of involution of the distal part of the right dorsal aorta.

4. *Retroesophageal Right Subclavian Artery*

This is the most common anomaly of the aortic arch system. The artery originates from the posterior wall of the aorta distal (close) to the origin of the left subclavian artery, and passes posterior to the esophagus and may cause difficulty in swallowing. Loss of a long segment of the right dorsal aorta may lead to this anomaly.

D. Anomalies of Venous Return

1. Double inferior vena cava at the lumbar level caused by the persistent connection between the left posterior cardinal vein and the left subcardinal vein.
2. Absence of the inferior vena cava.
3. Left superior vena cava is caused by persistence of the left anterior cardinal vein. It opens into the coronary sinus. It may exist along with right superior vena cava (double superior vena cava) or may exist alone, in which case the right common cardinal and right anterior cardinal veins are obliterated.
4. Anomalies of the pulmonary veins:
   a. Total anomalous pulmonary venous connection. None of the pulmonary veins connect with the left atrium, but open into the right atrium.
   b. Partial anomalous pulmonary venous connection. One or more of the pulmonary veins are connected to the right atrium.

TABLE 9.1. Cardiovascular Anomalies Associated with Chromosomal Abnormalities

	Frequency (%)
Trisomy 21	
Atrioventricular septal (endocardial cushion) defect	40%
Interventricular septal defect	5
Interatrial septal defect	5
Without cardiac anomalies	50
Trisomy 18	
Interventricular septal defect	70
Preductal coarctation of aorta	10
Interatrial septal defect, secundum type	10
Without cardiac defects	10
Trisomy 13	
Interventricular septal defect without tetralogy	35
Interatrial septal defect	35
Preductal coarctation	15
Persistent left superior vena cava	10
Without cardiac defects	5
Turner's syndrome (45,XO)	
Preductal and postductal coarctation	70
Aortic stenosis	10
Persistent left superior vena cava	5
Without cardiac defects	15

Frequencies were averaged from three different studies.

TABLE 9.2. Summary of Development of the Cardiovascular System

Age (days)	Embryonic structure
18	Cardiogenic plate
20	Endocardial plexus
22	Fusion of primitive heart tubes, appearance of first aortic arch
23	Conoventricular loop
26	Septum primum, foramen primum, atrioventricular canal, interventricular septum
28	Venous valves, endocardial cushions and conotruncal ridges
32	Endocardial cushions approximated
38	Septum secundum, foramen secundum and absorption of bulbus cordis
42	Obliteration of foramen primum, division of truncus arteriosus, and appearance of ductus venosus
44	Fusion of endocardial cushion
48	Completion of intraatrial and interventricular septa, formation of foramen ovale, absorption of pulmonary vein and beginning of superior and inferior venae cavae
56	Absorption of sinus venosus, definitive plan of main blood circulation

REVIEW QUESTIONS — Chapter 9

1. In the middle of the third week, angioblasts (blood islands) are formed in all of the following except the:
   a. yolk sac
   b. chorion
   c. connecting stalk
   d. intraembryonic mesoderm
   e. amnion
2. The heart (endocardial) tube develops in the:
   a. splanchnic mesoderm
   b. somatic mesoderm
   c. paraxial mesoderm
   d. intermediate mesoderm
   e. septum transversum
3. The inflow of heart tube is the:
   a. bulbus cordis
   b. ventricle
   c. atrium
   d. sinus venosus
4. During early development of the heart, four layers become apparent. Of these, the cardiac jelly represents:
   a. connective tissue on the surface of the heart
   b. a primitive muscular layer
   c. an inner elastic layer of the heart wall
   d. a thick acellular matrix
   e. the innermost lining of the heart
5. The sinoatrial (SA) node develops in the:
   a. right common cardinal vein
   b. primitive atrium
   c. primitive ventricle
   d. bulbus cordis
   e. sinus venosus
6. The cresta terminalis that separates the trabeculated and smooth part of the right atrium is derived from the:
   a. left venous valve on the sinoatrial opening
   b. right venous valve on the sinoatrial opening
   c. septum secundum
   d. septum primum
   e. valve to inferior vena cava

7. The vitelline circulation:
   a. develops in the wall of the yolk sac
   b. gives rise to umbilical vessels
   c. brings blood from the placenta
   d. delivers blood to the placenta
   e. all of the above
8. Differentiation of the heart starts at the:
   a. truncus arteriosus
   b. conus cordis
   c. ventricle
   d. atrium
   e. sinus venosus
9. As a result of cephalic folding of the embryo, the oropharyngeal membrane and the structures located cephalic to it rotate 180°, so that the:
   a. pericardial cavity becomes dorsal to the cardiogenic (heart) tube
   b. common atrium migrates dorsal to the ventricles
   c. septum transversum becomes ventral to the foregut
   d. dorsal aortae come together and fuse with each other
   e. sinus venosus shifts ventrocephalically
10. The cardiogenic plate is induced by the:
    a. underlying endoderm
    b. overlying ectoderm
    c. notochordal plate
    d. oropharyngeal membrane
    e. intraembryonic angiogenetic cords
11. The inferior vena cava develops from the:
    a. right vitelline vein
    b. left umbilical vein
    c. right common cardinal vein
    d. right precardinal (anterior cardinal) vein
    e. all of the above contribute to formation of inferior vena cava
12. The intraembryonic coelom located cranial to the prochordal plate (oropharyngeal membrane) gives rise to the:
    a. oral cavity
    b. pericardial cavity
    c. septum transversum
    d. thoracic cavity
    e. peritoneal cavity
13. The proximal part (closer to the entrance to the atrium) of the superior vena cava is derived from the:
    a. left common cardinal vein
    b. right common cardinal vein
    c. right vitelline vein
    d. right supracardinal vein
    e. right sinus venosus

14. The derivatives of the precardinal (anterior cardinal) veins include the:
    a. brachiocephalic veins
    b. subcardinal veins
    c. supracardinal veins
    d. azygos vein
    e. all of the above
15. During the bending of the endocardial (heart) tube, the atrium moves:
    a. inferiorly
    b. ventrally and to the right
    c. to the left
    d. posteriorly and superiorly
    e. ventrally and to the left
16. Bending of the endocardial (heart) tube is caused by:
    a. active migration of the myocytes
    b. a change in the shape of cells
    c. differential growth
    d. slow growth of pericardial cavity
    e. all of the above
17. Differentiation of the heart starts at the:
    a. sinus venosus
    b. atrium
    c. ventricle
    d. bulbus cordis
    e. aortic arches
18. At the completion of division of truncus arteriosus, one part joins the left fourth aortic arch and the other part joins the:
    a. third aortic arch on both sides
    b. seventh segmental artery
    c. right fourth aortic arch
    d. sixth aortic arch
    e. descending aorta
19. The neural crest cells migrate to the developing heart and contribute to the:
    a. atrial myocardium
    b. ventricular myocardium
    c. cellular components to the wall of the truncus arteriosus
    d. endothelial components to the outflow tract of the heart
    e. all of the above
20. A child was born with difficulty breathing, a raised sternal angle, flared nostrils and excessive sweating. Further investigation found a defect close to the entrance of the superior vena cava to the right atrium. The anomaly is (questions 20–23 pertain to this case):
    a. an interventricular septal defect
    b. an osteum secundum defect
    c. an osteum primum defect
    d. a common atrium
    e. a persistent ductus arteriosus

21. The general symptoms were caused by:
    a. a left-to-right shunt
    b. a right-to-left shunt
    c. no shunting of blood
    d. any of the above
    e. none of the above

22. The cause of this malformation could be:
    a. increased programmed cell death in the conotruncal region that delays fusion of the truncal region
    b. failure of neural crest cell migration to the conotruncal region
    c. a defect in atrioventricular septation
    d. excessive resorption of the septum primum
    e. delayed postnatal closure

23. Other findings included:
    a. cyanosis
    b. two pulmonary veins opened in right atrium
    c. trisomy 21
    d. maternal rubella infection during early pregnancy
    e. all of the above

Questions 24–26:
Soon after birth, a female child began to have difficulty in breathing. There was no cyanosis, and no septal defect was found. Her pulmonary circulation was increased. Her mother indicated that she felt sick in her early pregnancy and was treated for rubella infection.

24. The symptoms result from:
    a. a left-to-right shunt
    b. a right-to-left shunt
    c. a right-to-left shunt without cyanosis
    d. these symptoms are not related to shunt

25. The most appropriate anomaly would be:
    a. an overriding pulmonary artery
    b. an overriding aorta
    c. a persistent ductus arteriosus
    d. a patent foramen ovale
    e. an endocardial cushion defect

26. Before any intervention, which of the following should be evaluated?
    a. size of defect
    b. the amount of blood passing through the defect
    c. the pressure difference between the pulmonary trunk and aorta
    d. systemic and pulmonary resistance
    e. all of the above

27. All of the following produce cyanosis except:
    a. a preductal coarctation of aorta
    b. a tetralogy of Fallot
    c. a persistent truncus arteriosus
    d. an atrioventricular septal defect
    e. a complete transposition of the great arteries

28. Aortic stenosis or aortic atresia
    a. leads to right ventricular hypertrophy
    b. results in left ventricular hypoplasia
    c. may be caused by unequal division of the truncus arteriosus
    d. associated with decreased coronary circulation
    e. all of the above are correct

29. Prenatal closure of the ductus arteriosus may:
    a. cause right-to-left shunt
    b. be caused by maternal ingestion of aspirin during pregnancy
    c. be caused by in utero administration of prostaglandins
    d. cause an excessive load on the left ventricle
    e. increase circulation through the aorta

30. Regarding coarctation of aorta:
    a. the ductus arteriosus typically closes after birth in preductal coarctation
    b. the blood pressure is higher in the upper limbs than in the lower limbs in postductal coarctation
    c. preductal coarctation is found more frequently than postductal coarctation in live births
    d. a collateral circulation is established in both types of coarctation
    e. rib notching is found in both types of coarctation

31. The truncal ridges may fail to spiral, resulting in:
    a. tetralogy of Fallot
    b. a ventricular septal defect
    c. transposition of great vessels
    d. a common atrium
    e. all of the above

32. Tetralogy of Fallot is characterized by:
    a. aortic stenosis
    b. an atrial septal defect
    c. right ventricular hypertrophy
    d. overload of the pulmonary circulation
    e. all of the above

ANSWERS TO REVIEW QUESTIONS

1. e
2. d
3. d
4. d
5. e
6. b
7. a
8. b
9. c
10. a
11. a
12. b
13. b
14. e
15. d
16. e
17. a
18. d
19. c
20. b
21. a
22. d
23. b
24. a
25. c
26. e
27. a
28. e
29. b
30. b
31. c
32. c

CHAPTER

10

RESPIRATORY SYSTEM

I. BODY CAVITY[1]

A. Closure of the Intraembryonic Coelom

The early intraembryonic coelom appears as a horseshoe-shaped cavity. The curve of the horseshoe represents the future pericardial cavity, and the lateral limbs represent the future pleural and peritoneal cavities (Fig. 10.1). Communication between the extraembryonic and intraembryonic coeloma becomes occluded during lateral folding of the embryonic disk (Fig. 10.2). The lateral limbs of the intraembryonic coelom approach each other and fuse on the ventral aspect of the embryo. This folding of the embryo transforms the intraembryonic coelom into a closed cavity. The ventral mesentery thus formed disappears except where it is attached to the lower part of the foregut, resulting in a large cavity that extends from the thoracic to the pelvic regions. Four well-defined parts of the coelomic cavity can be recognized:

1. a large pericardial cavity;

2,3. pericardio-peritoneal canals (pleural canals), which are situated dorsal to the septum transversum on each side of the foregut, connecting the pericardial and peritoneal cavities;

4. a large peritoneal cavity.

---

[1]Review body folding, Chapter 4.

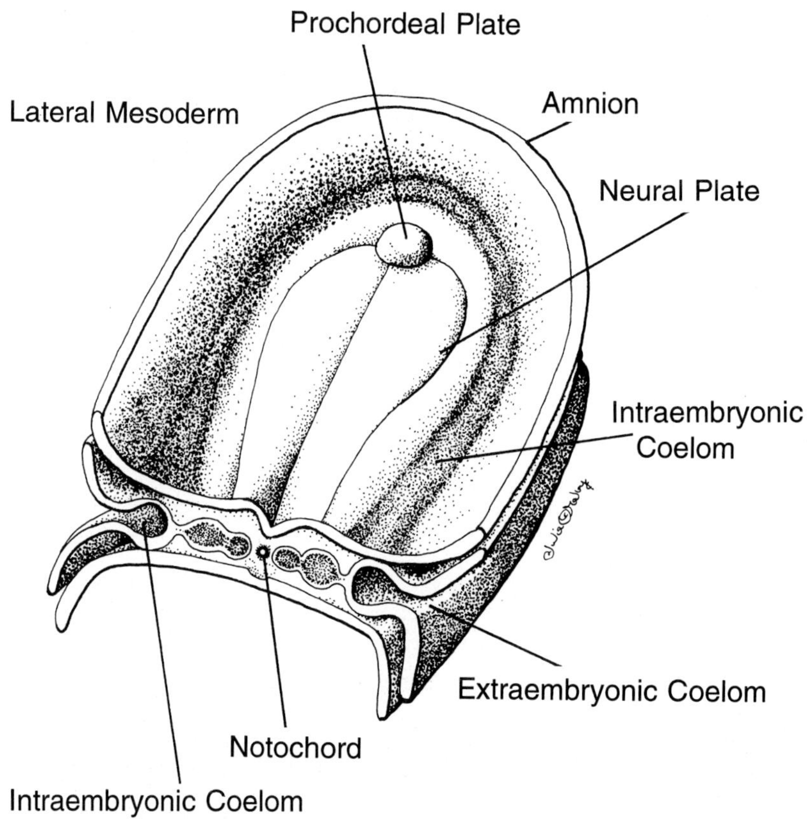

Figure 10.1
Primitive body cavity or coelom of an early embryo.

The cells of the somatic mesoderm lining these cavities become the mesothelium and will later develop into the parietal serous membranes. The cells from the splanchnic mesoderm will give rise to the visceral serous membranes.

II. DEVELOPMENT OF THE LUNGS

A. Lung Buds (Lung Primordium) Formation (26 days to 6 weeks)

At approximately day 26, the median laryngotracheal (pharyngeal) groove appears in the ventral wall of the primitive pharynx. Interaction between the epithelium and mesenchyme is necessary for normal development. The surrounding mesenchyme determines the pattern of growth and the branching of the endodermal components (Fig. 10.3).

The laryngotracheal groove deepens to form the respiratory (laryngotracheal) diverticulum, which grows caudally and becomes separated from the pharynx by the tracheoesophageal septum (Fig. 10.2). The septum appears as epitheliomesenchymal ridges that extend inward from the sides of the foregut. These ridges

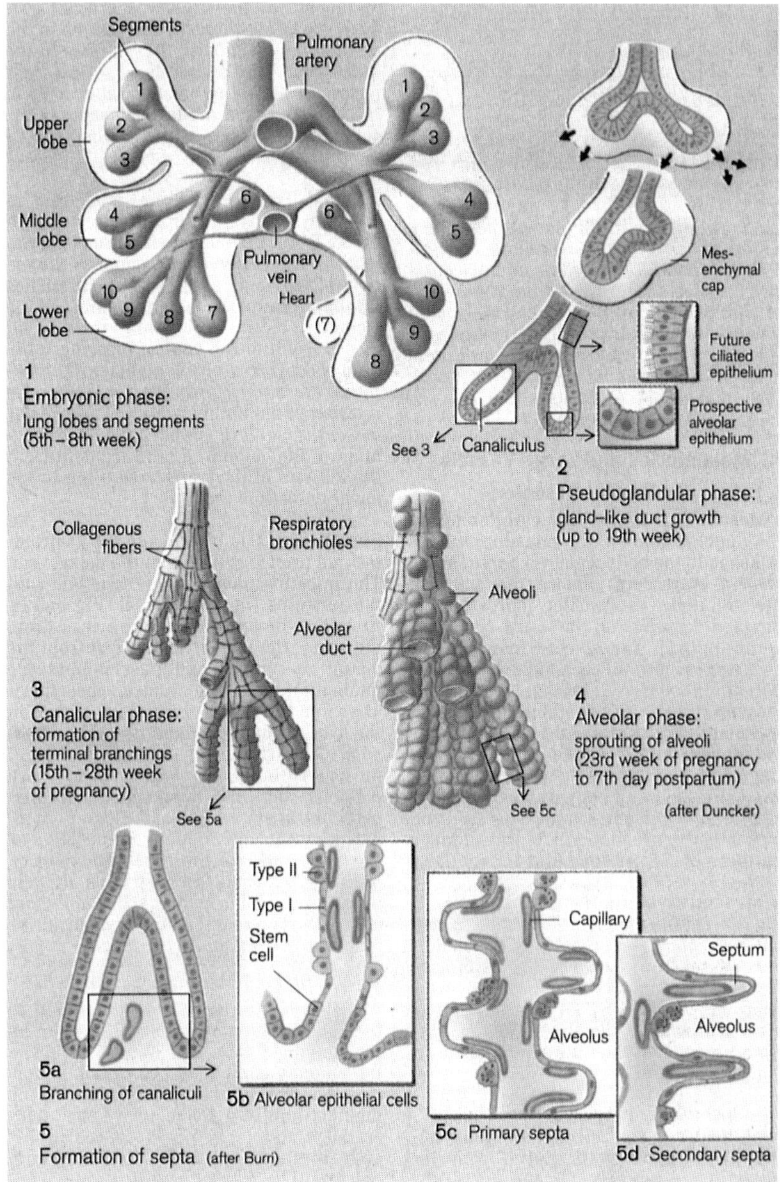

Figure 10.2

Differentiation of the lungs (see Color Plate 10.2). Reproduced with permission from Thieme Medical Publishers Inc., New York, 1995, *Color Atlas of Embryology,* Ulrich Drews, Chapter 7: Gastrointestinal Tract.

gradually grow toward each other and fuse. This septum divides the primitive pharynx into the laryngotracheal tube and the esophagus. The distal end of the respiratory diverticulum divides to form the right and left lung buds. The undivided proximal part forms the trachea. At the proximal end of the trachea, the mesenchyme derived from the fourth, fifth and sixth branchial arches proliferates

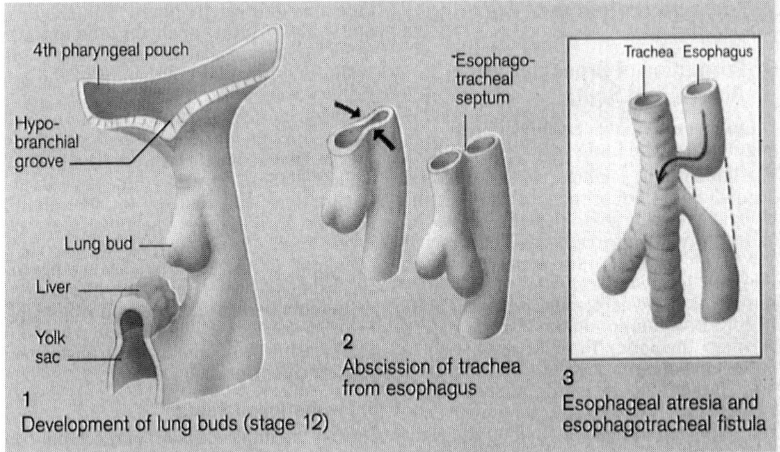

Development of lung buds

Cardinal veins
Pleuropericardial fold
Lung bud
Coelomic canal
See 2
Diaphragm
Pericardium
Pleuro-
peritoneal
fold
Liver
bud
Peritoneal
cavity
Septum transversum
1
Demarcation of pleural cavities
Aorta
Parietal
visceral
pleura
Trachea
Hilus of
lung
Pulmonary artery
Pulmonary vein
Phrenic nerve
2
Development of hilus of lung

Growth of lungs into the coelomic canals

Figure 10.3

Lung buds (see Color Plate 10.3). Reproduced with permission from Thieme Medical Publishers Inc., New York, 1995, *Color Atlas of Embryology*, Ulrich Drews, Chapter 7: Gastrointestinal Tract.

and produces the arytenoid swellings, which give rise to laryngeal cartilages and laryngeal muscles.

The lung buds grow first dorsally around the esophagus, and then caudolaterally, thereby penetrating into the pericardioperitoneal canal, which is the lateral limb of the U-shaped coelomic cavity (Fig. 10.3). The lung buds expand, rapidly filling the pericardioperitoneal canal, now referred to as pleural cavities. The splanchnic mesoderm, which was the medial lining of the pleural cavities, develops into visceral pleura; the somatic mesoderm, which was the lateral lining of the pleural cavities, develops into the parietal pleura.

The right lung bud divides into three main bronchii, and the left bud divides into two main bronchii.

Changes in the bronchial tree are brought about by mitosis, cell death and cell migration; its division and growth occur in areas of low resistance in the mesenchyme.

B. Development of the Lower Respiratory Tract

Further development of the lungs is divided into the following periods (Fig. 10.2):

1. Pseudoglandular period (7 weeks to 16 weeks)
   At seven and eight weeks, respectively, segmental bronchi and bronchopulmonary segments are present. The lungs resemble exocrine glands. About 16–25 generations of bronchi and bronchioles are completed.
2. Canalicular period (13 weeks to 25 weeks)
   At 13 weeks, the terminal bronchioles divide into respiratory bronchioles. Canalization, the invasion of lung epithelium by capillaries, begins at about 17 weeks. Each respiratory bronchiole gives rise to three to six alveolar ducts. Each duct ends in a terminal (alveolar) sac of cuboidal epithelium. Capillaries develop rapidly and grow closer to the epithelium, and a blood-air barrier is established.
3. Terminal sac period (24 weeks to birth)
   At 24 weeks, the formation of terminal sacs and capillary network is completed. The epithelial lining in sacs is continuous.
4. Alveolar period (28 weeks to postnatal life)
   At 28 weeks, the epithelium in the terminal sac is greatly changed. The epithelial cells differentiate into the following cell types:
   a. Type I cells, or pneumocytes
      These are cuboidal cells with brush borders.
   b. Type II cells
      These contain osmiophilic granules (mainly lecithin) that are the precursors of surfactant. At this stage the fetus is capable of respiration. After this stage, new alveoli are formed as small buds on the existing alveoli, and the parent alveolus is transformed into an alveolar duct. The alveoli increase greatly in number until age 3 years and continue to grow at a slower rate up to 8 years of age.

Thyroxin is a potent stimulant of surfactant production. Suprarenal steroids also influence the maintenance of type II cells and the production of surfactant during intrauterine life. Fetal thoracic movements become marked during the third trimester, and are important for development of the airways. These movements are brought about by contraction of the diaphragm and abdominal wall muscles.

The endodermal lining of the respiratory diverticulum gives rise to the following:

1. the epithelial lining of the tracheobronchial tree and pulmonary alveoli,
2. glands of the larynx, trachea and bronchi.

The splanchnic mesoderm surrounding the lung bud gives rise to the following:

1. connective tissue
2. cartilages
3. smooth muscle
4. capillaries

The capability of breathing after birth depends on the functionally ready respiratory units and a sufficient amount of surfactant. Development of an adequate pulmonary vasculature is critical to the survival of premature infants.

C. Changes at Birth

The lungs at birth are about half-filled with fluid derived from the lungs, amniotic cavity and tracheal glands. During and after birth, this fluid should be replaced for lungs to function properly. The fluid is reduced as follows:

1. by expulsion caused by pressure on the thoracic cavity;
2. by absorption by lymphatic and vascular network.

The cold, dry extrauterine environment, little anoxia and tying of the umbilical cord stimulate the first breath, which is followed by rhythmic respiration.

III. DIVISION OF THE BODY CAVITY

A. Separation of the Pericardial Cavity from the Pleural Cavities

The mesenchyme, containing the common cardinal veins and the phrenic nerve, bulges out from the body wall as pleuropericardial membranes (Fig. 10.4). Subsequently, the lung buds grow into the medial wall of the pericardioperitoneal canals or primitive pleural cavities. These cavities expand ventrally around the heart by splitting the mesenchyme of the body wall. In this manner, the pleuropericardial membranes become free dorsally and project into the cephalic end of the pericardioperitoneal canals. The free ends of the pleuropericardial membranes extend dorsally, approach each other and finally fuse with each other and with the root of the lungs, separating the pericardial cavity from the pleural cavities. Because of the large size of the right common cardinal vein, the membrane on the right side is larger, and the right opening closes slightly earlier than the left one. The pleuropericardial membranes form the adult fibrous pericardium.

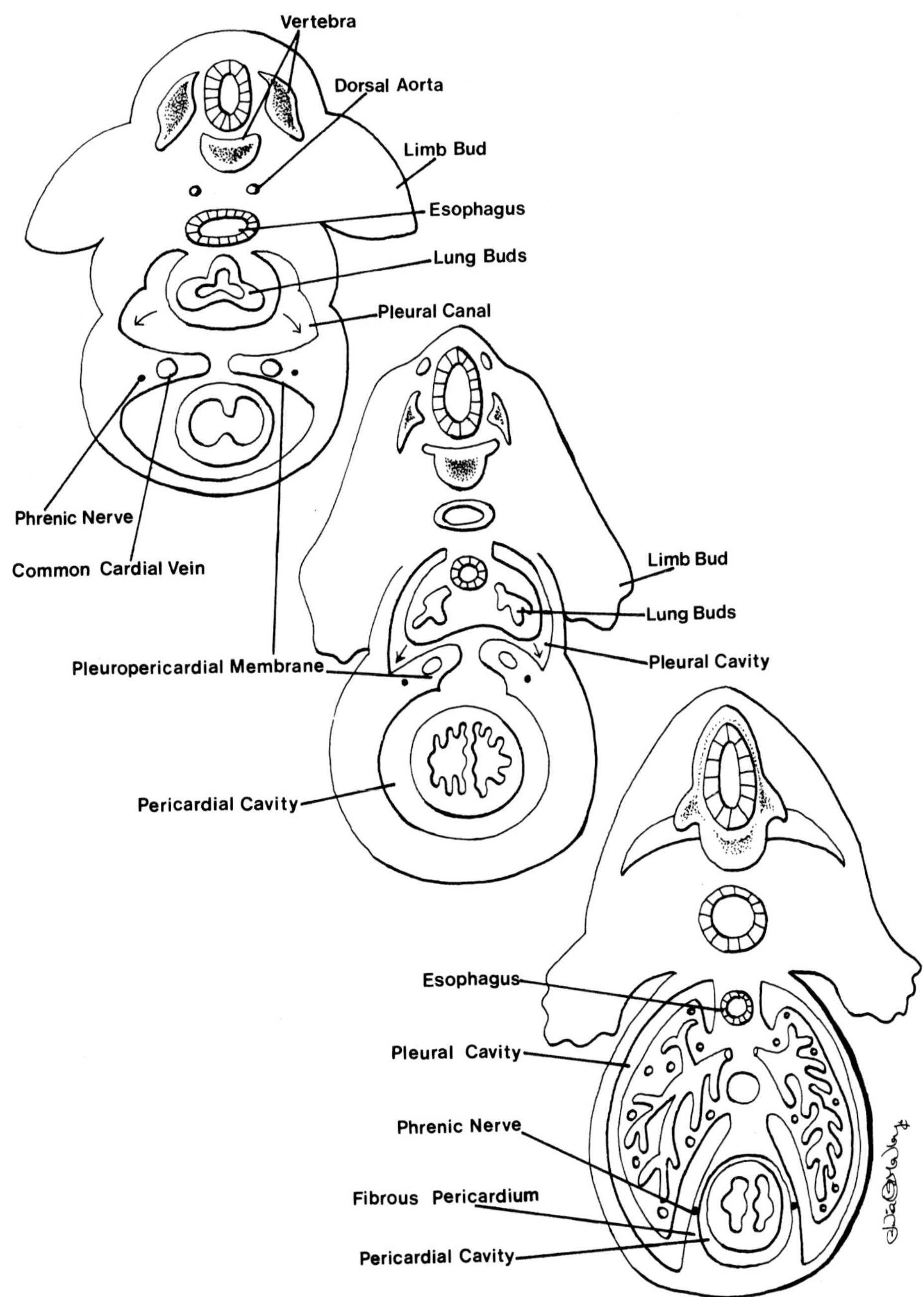

Figure 10.4
Separation of pericardial cavity from the pleural cavities.

B. Separation of the Pleural Cavities from the Peritoneal Cavities

At the caudal ends of the pericardioperitoneal canals, a pair of crescent-shaped folds appears as pleuroperitoneal membranes. These are attached dorsolaterally to the body wall, and their free edges project into the caudal ends of the pericardioperitoneal canals (Fig. 10.5). The pleuroperitoneal membranes grow medioventrally, and their free ends fuse with the dorsal mesentery of the esophagus and with the septum transversum. Although fusion of the pleuroperitoneal membranes separates the pleural cavities from the peritoneal cavity, closure of the openings between these cavities results from the rapid growth of the liver and invasion of the muscular tissue into the pleuroperitoneal membranes. The opening on the right side closes slightly earlier than the one on the left.

IV. DEVELOPMENT OF THE DIAPHRAGM

The diaphragm develops from four structures:

A. The septum transversum forms a partial partition between the pericardial and peritoneal cavities, leaving a large opening on both sides of the foregut (the pericardioperitoneal canals). The septum fuses with the ventral mesoderm of the esophagus and later with the pleuroperitoneal membranes. The septum forms the central tendon of the adult diaphragm.

B. The pleuroperitoneal membranes, as mentioned earlier, fuse with the dorsal mesentery of the esophagus and with the dorsal portion of the septum transversum. Although these membranes form a large portion of the primitive diaphragm, they represent a relatively small intermediate portion of the adult diaphragm (Fig. 10.5).

C. The dorsal mesentery of the esophagus forms the median portion of the diaphragm in which the crura of the diaphragm develop from extensions of muscle fibers into this mesentery.

D. The muscular components from the lateral and dorsal body wall contribute the peripheral rim of the diaphragm, external to the portion derived from the pleuroperitoneal membranes. These muscular components are therefore innervated by the intercostal nerves. Muscular tissues later spread over the pleuroperitoneal membrane.

Further expansion of the pleural cavities into the body wall forms the costodiaphragmatic recesses.

V. CONGENITAL MALFORMATIONS OF THE RESPIRATORY TRACT

A. Tracheoesophageal fistula
Abnormal communication between the trachea and esophagus results from incomplete division of the foregut into the respiratory and digestive portions, which may have been caused by incomplete fusion of the epithelial ridges or by

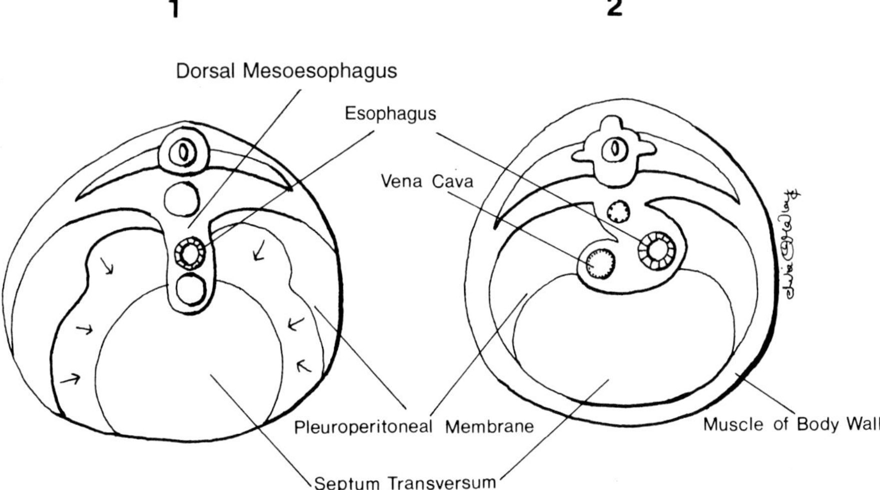

Figure 10.5
Development of the diaphragm.

deviation (misdirection) of the tracheoesophageal septum. Rarely does a tracheoesophageal fistula occur without esophageal atresia. Development of an abnormal epithelial connection between already separated tubes is also suggested. There are four main types:

1. Esophageal atresia with the fistula between the trachea and the distal portion of the esophagus is most common (about 90% of cases) (Fig. 10.6) and may cause chemical pneumonia.
2. Esophageal atresia without fistula. Either the fistulous canal becomes ligamentous or the distal portion of the esophagus is not present. Saliva in lungs may cause pneumonia. This is not emphysema, because there is no destruction of lung tissue. The alveoli are distended up to three to ten times normal (Fig. 10.6b).
3. Esophageal atresia with a fistula between the proximal part of the esophagus and trachea.
4. Esophageal atresia with the fistula connecting both the proximal and distal ends of the esophagus with the trachea (Fig. 10.6). Tracheo-esophageal fistula may be associated with vertebral anomalies, anorectal malformation and persistent ductus arteriosus.

B. Tracheal stenosis or atresia may also be caused by unequal division of the foregut.

C. Trachea diverticulum may give rise to the tracheal lobe.

D. Congenital bronchial cysts may result from abnormal dilatation of the terminal bronchi.

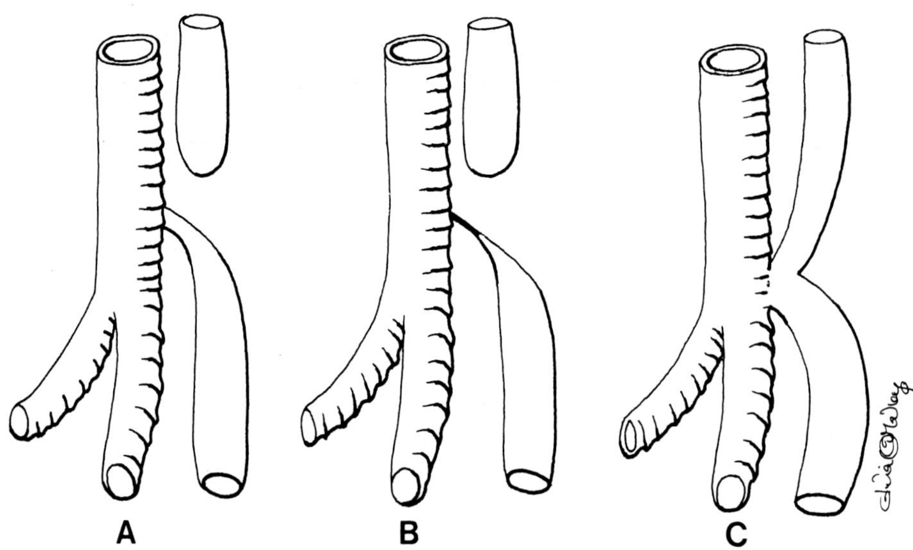

Figure 10.6

Malformations of esophagus and trachea: (A) esophageal atresia with esophageal-tracheal fistula; (B) esophageal atresia without fistula; (C) esophageal atresia with fistula connecting both proximal and distal esophagus.

E. Agenesis of the lungs
Absence of one or both lungs results from failure of one or both lung buds to develop. Bilateral agenesis is rare. Unilateral agenesis may be compatible with normal life.

F. Congenital lobar overinflation (emphysema)
This has been attributed to a wide variety of abnormalities, including external bronchial compression, bronchiogenic cysts and interbronchial flap valves. Histologically, bronchial walls contain a decreased amount of cartilage. Shortly after birth, the neonate develops marked respiratory distress. There is overexpansion of part of the lung. The left upper lobe is mostly involved.

G. Respiratory distress syndrome (hyaline membrane disease)
Absence or deficiency of surfactant is probably the major cause of this disease. Prolonged intrauterine asphyxia may produce irreversible changes in surfactant-producing cells. A membrane-like substance is formed from the injured pulmonary cells. It results in failure of the alveoli to dilate. This accounts for 30% of all neonatal deaths and 50 to 70% of deaths in premature infants.

H. Diaphragmatic hernia:

1. Posterolateral defect
This results from defective formation and/or fusion of the pleuroperitoneal membranes. The defect, usually unilateral, consists of a large opening often known as the foramen of Bochdalek. It is five times more common on the left side than the right side and may be caused by early

closure of the right pleuroperitoneal opening. The abdominal viscera, which pass through the defect into the pleural cavities, push the heart anteriorly or to the right, and the lungs may be hypoplastic. Varying degrees of hypoplasia are present. Initiation of respiration at birth is difficult (Chapter 13, Fig. 13.7).

2. Congenital eventration of the diaphragm
   In this condition, half of the diaphragm has defective musculature, because muscular tissue failed to extend into the pleuroperitoneal membrane of the affected side.

3. Parasternal or retrosternal hernia of Morgagni
   This is located between the sternal and costosternal portions of the diaphragm. Presumably, a small part of the muscle failed to develop from the septum transversum.

4. Esophageal hiatal hernia
   An excessively large esophageal opening may be caused by weakness of the muscle of the right crus. The positive intraabdominal pressure and negative intrathoracic pressure push the cardia of the stomach through the defect. A child born with this defect vomits when laid on its back because the stomach contents move up toward the esophagus. Nursing of the child must be done in an upright position.

I. Esophageal hernia
   Caused by congenital shortness of the esophagus. The cardia of the stomach is retained in the thoracic cavity.

J. Congenital pericardial defects
   Usually found on the left side. A portion of the heart, usually an auricle, herniates into the pleural cavity at each heartbeat.

K. Pulmonary hypoplasia and immaturity
   Various aspects of lung growth retardation, cellular maturity, tissue maturity and the complex nature of lung tissue are important. One or all of these may influence growth of the lung. To consider hypoplasia, the lung must be too small, it must have markedly decreased airways, and it must have too few alveoli and too few cells. Decreased cellular maturity is assessed by reduced glycogen and the appearance of osmiophilic lamellar bodies in pulmonary cells. This could be measured either in aspirated amniotic fluid or directly in the lungs.

Hypoplastic lungs caused by diaphragmatic hernia is associated with hyaline membrane formation, suggesting that there is a type II cell immaturity in this hypoplasia. This has been a model to study pulmonary hypoplasia. In some cases, maternal CO_2 breathing results in increased cellular and tissue maturity, leading to hyperplasia. The following factors have been found to be associated with hypoplasia:

A. Lung compression
Aside from diaphragmatic hernia, any chest wall anomaly (e.g., thoracic dystrophies, osteogenesis imperfecta, plural effusion, ascites or prune belly syndrome) may result in lung compression that leads to hypoplasia.

B. Oliogohydramnios
The amount of liquid in the lung is a major determinant of lung growth. Reduction in the amount of amniotic fluid, which may have been caused by prolonged leakage, causes hypoplasia of lung. This may have resulted indirectly from compression of the fetus in general and compression on the lungs in particular.

C. Renal anomalies
The urinary and respiratory systems develop at the same time. The strong association of lung hypoplasia with renal anomalies suggests that these may be linked to a common developmental defect. The common factor may be abnormal synthesis of proline, which is essential for collagen synthesis, thought to be involved in bronchial branching. Although rare, hypoplasia has been reported in newborns having renal anomalies without oligohydramnios.

D. Fetal lung movements
There is now strong evidence that fetal lung movements are critical for normal lung development and growth. Experimental denervation of the diaphragm in a sheep resulted in pulmonary hypoplasia, decreased type II cell maturity and decreased tissue maturity.

E. Hormonal effects
Glucocorticoid administration may accelerate lung maturation or may affect lung growth. Depending on the dose, these may induce type II cell maturation or inhibit DNA synthesis, which may lead to hypoplasia. Similarly, the administration of triamcinolone acetonide (TCA) increases lung maturity. Depending on gestational age, it may lead to hypoplasia. It also produces oligohydramnios.

Maternal diabetes in the human is associated with an increase in frequency of respiratory distress syndrome resulting from lung immaturity. Administration of epidermal growth factor results in increased type II cell maturation and DNA synthesis.

REVIEW QUESTIONS — Chapter 10

1. The coelomic cavity consists of four parts. The lung buds grow into the:
   a. pericardial cavity
   b. pericardio-peritoneal canals
   c. peritoneal cavity
   d. septum transversum
   e. amniotic cavity

2. The laryngotracheal groove:
   a. is an epitheliomesenchymal ridge that gradually grows and fuses on the opposite wall
   b. deepens to form a respiratory diverticulum (lung bud)
   c. is part of the splanchnic mesoderm that will line the pleural cavity
   d. is a part of the somatic mesoderm that will develop into the visceral pleura

3. The epithelial component of a lung bud (respiratory diverticulum) gives rise to all of the following EXCEPT the
   a. type I pulmonocytes (pneumocytes)
   b. type II pulmonocytes (pneumocytes)
   c. clara cells
   d. endothelial cells
   e. goblet cells

4. The primitive pharynx is divided into the trachea and esophagus. The area demarcating this division is represented by the:
   a. root of the tongue
   b. foramen cecum
   c. epiglottis
   d. first tracheal ring
   e. none of the above

5. The mesenchyme surrounding the lung buds determines the pattern of their growth and gives rise to the:
   a. smooth muscles
   b. pseudostratified columnar epithelium
   c. epithelial lining of the alveolar sac
   d. type II cells
   e. all of the above

6. The pneumocytes differentiate during the:
   a. pseudoglandular period
   b. canalicular period
   c. terminal sac period
   d. alveolar period

7. The factor that could affect normal development of the lungs is:
   a. oligohydramnios
   b. fetal lung movements
   c. hormones
   d. lung compression
   e. all of the above

8. The stimulus for surfactant production during intrauterine life is:
   a. thyroxin
   b. lecithin
   c. cholesterol
   d. thyroid-stimulating hormone
   e. all of the above

9. The maximum number of alveoli in the lungs is reached at the:
   a. canalicular period
   b. terminal sac period
   c. alveolar period
   d. age of 1 year
   e. age of 8 years

10. Which adult structure is derived from the pleuroperitoneal membrane?
    a. lungs
    b. heart
    c. fibrous pericardium
    d. diaphragm
    e. thymus

11. The central tendon of the diaphragm is formed from the:
    a. muscles of the dorsal and lateral body wall
    b. dorsal mesentery of the esophagus
    c. pleuroperitoneal membrane
    d. septum transversum

12. The pleuroperitoneal membrane of the diaphragm is replaced by the:
    a. central tendon of the diaphragm
    b. dorsal mesentery of the esophagus
    c. septum transversum
    d. muscles of the body wall

Questions 13–16:
A child was born 4 weeks before the due date. Soon after birth, he became cyanotic, could not breathe and died. His chest was raised and his abdomen flat. Chest X-rays showed large air shadows.

13. What condition did he have?
    a. respiratory distress syndrome
    b. congenital diaphragmatic hernia
    c. tracheoesophageal fistula
    d. congenital shortening of the esophagus
    e. congenital overinflation

14. The major cause for difficulty in breathing is:
    a. absence or deficiency of surfactant
    b. hypoplastic lungs
    c. prolonged intrauterine asphyxia
    d. esophageal atresia
    e. external bronchial compression

15. The cells most directly involved in this condition are:
    a. type I pneumocytes
    b. type II pneumocytes
    c. both
    d. neither

16. The most significant prenatal test would be:
    a. measurement of the lecithin/sphingomyelin ratio in the amniotic fluid
    b. karyotype analysis
    c. measure of a-fetoprotein in the maternal blood
    d. ultrasonography
    e. all of the above

REVIEW QUESTIONS — Chapter 10

1. b 2. a 3. d 4. e
5. a 6. d 7. e 8. a
9. e 10. c 11. d 12. d
13. b 14. b 15. d 16. d

CHAPTER

11

PHARYNGEAL APPARATUS

The embryonic pharynx is the rostral-most part of the foregut and is bounded laterally and ventrally by the pharyngeal apparatus. The pharyngeal apparatus consists of the pharyngeal arches, the pharyngeal groove (clefts), the pharyngeal pouches, and the membranes formed by the opposing clefts and pouches (Fig. 11.1).

The surface ectoderm of the ventral and lateral aspects of the pharyngeal gut show oblique cleft-like invaginations, called pharyngeal grooves. Coinciding to the grooves, outpocketings from the endodermal lining of the pharynx develop as pharyngeal pouches. The grooves and pouches approach each other, dividing the pharyngeal mesoderm into segments called pharyngeal arches. The mesenchyme of each condenses to form a core of cartilage and striated muscles. Four pairs are visible externally at 4 weeks. Ectodermal thickenings of arches, the epipharyngeal disks, give rise to ganglia of cranial nerves V to X.

I. CRANIAL PHARYNGEAL ARCH COMPONENTS

The skeletal component of pharyngeal arches receives midbrain and hindbrain neural crest cells.

A. First pharyngeal arch (mandibular arch) (Table 11.1)

1. Derivatives of Meckel's cartilage

   a. The head of malleus, and body and short crus of incus (two middle ear bones).

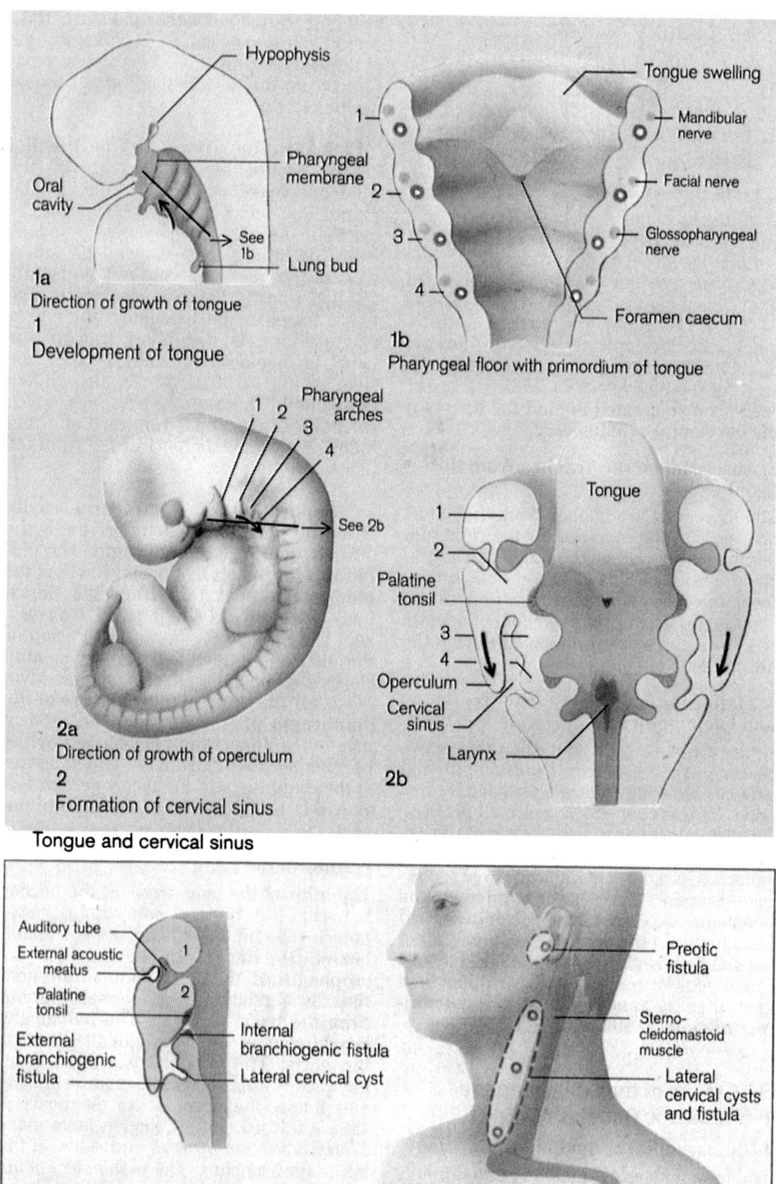

Figure 11.1

Cervical sinus and cervical cysts (see Color Plate 11.1). Reproduced with permission from Thieme Medical Publishers Inc., New York, 1995, *Color Atlas of Embryology,* Ulrich Drews, Chapter 7: Gastrointestinal Tract.

b. Anterior ligament of malleus and sphenomandibular ligament. The ventral portion of this cartilage largely disappears. The mandible is formed secondarily around this portion by intramembranous ossification.

TABLE 11.1. Pharyngeal Arch Components

Arch	Skeletal structure	Musculature	Innervation
I	Head of malleus Body and short crus of incus Ant. lig. of malleus Sphenomandibular lig. Maxilla, zygomatic, mandible Squamous portion of temporal	Muscles of mastication Mylohyoid Ant. belly of digastric Tensor veli palatini Tensor tympani	Mandibular V_3
II	Handle of malleus Long crus of incus Stapes Styloid process Stylohyoid lig. Lesser horn of hyoid	Muscles of facial expression Stylohyoid Post. belly of digastric Stapedius	Facial VII
III	Body and greater horn of hyoid	Stylopharyngeus	Glosso-pharyngeal IX
IV	Laryngeal cartilages	Pharyngeal constrictors Cricothyroid Levator veli palatini	Superior laryn-geal nerve X
VI	Laryngeal cartilages	Intrinsic muscles of larynx	Recurrent laryn-geal nerve X

Components of Pharyngeal Pouches

Pouch	Components
I	Tympanic cavity and antrum Pharyngeotympanic (eustachian) tube Medial side of tympanic membrane
II	Tonsillar fossa Crypts of palatine tonsil
III	Inferior parathyroid Thymus
IV	Superior parathyroid
V	Ultimobranchial body

c. Facial bones: the maxilla, zygomatic, temporal and mandible are discussed with development of the face.

2. Musculature of the first arch

    a. Muscles of mastication

    b. Anterior belly of digastric

    c. Mylohyoid

    d. Tensor tympani

    e. Tensor palatini

3. The nerve for this arch is the trigeminal (V), especially its mandibular branch. The ophthalmic branch does not supply any components.
4. The ectoderm covering the first arch forms the epidermis of the face, the lining of the vestibule of the mouth, the palate and two-thirds of the anterior tongue, and the parotid gland.

B. Second pharyngeal arch (hyoid arch) fuses with cardiac swelling
1. Derivatives of Reichert's cartilage
   a. Complete malleus and incus
   b. Stapes (middle ear bone)
   c. Styloid process of the temporal bone
   d. Lesser horn of the hyoid bone
   e. Stylohyoid ligament
2. Muscles of the second arch
   a. Stapedius
   b. Stylohyoid
   c. Posterior belly of the digastric
   d. Muscles of facial expression
3. The nerve of the second arch is the facial (VII).
4. The ectoderm covering the second arch forms the epidermis over the dorsal aspect of the auricle and upper neck.

C. Third pharyngeal arch
1. The cartilage of this arch gives rise to the greater horn and the body of the hyoid bone.
2. The muscles of the third arch are the stylopharyngeus and upper pharyngeal constrictors.
3. The nerve for this arch is the glossopharyngeal (IX).
4. The ectoderm covering the third arch forms the epidermis over the lower neck.

D,E. The fourth and fifth pharyngeal arches (fifth arch is temporary)
1. The cartilages of these arches fuse to form the thyroid, arytenoid, corniculate, and cuneiform cartilages.
2. Muscles of the fourth arch:
   a. Levator palatini
   b. Lower pharyngeal constrictors

c. Cricothyroid

3. The superior laryngeal branch of the vagus is the nerve of the fourth arch.

F. Sixth pharyngeal arch

1. The cartilage gives rise to the cricoid cartilage and upper rings of the trachea.

2. Muscles include the remaining intrinsic muscles of the larynx.

3. The recurrent laryngeal branch of the vagus is the nerve of the sixth arch.

Each arch receives an artery, the fate of which is discussed in the chapter on the cardiovascular system (Chapter 9). It is known that neural crest cells migrate into the pharyngeal arches, and the mesectoderm subsequently helps in the formation of arch cartilage. Although the nerves of the second to sixth arches have little cutaneous distribution, they supply the mucous membranes of the tongue, pharynx and larynx. The muscles of each arch do not always confine to bone and cartilages of their own arch, but their origin can easily be determined by their nerve supply.

II. PHARYNGEAL POUCHES

A. First pharyngeal pouch (tubotympanic recess) derivatives (Fig. 11.1, Table 11.1)

1. Tympanic cavity and antrum

2. Pharyngeotympanic (eustachian) tube

3. Contributes to the formation of medial side of the tympanic membrane

B. The second pouch remains as the tonsillar fossa. Its endoderm forms the surface epithelium and the lining of the crypts of the palatine tonsil.

C. Third pouch derivatives

1. Dorsal bulbar portion differentiates into the inferior parathyroids (parathyroid III). Some studies indicate that all parathyroids may be ectodermal in origin.

2. The ventral elongated portion differentiates into the thymus. Both the inferior parathyroid and thymus separate from the pouch and migrate caudally.

D. Fourth pouch develops into a superior parathyroid (parathyroid IV). These separate from the pouch and descend.

E,F. Fifth pouch gives rise to the ultimopharyngeal body, which is incorporated into the thyroid gland and possibly gives rise to parafollicular cells.

III. PHARYNGEAL GROOVES (CLEFTS)

A. The first groove remains as the external auditory meatus. The epithelial lining contributes to the formation of the lateral side of the tympanic membrane (Fig. 11.1).

B. The remaining pharyngeal grooves become buried by the rapidly growing second arch, which enlarges and grows caudally and finally fuses with the epicardial ridge in the lower part of the neck. As a result of this, the second, third and fourth grooves form the lining of a cavity, the cervical sinus. The cervical sinus later becomes obliterated. These changes result in the smooth contour of the neck.

IV. STRUCTURES ORIGINATING FROM THE FLOOR OF THE PHARYNX

A. The tongue develops from pharyngeal arches 1, 3 and 4 and from occipital somites.

1. The mesoderm in the ventromedial part of the first pharyngeal arch gives rise to one median tongue bud, the tuberculum impar, and two distal tongue buds, the lateral lingual swellings (Fig. 11.1). The lateral lingual swellings enlarge rapidly, overgrow the tuberculum impar, and fuse with each other to form the anterior two-thirds, or the body of the tongue. The plane of fusion is indicated by the median sulcus. The tuberculum impar does not form a significant part of the adult tongue. Thus, the mucosa covering the body of the tongue is innervated by the mandibular division of the trigeminal nerve (the first pharyngeal arch nerve).

2. The mesoderm from the second arch gives rise to another median elevation, the copula. The mesoderm from the third and fourth arches, the hypopharyngeal eminence, overgrows the copula and gives rise to the posterior one-third, or the root of the tongue. Thus, this part of the tongue is innervated by the glossopharyngeal, the third arch nerve (Fig. 11.1).

3. The third median swelling formed by the posterior part of the fourth arch marks the development of the epiglottis. The anterior two-thirds of the tongue is separated from the posterior one-third by the terminal sulcus. The circumvallate papillae are situated anterior to the terminal sulcus, but their taste buds are innervated by the glossopharyngeal nerve. The reason usually given for this is that the posterior third of the tongue becomes pulled slightly forward during development of the mucosa. The mesoderm of the pharyngeal arches gives rise to connective tissue, lymphatics, and the blood vessels, and perhaps some muscle fibers of the tongue.

4. Most of the muscles of the tongue are derived from the myotomes of the occipital somites (Chapter 8, Fig. 8.3). The muscles of the tongue are supplied by the hypoglossal nerve, which accompanies the myoblasts from the occipital myotomes.

B. Thyroid gland
The thyroid gland begins to develop as an endodermal diverticulum (thyroid diverticulum), which descends from the floor of the primitive pharynx between the median tongue bud and copula (Fig. 11.1). The thyroid descends in front of the neck and remains connected to the floor of the pharynx by the thyroglossal duct; its opening in the tongue is called the foramen cecum. The duct later disappears.

C. Salivary glands
The parotid gland develops from the epithelial buds, which arise from the ectodermal lining of the primitive mouth. The submandibular and sublingual glands are formed from the endoderm in the floor of the mouth.

V. CONGENITAL MALFORMATIONS

A. Pharyngeal cysts
These are remnants of the cervical sinus. They are lined by a stratified squamous epithelium with fluid-containing cholesterol crystals. They gradually increase in size and readily become infected. Rarely, the epithelial lining undergoes malignant changes to give rise to squamous cell carcinoma.

B. Pharyngeal sinus
This is formed when a portion of the groove persists. It opens on the skin surface especially in the lower part of the neck. Occasionally there is mucus discharge from the opening.

C. Pharyngeal fistula
This is a tract that extends from the pharynx to open on the side of the neck. It is formed by the breakdown of the membrane between the pharyngeal groove and the pouch. It passes between the internal and external carotid arteries and opens in the tonsillar fossa. The fistula continually drains saliva and repeatedly becomes infected.

D. Pharyngeal vestige
These are cartilaginous or bony remnants that appear under the skin on the side of the neck.

E. Mandibulofacial dysostosis (Treacher–Collins syndrome)
This results from a deficient mesenchyme with an altered intercellular matrix in the first, and sometimes in the second, arch. The underlying cause may be interference with the migration of neural crest cells and/or excessive cell death. It manifests as an underdeveloped lower face, a small mandible (micrognathia),

a cleft of the lower eyelid, a down-slanting palpebral fissure, a high cleft palate, and abnormal external ears. Similar conditions have been produced in experimental animals by administration of the oral anti-acne drug isotretinoin (Accutane, vitamin A analogues). Dominant trait is frequent in this condition.

F. Pierre Robin sequence
This also involves first arch derivatives and includes micrognathia (small mandible) and glossoptosis (displacement of tongue). This produces breathing difficulties in newborns. If the mandible fails to develop, the tongue may remain between the maxillary shelves and keep the palate cleft.

G. Congenital thymic aplasia and absence of parathyroid glands (DiGeorge's syndrome)
This syndrome results when the third and fourth pharyngeal pouches do not develop. Facial abnormalities are related to abnormal development of the first pharyngeal arch. This condition may result from teratogenic effects during the 4th and 6th weeks of development. The infant is born with congenital hypoparathyroidism, malformation of the mouth (fish mouth), low set notched ears, a nasal cleft, thyroid hypoplasia, cardiac abnormalities (persistent truncus), and increased susceptibility to infections. It is believed that this is also caused by interference in migration of neural crest cells, which may have led to delayed cell proliferation. Incidence is high in alcoholic mothers. (See Chapter 16, Endocrine System).

H. Ectopic and accessory thymus.

I. Ectopic and supernumerary parathyroids.

J. Thyroglossal duct cysts and fistula
Cysts may be found anywhere along the migratory path followed by the thyroid gland. Sometimes a cyst becomes infected and forms a sinus, the thyroglossal fistula, which opens through the skin.

K. Ectopic and accessory thyroid
Failure of thyroid gland migration results in lingual thyroid. Incomplete descent results in cervical thyroid. Remnants of the thyroglossal duct may give rise to an accessory thyroid that usually is functional.

L. Congenital hypoplasia or aplasia of the thyroid.

M. Ankyloglossia (tongue tie)
The tip of the tongue is tied to the floor of the mouth. Normally, the cells from the edges of the tongue proliferate and grow into the underlying mesoderm. Later in development, these cells degenerate except for a few, which give rise to the frenulum. Degeneration of these cells causes the tongue to become free. If for some reason the degeneration is arrested, the tongue is not freed and the frenulum is short. This may cause speech impairment and difficulty in mastication.

N. Macroglossia
An excessively large tongue.

O. Microglossia
An abnormally small tongue that is often associated with micrognathia.

P. An excessively long frenulum results in an excessively mobile tongue, which may fall back and cause obstruction to the air passage, leading to suffocation.

Q. Cleft tongue
Incomplete fusion of the distal tongue buds.

R. Bifid tongue
Complete failure of fusion of lateral lingual swellings.

S. Ranula is a transparent cystic swelling that may be found under the tongue and is caused by noncanalization of the sublingual gland duct.

REVIEW QUESTIONS — Chapter 11

Select the *most appropriate* answer.

1. Which of the following develops from the first pharyngeal cleft (groove)?
   a. the thyroglossal duct
   b. the malleus
   c. the mylohyoid muscle
   d. the external auditory meatus
   e. all of the above

2. A small blind pit at the anterior border of the sternomastoid muscle that drips mucus is most likely the:
   a. persistent thyroglossal duct
   b. ectopic thyroid
   c. persistent cervical sinus
   d. cervical cyst
   e. ectopic thymus

3. All of the following develop from the first pharyngeal pouch EXCEPT the:
   a. tympanic membrane
   b. tympanic cavity
   c. tympanic antrum
   d. pharyngeotympanic tube
   e. epithelial covering of the middle ear ossicles

4. Which of the following develops in the third pharyngeal arch?
   a. the malleus
   b. facial nerves
   c. the anterior belly of the digastric
   d. the posterior belly of the digastric
   e. the common carotid artery

5. Absence of the lower jaw (agnathia) results from malformations of the:
   a. first pharyngeal arch
   b. first pharyngeal pouch
   c. second pharyngeal arch
   d. third pharyngeal arch
   e. all of the above contribute to the formation of the lower jaw

6. The foramen cecum of the tongue:
   a. marks the division between the second and third pharyngeal pouches
   b. is a remnant of the cupula
   c. marks the point of evagination of the thymus diverticulum
   d. marks the point of evagination of the thyroid diverticulum
   e. is formed by the aryepiglottic folds

7. The muscle that is derived from the first pharyngeal arch is the:
   a. mylohyoid
   b. stapedius
   c. stylohyoid
   d. stylopharyngeus
   e. levator vili palatini

8. Interference with migration of neural crest cells may result in malformation of structures derived from the:
   a. first pharyngeal arch
   b. second pharyngeal arch
   c. third pharyngeal arch
   d. fourth pharyngeal arch
   e. all of the above

9. A 2-week-old child became increasingly susceptible to infections. She also had low calcium levels in the blood. This may be associated with malformation of the:
   a. first pharyngeal cleft
   b. first pharyngeal arch
   c. second pharyngeal arch
   d. third pharyngeal pouch
   e. fourth pharyngeal pouch

10. The abnormalities described in the child mentioned above could be attributed to:
    a. failed development of the pharyngeal arches
    b. failed migration of the thymus
    c. interference with migration of neural crest cells in the third arch area
    d. failure of migration of the parathyroids
    e. all of the above

11. The pharyngeal cyst, a remnant of the cervical sinus, is lined by the:
    a. ectoderm
    b. endoderm
    c. somatic mesoderm
    d. splanchnic mesoderm
    e. neural crest cells

12. Thymic lymphocytes are derived from the:
    a. ectoderm
    b. endoderm
    c. mesoderm
    d. neuroectoderm
    e. neural crest cells

13. The mesenchyme situated between the first pharyngeal groove and the first pharyngeal pouch gives rise to the:
    a. external auditory meatus
    b. internal auditory meatus
    c. core of the tympanic membrane
    d. tympanic cavity
    e. pharyngotympanic (eustachian) tube

14. The contents of each pharyngeal arch include all of the following EXCEPT:
    a. cells from the ectodermal neural crest
    b. an aortic arch artery
    c. a cranial nerve (or its branches)
    d. a mesodermal core from the paraxial mesoderm
    e. the whole or part of a bone

15. Abnormal development of the second branchial arch would have the most profound effect on the development of the:
    a. aorta
    b. internal carotid artery
    c. mandible
    d. middle ear ossicles
    e. permanent teeth

16. Which of the following nerves is not associated with the pharyngeal arches?
    a. the trigeminal nerve
    b. the facial nerve
    c. the glossopharyngeal nerve
    d. the vagus nerve
    e. the hypoglossal nerve

17. All of the following are derived from the second pharyngeal arch EXCEPT the:
    a. tonsillar fossa
    b. muscles of facial expression
    c. posterior belly of the digastric
    d. stapes
    e. stylohyoid muscle

18. The pharyngeal arch mesoderm gives rise to all of the following EXCEPT the:
    a. temporalis
    b. muscles of facial expression
    c. muscles of mastication
    d. muscles for movement of the tongue
    e. tensor tympani

19. The fact that general sensation from the anterior two-thirds of the tongue is carried by the trigeminal nerve indicates that this portion of the tongue is derived from the:
    a. first pharyngeal arch
    b. second pharyngeal arch
    c. third pharyngeal arch
    d. fourth pharyngeal arch
    e. floor of the stomodeum

20. The cells from the fifth/sixth pharyngeal pouch differentiate into the:
    a. parathyroid oxyphils
    b. parafollicular cells of the thyroid
    c. lingual tonsil
    d. epithelial cells of the thymus
    e. all of the above

21. The first pharyngeal groove:
    a. disappears
    b. forms the lining of the external auditory canal
    c. forms the lining of the cervical sinus
    d. gives rise to the thyroglossal duct
    e. continues with the pharyngeotympanic tube

22. All of the following muscles are derived from the first pharyngeal arch EXCEPT the:
    a. levator vili palatini
    b. tensor vili palatini
    c. tensor tympani
    d. mylohyoid
    e. anterior belly of the digastric

23. The structure that is derived from the first pharyngeal pouch is the:
    a. incus
    b. malleus
    c. stapes
    d. pharyngeotympanic tube
    e. tonsillar fossa

24. A developmental anomaly of the midline endoderm is:
    a. spina bifida
    b. dermoid cyst
    c. pilonidal cyst
    d. thyroglossal duct
    e. cyst of Rathke's pouch

ANSWERS TO REVIEW QUESTIONS

---

1. d	2. c	3. a	4. e
5. a	6. d	7. a	8. e
9. d	10. c	11. a	12. c
13. c	14. d	15. d	16. e
17. a	18. d	19. a	20. b
21. b	22. a	23. e	24. d

CHAPTER

12

CRANIOFACIAL DEVELOPMENT

The origin of the head skeleton (skull) and other tissue constituting the face is complex. For that reason, the development of the skull and face are discussed here rather than in the chapter on the skeletomuscular system (Chapter 8). In the present chapter, the formation of cranial bones that make up the roof and base of the skull as well as the formation of facial bones will be discussed.

The cranial bones are mesodermal in origin, and the facial bones arise from mesenchyme derived from the ectodermal neural crest. The facial region forms by the fusion of five different processes that develop around the stomodeum (opening of the mouth). Four of these processes arise from the first pharyngeal arch (Chapter 11), and the fifth process arises from the cranial mesoderm. Aside from the formation of upper and lower jaws, formation of the nose and palate will be discussed here. Review the pharyngeal arches components for further clarity.

I. DEVELOPMENT OF THE SKULL

The skull consists of two major entities:

A. The neurocranium, a protective case around the brain.

B. The viscerocranium, the main skeleton of the face derived from the pharyngeal arches.

A. The Neurocranium

The mesenchyme involved in the formation of neurocranial bones is influenced by overlying epithelium. The inductive interaction is typically matrix-mediated. The neurocranium is further subdivided, according to mode of ossification, into the membranous neurocranium and the cartilaginous neurocranium or chondrocranium.

1. Chondrocranium
   The cartilaginous base of the embryonic and fetal skull is laid down in mesenchymal condensation. Some studies have shown the presence of type II collagen and accumulation of a cartilage-specific proteoglycan in the areas of these mesenchymal condensations. The following cartilaginous precursors are formed (Fig. 12.1):

   a. The parachordal plate forms around the cephalic end of the notochord. It fuses with the sclerotomes of occipital somites and gives rise to the base of the occipital bone. This component later extends dorsally around the neural tube to form the edges of the foramen magnum.

   b. The hypophyseal cartilage (sphenoid complex) forms around the developing hypophysis (the pituitary gland).

   c. The precordal cartilage (trabeculae cranii), a precursor of the ethmoid complex, forms around the olfactory capsule.

Fusion of these three components gives rise to an elongated median plate that undergoes endochondral ossification to form the base of the skull. Other paired cartilages appear on either side of the median plate that are associated with the epithelial primordia of the sense organs, which also undergo endochondral ossification. These are:

   d. The ala orbitalis, which is associated with the optic capsule and gives rise to the lesser wings of the sphenoid.

   e. The ala temporalis, also associated with the optic capsule, gives rise to the greater wing of the sphenoid.

   f. The otic capsules (periotic capsule), associated with auditory organs, form around the membranous labyrinth as the petrous temporal.

By its interstitial growth, the chondrocranium allows expansion of the neurocranium while the brain grows. Ossification of the chondrocranium begins early in the third month, and most of the cartilages have two or more centers of ossification.

2. Membranous neurocranium
   The components of the membranous neurocranium arise as bony trabeculae from the mesenchyme investing the brain. The components are induced by the specific part of the developing brain to which they are related. The bones thus formed, the paired frontal and parietal and squamous part of the occipital region, undergo intramembranous ossification, stay flat, giving rise to the cranial vault (calvaria). All flat bones, in fetal life and after birth, remain separated by dense connective tissue

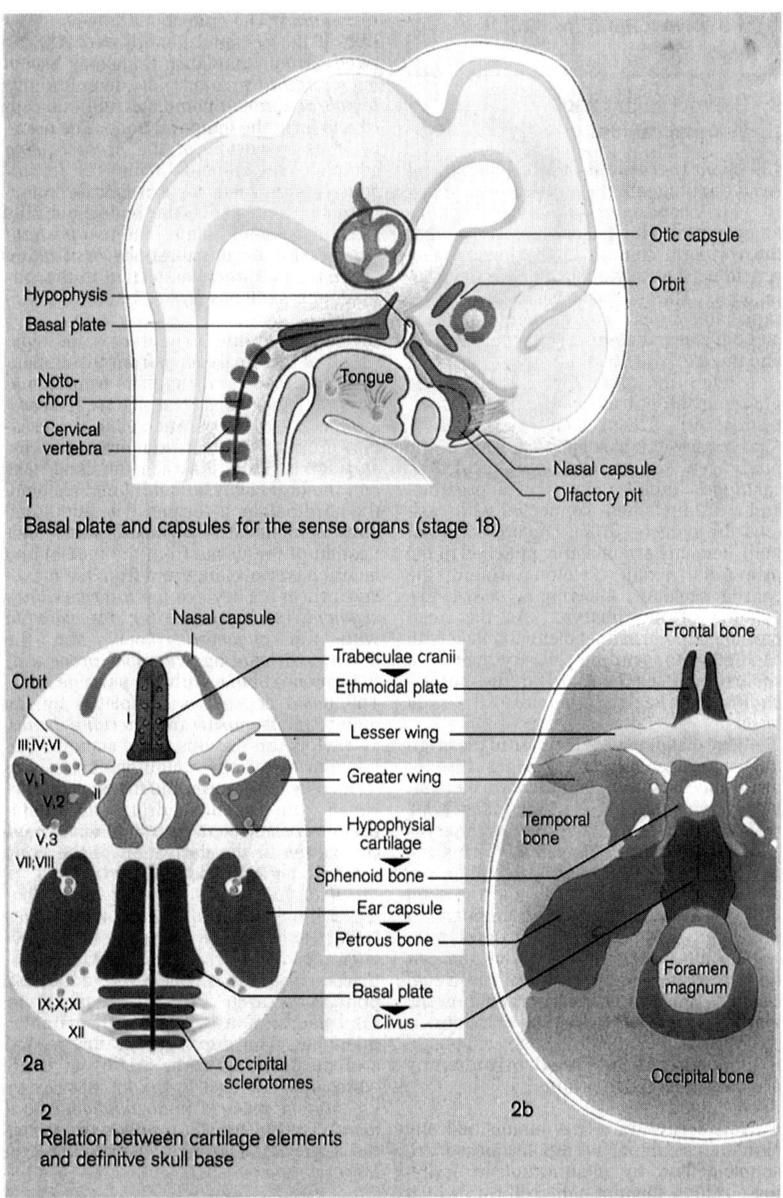

Figure 12.1
Skull base (see Color Plate 12.1). Reproduced with permission from Thieme Medical Publishers Inc., New York, 1995, *Color Atlas of Embryology,* Ulrich Drews, Chapter 9: Head.

membranes forming fibrous joints, the sutures. The sutures enable the skull to undergo changes in shape during childbirth and early childhood. At intersection of sutures, broader areas, called fontanelles, are occupied by connective tissue membranes. Six fontanelles are present at birth. The most prominent fontanelles are the anterior fontanelle, located at the intersection of two frontal and two parietal bones; and the posterior fontanelle, located at the intersection of the parietal and occipital bones. Pulsation in the anterior fontanelle, which can be seen or felt, is due to underlying cerebral arteries. Palpation of fontanelles enables the physician to estimate intracranial pressure, to assess the degree of cranial growth to diagnose premature closure of the sutures and, if needed, to obtain a blood sample.

B. Viscerocranium

The viscerocranium (like the neurocranium) consists of a cartilaginous part and a membranous part according to the mode of ossification. It is formed mainly by the mesenchyme from pharyngeal arches 1, 2 and 3. This mesenchyme is derived in turn from the neural crest cells.

1. The cartilaginous viscerocranium (see also the skeletal part of the pharyngeal arches).
   a. The first arch cartilages include the malleus and the body, the short crus of incus and Meckel's cartilage.
   b. The second arch cartilages include the long crus of incus, the stapes, the styloid process and the lesser horn of hyoid.
   c. The third arch cartilages include the body and the greater horn of hyoid.

All of these cartilages, except Meckel's cartilage, undergo endochondral ossification. Meckel's cartilage regresses and persists only as the sphenomandibular ligament.

2. The membranous viscerocranium
   The membranous viscerocranium consists of the bones that are associated with the upper and lower jaws and the ears. The mesenchyme forming these bones is again derived from the neural crest cells by epitheliomesenchymal interaction. These bones undergo intramembranous ossification.
   a. The maxilla, including the secondary palate, the zygomatic squamous part of the temporal bone and the mandible, arises from the first arch mesenchyme. The cellular proliferation that is accompanied by change and accumulation of extracellular matrix causes mandibular growth spurts. A secondary cartilage in the condylar processes leads to later growth in the length of the mandible.
   b. The premaxilla, nasal, lacrimal and vomar bones arise from the mesenchyme of the frontonasal process.

II. DEVELOPMENT OF THE FACIAL REGION

Very early the rostral-most part of the neural tube grows faster and causes the cephalic fold, which shifts the oropharyngeal membrane ventrally and caudally. The face beneath the developing brain at this stage is represented by the stomodeum. By the fourth week, the oropharyngeal membrane breaks down and the stomodeum communicates with the foregut. Five facial primordia appear around the stomodeum (Fig. 12.2):

A. The single frontonasal prominence (process) begins as a proliferation of mesenchyme on the ventral surface of the developing brain.

B. The paired maxillary prominences (processes) grow out from the upper end of each first pharyngeal arch.

C. The paired mandibular prominences (processes) grow medioventrally from lower part of the first pharyngeal arch.

The growth and fusion of these five primordia form the facial regions and the palate. It is suggested that neural crest cells migrating to these primordia have received specific morphogenetic instruction. Information regarding patterns of retinoic acid-binding protein and retinoic acid receptors is also essential for normal as well as anomalous growth of the facial region.

A. Formation of the Face

By the end of the fourth week, localized thickenings of the surface ectoderm, on each side of the lower part of the nasofrontal prominence, appear as nasal placodes (Fig. 12.2). The mesenchyme at the margins of the nasal placodes proliferates and changes it to horseshoe shaped processes. The medial limb of the placode is the medial nasal process, and the lateral limb is the lateral nasal process, and the depressions in between the nasal processes become deep and form the nasal pits. The maxillary prominence grows rapidly in the medial direction, bringing the two medial nasal processes closer together. The medial nasal processes later merge to give rise to the philtrum of the upper lip and primary palate (premaxilla). The merging and fusion to maxillary prominences with the periphery of the medial nasal processes completes formation of the upper lip. Hence, the upper lip is formed by both maxillary prominences and both medial nasal processes. During their growth, the maxillary prominences also merge with the lateral nasal processes. A nasolacrimal groove marks the line of fusion of these processes. The nasolacrimal groove extends from the developing eye to the upper lip. The ectoderm in the floor of these grooves thickens to form solid epithelial cords. The epithelial cords detach from the surface ectoderm and sink into the underlying mesenchyme. Canalization of the cords forms the nasolacrimal ducts. Merging of the maxillary prominence

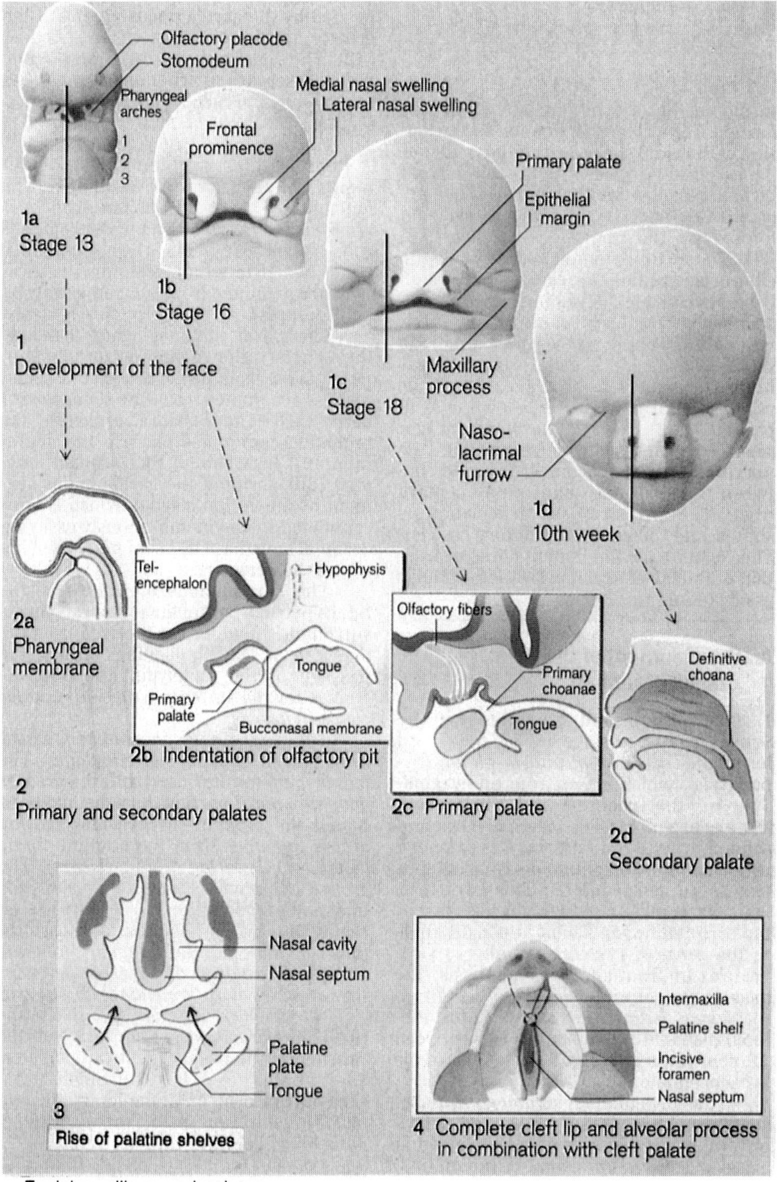

Figure 12.2

Facial swellings (see Color Plate 12.2). Reproduced with permission from Thieme Medical Publishers Inc., New York, 1995, *Color Atlas of Embryology*, Ulrich Drews, Chapter 9: Head.

with the lateral nasal process reduces the groove. The lateral nasal processes remain as the alar region of the nose. The two mandibular processes also grow medioventrally, come closer and merge, forming the lower lip and lower jaw. The mandibular

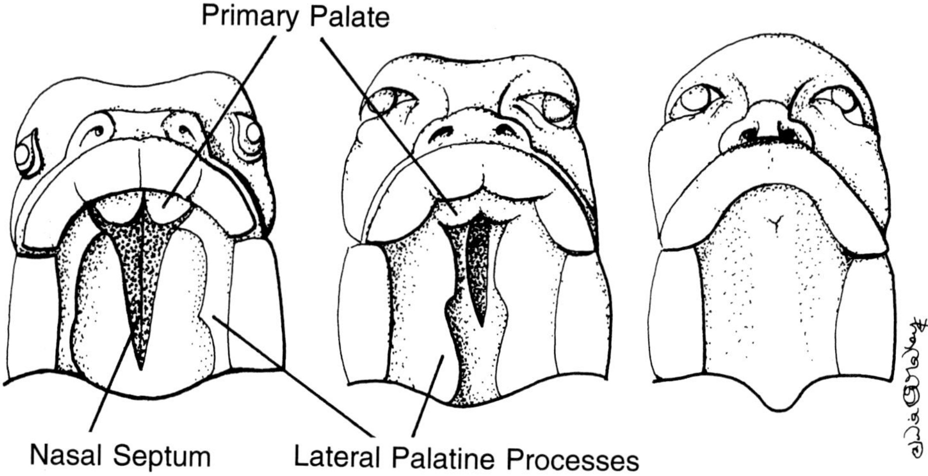

Figure 12.3
Ventral view showing the development of the palate.

prominences partly fuse with the maxillary prominences to limit the opening of the mouth.

B. Formation of the Palate

The palate forms from three primordia:

1. An unpaired median palatine segment that is an ingrowth of fused medial nasal processes (see above). This gives rise to the premaxilla (primary palate), which carries four upper incisor teeth.
2. A pair of lateral palatine shelves that are an inward growth of the maxillary prominence. These fuse to give rise to the secondary palate. The right and left palatine shelves grow vertically on each side of the tongue (Fig. 12.3). Active differential craniofacial growth and rapidly increasing length of pharyngeal cartilage I tend to lower the tongue. This withdrawal of the tongue causes the palatine shelves to elevate and become horizontal above the tongue. Many other factors like intrinsic shelf force are proposed in the elevation of palatine shelves. Two palatine shelves come close together, and the epithelial lining undergoes alteration in its basal lamina and programmed cell death. At the point of contact between the opposite shelves, the epithelial layer is completely lost and the mesenchyme of the shelves merge with each other. Hence, this process includes mesenchymal epithelial interaction, rapid mesenchymal proliferation, cell adhesion and programmed cell death. Many studies have indicated hormonal dependency and increased sensitivity to teratogens. Epithelial growth factor stimulates mesenchymal cell proliferation and prevents terminal differentiation of the median epithelium.

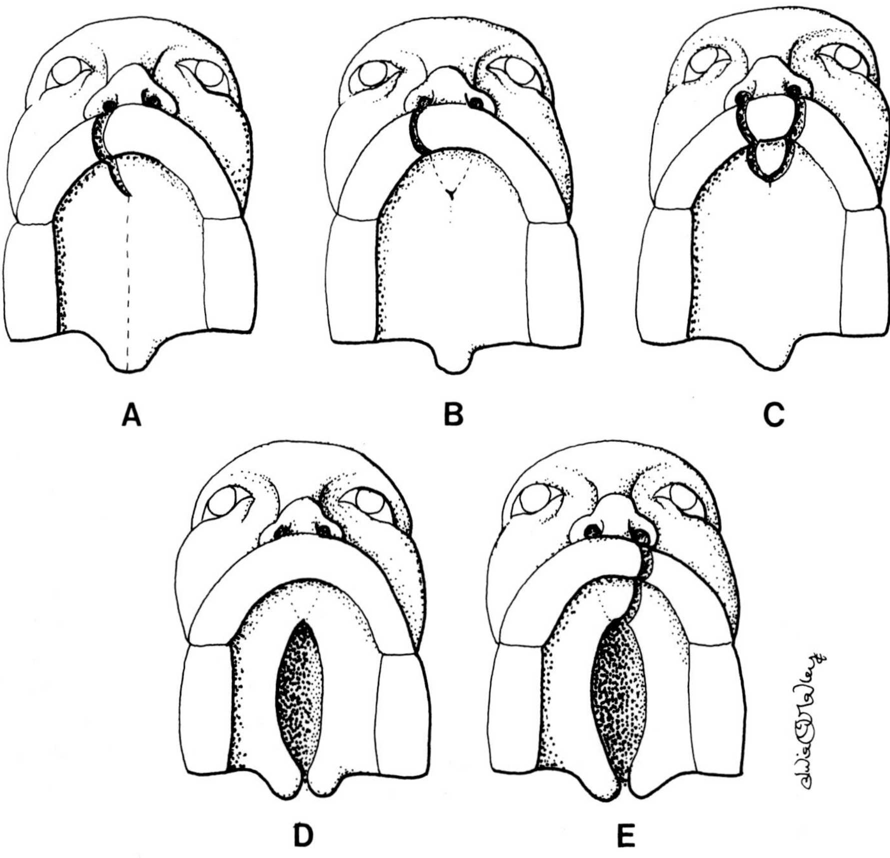

Figure 12.4

Ventral view showing anomalies of face and palate: (A) unilateral cleft lip and cleft primary palate; (B) unilateral cleft lip; (C) bilateral cleft lip; (D) cleft of the secondary palate; (E) cleft of the primary and secondary palates.

The palatine shelves similarly fuse with the premaxilla and then with the nasal septum, which is a downgrowth from the frontonasal prominence. The soft palate and uvula probably develop by epithelial displacement. Fusion of shelves begins anteriorly and proceeds posteriorly, thus separating the nasal cavity from the oral cavity.

3. Formation of the nasal cavity
   The depression in the nasal placode, the nasal pits, continues to deepen toward the pharynx because of rapid growth and displacement of the lateral and medial nasal processes. The nasal pits expand into substantial nasal cavities that are separated from the oral cavity by a thin oronasal membrane. The oronasal membrane disappears just before elevation of the palatine shelves. The fusion of palatine shelves to each other and the nasal septum permanently separates the nasal cavities from each other and from the oral cavity. The dorsal-most epithelium of the nasal pits becomes the highly specialized olfactory epithelium.

In summary, the medial nasal processes give rise to the philtrum of the upper lip, the premaxilla and the median portion of the nose. The lateral nasal process forms the alae of the nose. The maxillary prominences complete the upper lip and give rise to the secondary palate and upper jaw. The nasolacrimal duct forms the ectodermal thickening between the maxillary prominences and lateral nasal processes. The mandibular prominences give rise to the lower jaw.

III. CONGENITAL MALFORMATIONS OF THE CRANIOFACIAL REGION

It has been estimated that various types of craniofacial abnormalities account for a third of all congenital defects. These include malformation of the frontonasal prominence, clefting defects, malformations associated with the skull and abnormalities associated with pharyngeal apparatus derivatives. The etiology of most of these is multifactorial, but some could be demonstrated as having genetic hosts. Many teratogens are known to cause craniofacial anomalies. The most clinically significant of these are: 1) alcohol, especially when consumed during the first trimester of pregnancy, 2) hydantoin, 3) vitamin A analogues (Accutane), 4) toluene, 5) ionizing radiation, 6) hyperthermia and 7) cigarette smoking. The best studied of these is alcohol. Fetal alcohol syndrome is known to affect as many as 2:1000 live births. Although many of these are truly congenital in nature, many can be attributed to mechanical stress during intrauterine life, and yet others are secondary to malformation of the brain.

A. Craniosynostosis

These malformations result from premature closure of certain sutures in the neurocranium. Several anomalies of the outer and middle ear are also associated with craniosynostosis.

1. Scaphocephaly: premature closure of the sagittal suture, between the two parietal bones, results in a long, narrow, wedge-shaped skull.
2. Acrocephaly or oxycephaly: premature closure of the coronal suture, between the frontal and parietal bones, results in a high tower-like skull. This also is seen in Apert's syndrome in association with syndactyly of the fingers and toes, and hypertelorism.
3. Plagiocephaly: unilateral premature closure of the coronal and lambdoid sutures results in a twisted asymmetrical skull.
4. Crouzon's syndrome: premature closure of the sagittal and coronal sutures results in wide-set eyes (hypertelorism), small maxilla and low-set ears. The hypoplastic maxilla and zygomatic bone cause a shallow orbit and exophthalmia. Although the anterior fontanelle closes early, mental development is usually normal. This is a dominant genetic condition.

5. Microcephaly (and microencephaly): premature closure of all the sutures and fontanelles results in a small skull. As the calvarium grows congruent with the brain, the brain remains small, which results in severe mental retardation. It may present as generalized intrauterine growth retardation.

B. Holoprosencephaly

This results from a disturbance in the early development of the forebrain. It has a severe to mild spectrum of anomalies and a high association with fetal alcohol syndrome. Alcohol-induced holoprosencephaly is the most common cause of congenital mental retardation. It has been reported that one incidence of overconsumption or chronic consumption of even small amounts of alcohol may result in some degree of mental retardation with minor physical defects. The frequency in the Western world is 2 in 1000 live births. Inhalation of toluene during the critical period of forebrain induction increases its risk. Children of diabetic mothers are also at high risk. Because of its influence on the surrounding structures, the malformations of forebrain manifest externally as typical facial anomalies like a long upper lip with no philtrum, generalized reduction in size of the nasofrontal prominence causing a high arched palate and a short upturned nose, a short retracted lower jaw, and an overall small skull with maldevelopment of other parts of the brain. In its severe form it is represented: 1) as anencephaly (absence of cranial vault) and 2) cyclopia (single median eye primordia, single tubular nose (proboscis) situated above the eye primordia). Although many cases of holoprosencephaly appear to be multifactorial, it should be stressed again that maternal consumption of alcohol during pregnancy is regarded as a leading cause of this condition. Trisomy 13 and 18 are also associated with it.

C. Frontonasal Dysplasia

This includes nasal malformation caused by an excess of tissue in the frontonasal prominence. It results in a broad nasal bridge and hypertelorism (increased distance between the eyes). In some cases, a median nasal cleft is found.

D. Clefting Defects

Cleft lip and cleft palate are relatively common anomalies. Cleft lip with or without cleft palate is seen in 1:1000 Caucasian (higher in Oriental, lower in African) live births. It is more frequent in males. The frequency of cleft palate without cleft lip is 1:2500, and it is more common in females. Although clefting may occur alone, it is frequently combined with other craniofacial syndromes. The etiology of clefting is multifactorial. Factors that disturb the fusion of the facial prominences result in these defects. Both genetic and environmental factors contribute to most of the defects. A single mutant gene, chromosomal abnormalities and specific teratogens are also known to cause clefting. Some common drugs like anticonvulsants, phenytoin (Dilantin), vitamin A, vitamin A analogues, particularly the oral anti-acne drug isotretinoin (Accutane), and cortisone have been shown to induce clefting in experimental animals. Vitamin A analogues particularly retard proliferation of the frontonasal prominence mesenchyme.

1. Cleft Lip

Cleft lip results when the maxillary prominences and medial nasal processes fail to merge and fuse. It is more frequent unilaterally than bilaterally. It may extend to the primary palate. The underlying mechanism is hypoplasia of the maxillary prominence, resulting from either inadequate proliferation and migration of neural crest ectomesenchyme or excessive cell death. Abnormal or inadequate development of the lateral nasal processes has also been implicated (Fig. 12.4).

a. Unilateral cleft lip results from failure of the maxillary prominence to merge with the medial nasal process on the affected side and is more common on the left than the right.

b. Bilateral cleft lip results from failure of the maxillary prominences to merge with the medial nasal prominences on both sides. Such a defect when complete deforms the face because of discontinuity of the orbicularis oris muscle.

c. Median cleft lip results from failure of the medial nasal processes to merge, proliferate and push out the overlying epithelium. The intermaxillary segment (primary palate) may not form. It is associated with frontonasal dysplasia.

d. Oblique facial cleft results from failure of fusion of the lateral nasal process with the maxillary prominence. It may be caused by hypoplasia of either or both of the merging processes. A deep epithelial-lined fissure extends from the medial corner of the eye to the upper lip. Disruption in fusion may also be caused by an amniotic band.

2. Cleft Palate

This results from failure of the palatine shelves to fuse and merge with each other, with the nasal septum and with the primary palate. The defect may be partial or complete, may be unilateral, or bilateral, with or without cleft lip. It could arise in various ways: a) defective or inadequate growth of the palatine shelves, b) failure of the shelves to elevate and attain a horizontal position at correct time, c) failure of the shelves to contact each other and fuse, d) secondary rupture after fusion and e) the shape of the embryonic face (exceedingly wide head). After the threshold period (time for maximal attraction and adhesions), delayed movements of shelves would not allow fusion and merging. This may explain the higher incidence of cleft palate in females, as the fusion of shelves is one week later than in males. Retarded growth of palatine shelves has been ascribed to poor vascularization of this area. Incomplete dissolution of the covering epithelium is also stressed by some authors. Generally, children with clefting defects show an underdeveloped midface. In children with cleft palate, swallowing and speech are affected.

a. Anterior cleft: cleft of primary palate results when the palatine shelves fail to fuse with the primary palate. These are anterior to the incisive foramen (Fig. 12.4).

b. Posterior cleft: cleft secondary palate results when the palatine shelves fail to meet and fuse with each other. These are posterior to the incisive foramen.

c. Combined cleft: cleft of the primary and secondary palates results when the palatine shelves fail to fuse with each other and with the primary palate.

It should be noted that combined cleft lip and cleft palate defects involve two separate abnormalities.

E. Macrostomia

Incomplete fusion and merging of the maxillary prominences with the mandibular prominences results in a large mouth.

F. Microstomia

Excessive fusion and merging of the maxillary prominences with mandibular prominences results in a small mouth opening.

G. Nasolacrimal Duct Atresia

Failure of lacrimal duct canalization results in a blocked nasolacrimal duct. Usually, only a small portion of the duct is involved. Thickened matter is seen in the medial corner of the eye. Patency is established by probing the duct.

REVIEW QUESTIONS — Chapter 12

Multiple choice. Select the *most appropriate* answer.

1. A child has unilateral cleft lip and cleft palate. It results from nonfusion of the:
   a. medial nasal process
   b. medial nasal process and lateral nasal process
   c. medial nasal process and maxillary prominence
   d. maxillary prominence and lateral nasal process
   e. maxillary prominence and mandibular prominence

2. The facial region is formed by fusion of five different processes. Four of these processes arise from the:
   a. first pharyngeal arch
   b. first and second pharyngeal arches
   c. outer layers of cells from the ectomeninges
   d. mesenchyme related to the head region
   e. none of the above

3. The secondary palate is formed by:
   a. fusion of the medial nasal processes
   b. fusion of the medial and lateral nasal processes
   c. downward growth of the nasal septum
   d. fusion of the palatine shelves of the maxillary process
   e. all of the above

4. Which of the skull bones or parts of the skull bone develops around the notochord?
   a. the base (body) of the sphenoid bone
   b. the base (body) of the occipital bone
   c. the Petrous portion of the temporal bone
   d. the ethmoid bone

5. The muscles for movement of the tongue are derived from the:
   a. first pharyngeal arch
   b. second pharyngeal arch
   c. third pharyngeal arch
   d. occipital somites
   e. cervical segment somites

6. The middle portion of the upper lip (philtrum) arises from the:
   a. maxillary prominence
   b. mandible prominence
   c. palatine shelves
   d. frontonasal prominence
   e. ectodermal epithelial cord

7. During development of the face, tissues from one process merge with tissues of another process. A successful merging is accomplished by:
   a. mesenchymal epithelial interaction
   b. rapid mesenchymal proliferation
   c. cell adhesion
   d. programmed cell death (apoptosis).
   e. all of the above

8. Premature closure of sutures may result in:
   a. a long, narrow, wedge-shaped skull
   b. a high tower-like skull
   c. a small skull
   d. a twisted asymmetrical skull
   e. any of the above

9. During development of the face, factor(s) that could interfere with the fusion of facial prominences include:
   a. a single gene mutation
   b. chromosomal abnormalities
   c. environmental teratogens
   d. a vitamin A analogue
   e. any of the above

10. For normal development, a pair of lateral palatine shelves must fuse with each other and the:
    a. premaxilla
    b. nasal septum
    c. both
    d. none

11. Nasolacrimal duct atresia:
    a. results from failure of canalization
    b. may be caused by a hypoplastic nasolacrimal duct
    c. results from failure of fusion between the lateral nasal processes and the maxillary prominence
    d. is due to nonproliferation of the surface ectoderm in this area
    e. any of the above could cause this

12. Lateral nasal processes:
    a. merge together to form the nasolacrimal duct
    b. contribute to the formation of the upper lip
    c. form the alar region of the nose
    d. give rise to the philtrum
    e. all of the above are correct

13. Unossified areas at the intersection of the sutures in the fetal skull are known as:
    a. the membranous neurocranium
    b. the membranous viscerocranium
    c. the frontonasal prominence
    d. Meckel's cartilage
    e. the fontanelles

14. The ethmoid bone develops from the:
    a. trabeculae cranii (prechordal cartilage)
    b. ala orbitalis
    c. ala temporalis
    d. parachordal cartilage
    e. periotic capsule

15. The cartilaginous viscerocranium is derived from the:
    a. maxillary process
    b. mandibular process
    c. second pharyngeal arch
    d. third pharyngeal arch
    e. all of the above

16. The components of the membranous neurocranium are induced by the:
    a. ectoderm
    b. ectodermal neural crest
    c. ectodermal mesenchymal interaction
    d. developing brain
    e. all of the above

17. The hypophyseal cartilage fuses with the ala orbitalis and ala temporalis to form the:
    a. occipital bone
    b. temporal bone
    c. sphenoid bone
    d. zygomatic bone
    e. ethmoid bone

18. Growth and fusion of the maxillary prominences, mandibular prominences and frontonasal prominence give rise to all of the following except the:
    a. upper jaw
    b. lower jaw
    c. primary palate
    d. secondary palate
    e. hyoid bone

19. Craniosynostosis includes all of the following except:
    a. scaphocephaly
    b. holoprosencephaly
    c. Crouzon's syndrome
    d. microcephaly
    e. acrocephaly

20. The oblique facial cleft (fissure):
    a. is found between the maxillary prominence and mandibular prominence
    b. is a remnant of the primary palate
    c. represents failure of the maxillary prominence to merge with the medial nasal process
    d. may represent a disruption in fusion caused by an amniotic band
    e. none of the above is true

21. Premature closure of the fontanelles and sutures of the skull may lead to:
    a. scaphocephaly
    b. acrocephaly
    c. Crouzon's syndrome
    d. microcephaly
    e. acrania

22. Failure of the brain to develop may result in:
    a. plagiocephaly
    b. microcephaly
    c. anencephaly
    d. acrocephaly
    e. scaphocephaly

23. The most common congenital mental retardation induced by maternal overconsumption of alcohol is:
    a. microcephaly
    b. acrocephaly
    c. holoprosencephaly
    d. clefting defect
    e. all of the above

ANSWERS TO REVIEW QUESTIONS

1. c	2. a	3. d	4. b	5. d
6. d	7. e	8. e	9. e	10. c
11. a	12. c	13. e	14. a	15. c
16. e	17. c	18. e	19. b	20. e
21. b	22. c	23. c		

CHAPTER

13

DIGESTIVE SYSTEM

The embryonic foldings during the fourth week incorporate a portion of the yolk sac inside, creating a tubular embryo. The primitive gut thus formed is bounded cranially by the oropharyngeal (stomodeal) membrane and caudally by the cloacal membrane. Its cranial part, associated with the pericardial cavity and the septum transversum, is known as the foregut. Its part connected to the yolk sac is the midgut, and its caudal part is the hindgut. The primitive gut tube undergoes epithelial proliferation (which occludes the lumen), elongation and rotation. Later recanalization, cellular differentiation and functional maturation complete its development. The foregut includes the pharynx, the esophagus, the stomach and the proximal duodenum. The epithelial diverticula from the foregut give rise to the thyroid gland, the respiratory primordium (trachea and bronchial tree), the biliary primordium and the pancreatic buds. The midgut includes the distal duodenum and the remaining small intestines, cecum, ascending colon and proximal half of the transverse colon. The hindgut includes the distal half of the transverse colon, the descending colon, the sigmoid colon, the rectum and the cloaca. The allantois opens into the cloaca. The primitive gut suspends from the posterior body wall by the dorsal mesentery, which undergoes positional changes caused by gut rotation and divides the peritoneal cavity into the lesser and greater sacs. Some of the developmental anomalies related to the gastrointestinal tract are discussed at the end of this chapter.

I. STOMODEUM

During development of the face (Chapter 12), five prominences form around an ectodermal depression that is the outer part of the oropharyngeal membrane. Head folding and differential growth of these prominences shift the oropharyngeal membrane internally and deepen the depression as the stomodeum (oral cavity). The oropharyngeal membrane ruptures and the stomodeum becomes continuous with the foregut. The ectodermal–endodermal junction shifts further caudally. Ectodermal diverticula on each side give rise to parotid glands, whereas endodermal diverticula give rise to submandibular and sublingual glands.

At formation of the primitive gut, the arteries of the yolk sac, the vitelline plexus, anastomose with the branches of the dorsal aorta. The arterial supply to the gut develops through this consolidation. The vitelline branches supplying the thoracic gut form esophageal arteries. The abdominal gut is supplied by the celiac trunk, the superior mesenteric artery and the inferior mesenteric arteries. The boundaries of the abdominal foregut, the midgut and the hindgut are determined respectively by the portion of the gut supplied by these arteries.

II. FOREGUT

The foregut derivatives include: the pharynx and its derivatives, the respiratory tract, the esophagus, the stomach, the proximal part of the duodenum (as far as the point of entrance of the common bile duct), liver and biliary duct apparatus, and the pancreas. The pharyngeal derivatives and the respiratory system have already been discussed (see Chapter 11, Fig. 11.1).

A. Esophagus

After separation from the trachea, the esophagus elongates rapidly. The endoderm proliferates and almost obliterates the lumen. Later, recanalization occurs. The upper two-thirds of the esophagus is surrounded by mesoderm derived from the caudal pharyngeal arches. This mesoderm gives rise to striated muscle, which is therefore innervated by the vagus nerve. The splanchnic mesoderm around the lower one-third of the esophagus forms the smooth muscle, which is supplied by a visceral (splanchnic) plexus.

B. Stomach

The stomach appears as a fusiform dilatation of the foregut. The dorsal border grows faster than the ventral border, resulting in greater and lesser curvatures. The right and left vagus nerves are attached on its surface. Its asymmetrical growth causes it to rotate on both its longitudinal (90° clockwise) and transverse axes. The results of rotation of the stomach are: a) the lesser curvature moves to the right and pulls the ventral mesogastrium (mesentery) to the right, b) the greater curvature moves to the left and pulls the dorsal mesogastrium to the left, c) the right side becomes dorsal and carries the right vagus to the dorsal side, d) the left side becomes ventral and carries the left vagus to the ventral side, e) the caudal (pyloric)

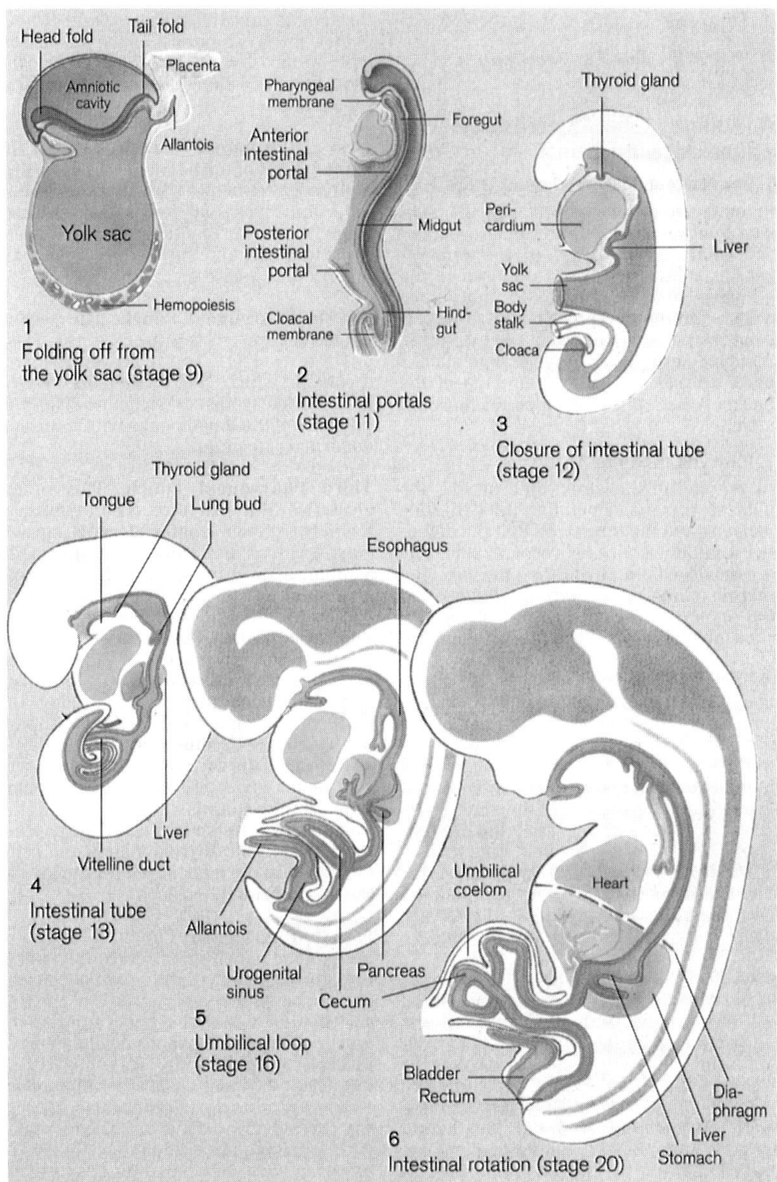

Figure 13.1

Overview: digestive tube (see Color Plate 13.1). Reproduced with permission from Thieme Medical Publishers Inc., New York, 1995, *Color Atlas of Embryology,* Ulrich Drews, Chapter 7: Gastrointestinal Tract.

end of the stomach moves upward and to the right and f) the cephalic (cardiac) end moves downward and to the left (Fig. 13.1). The gastric epithelium takes shape during the late embryonic period; differentiation of its cells into different types begins in the early fetal period.

C. Duodenum

The duodenum develops from the most caudal part of the foregut and the most cephalic part of the midgut. It grows rapidly and forms a C-shaped loop that projects ventrally. When the stomach rotates to the left, the duodenal loop is displaced to the right. The biliary opening migrates from the ventral aspect to the left side of the duodenum. Later, the midgut mesentery crosses the duodenum and fuses with it. The endoderm of the duodenum proliferates and obliterates the lumen, which later undergoes recanalization.

D. Liver and Biliary Duct Apparatus

Rapidly proliferating cell strands, the hepatic diverticulum, grow out from the caudal end of the foregut (Fig. 13.2) and extend into the septum transversum. It divides into a large cephalic and a small caudal part.

Development of the glandular epithelium (endoderm) is induced by the surrounding mesoderm, which is specific for each particular gland. This specificity may be due to the mode of vascularization of the mesoderm.

The columns of endodermal cells proliferate from the hepatic diverticulum, forming interlacing cords of hepatocytes. As the portal vein enters the developing liver, the mesoderm induces some of the hepatocyte to transform into the epithelial lining of the intrahepatic biliary apparatus. A phase of epithelial proliferation is followed by recanalization.

The paired vitelline veins and umbilical veins that pass through the septum transversum become broken by the liver cells to form the hepatic sinusoids. Some sinusoids arise from close vesicles, and join the others.

The connective tissue, hemopoietic tissue, and Kupffer cells are derived from the splanchnic mesoderm of the septum transversum. The origin of Kupffer cells is debated. The right lobe of the liver becomes larger than the left. The caudate and quadrate lobes develop from the left lobe.

The main hepatic diverticulum and its branches become canalized to form the common hepatic duct and the left and right hepatic ducts, which continue with small ducts. The canaliculi begin as invagination of adjacent cell membranes. Another solid column of cells grows out from the main hepatic diverticulum. The end of this outgrowth expands to form the gallbladder, whereas the narrow stem remains as the cystic duct.

The stalk connecting the hepatic and cystic ducts to the duodenum becomes the common bile duct. When the duodenum is displaced, it carries the entrance of the common bile duct to its dorsal aspect. Hemopoiesis begins in the liver at about the fifth week and peaks at the end of the second trimester and then rapidly declines before birth. This hemopoietic function is mainly responsible for the large size of the liver. At about the 10th week, the weight of the liver is about 10% of total body weight; at the time of birth, the weight of the liver is about 5% of total body weight. Hepatocyte growth factor is believed to be important in regeneration of the liver.

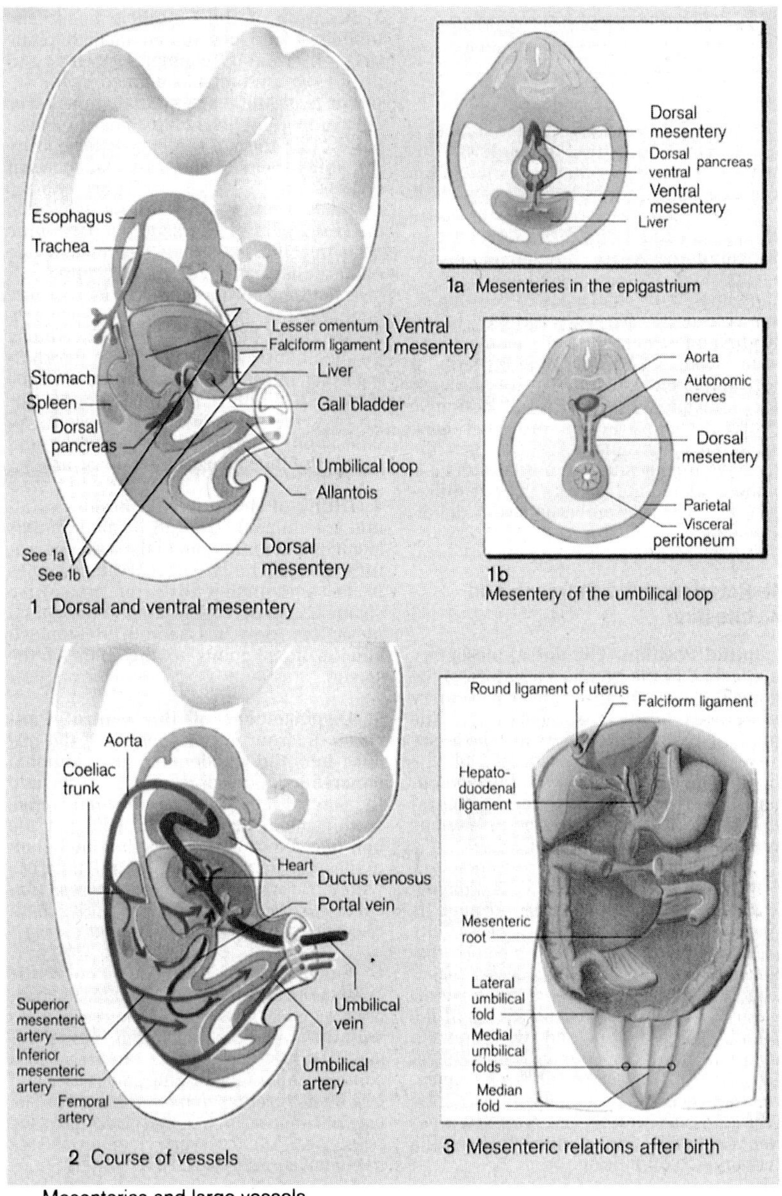

Figure 13.2

Overview: mesenteries (see Color Plate 13.2). Reproduced with permission from Thieme Medical Publishers Inc., New York, 1995, *Color Atlas of Embryology,* Ulrich Drews, Chapter 7: Gastrointestinal Tract.

E. Pancreas

The pancreas develops from the dorsal and ventral pancreatic buds, which arise from the endoderm of the most caudal part of the foregut. The large dorsal pancreatic bud appears opposite and slightly above the hepatic diverticulum and grows rapidly into the dorsal mesentery (Fig. 13.2). The ventral bud appears as an evagination from the common bile duct.

When the duodenum is displaced, the ventral bud is carried dorsally with the bile duct and it comes to be immediately below and behind the dorsal bud. The dorsal and ventral pancreatic buds fuse. The ventral bud forms the uncinate process and the inferior part of the head of the pancreas. The main pancreatic duct forms from the ducts of the ventral bud. The dorsal bud forms the bulk of the pancreas. The proximal part of its duct sometimes persists as the accessory pancreatic duct, which opens above the main duct. The main pancreatic duct, together with the common bile duct, enters the duodenum at the site of the major papilla. Secondary branching occurs in typical glandular arrangement. The islet cells arise as buds from the ducts. The first differentiating cells are α-cells.

III. MIDGUT

The midgut derivatives include: the distal duodenum, the jejunum and ileum, the cecum and appendix, the ascending colon, and the right half to two-thirds or proximal part of the transverse colon. The midgut derivatives are supplied by the superior mesenteric artery.

Initially, the midgut is suspended from the dorsal mesentery and is connected with the yolk sac by a wide opening, which reduces to a narrow vitelline duct. The midgut elongates rapidly and forms a primary intestinal loop. The vitelline duct is at the apex of the loop (Fig. 13.3).

Enlargement of the liver and kidneys fills the abdominal cavity and forces the intestinal loop to migrate through the umbilicus into the extraembryonic coelom (herniation). During its migration, the intestinal loop rotates 90° counterclockwise around an axis formed by the superior mesenteric artery and the vitelline duct (Fig. 13.3), as seen in front. The cranial limb moves to the right, and the caudal limb moves to the left. The cranial limb of the loop grows rapidly and forms intestinal coils. A cecal swelling appears in the caudal limb of the loop. At about the 10th week, the size of the liver and kidneys is relatively decreased, and the size of the abdominal cavity is increased. The intestine returns rapidly to the abdomen (reduction of the hernia).

The small intestines, which are formed from the cranial limb, return first and pass behind the superior mesenteric artery. As the intestines return to the abdominal cavity, they undergo an additional 180° counterclockwise rotation, making a total of 270°.

At first, the cecal swelling is conical; later, the upper part expands and forms the cecum, while the lower part elongates and forms the appendix. By the time the intestines return to the abdominal cavity, the vitelline duct is normally obliterated and loses its connection to the intestines.

The cranial limb of the loop forms the distal part of the duodenum, jejunum and proximal ileum. The caudal limb of the loop develops into the distal part of the ileum, cecum and appendix, ascending colon, and proximal two-thirds of the transverse colon.

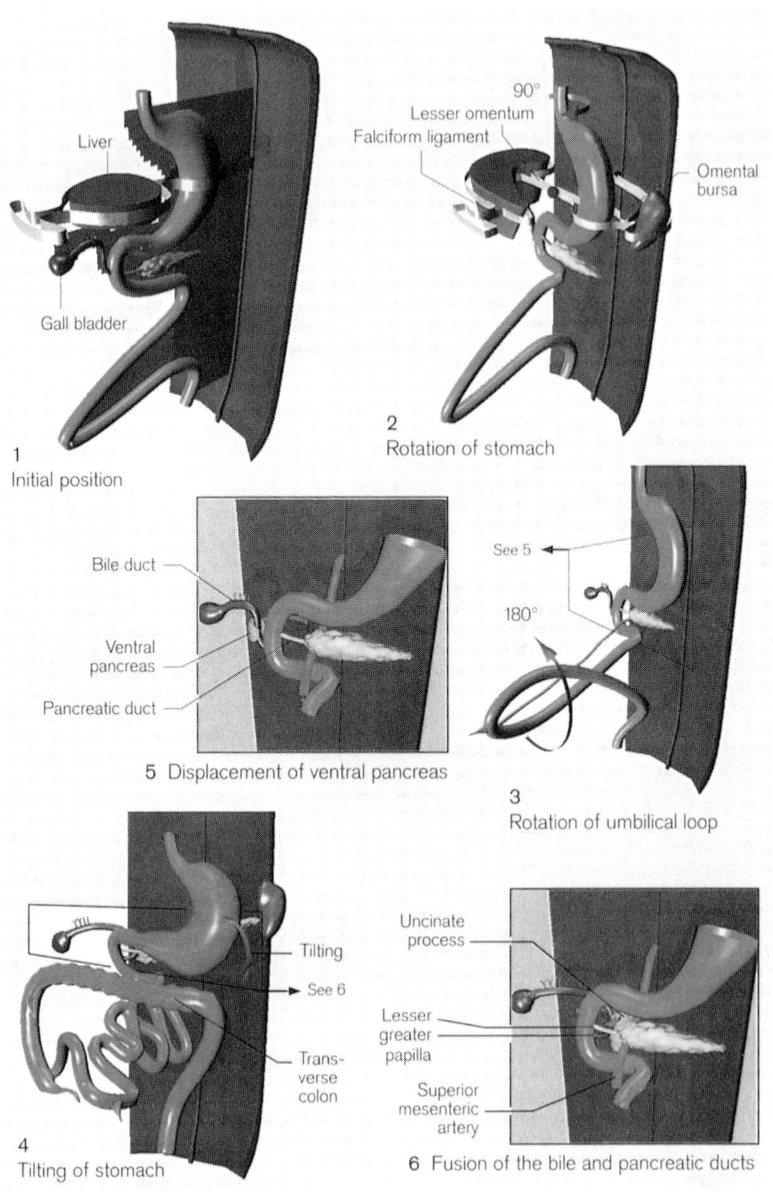

Figure 13.3

Gastrointestinal rotation (see Color Plate 13.3). Reproduced with permission from Thieme Medical Publishers Inc., New York, 1995, *Color Atlas of Embryology,* Ulrich Drews, Chapter 7: Gastrointestinal Tract.

Briefly, intestinal villi and peristalsis begin to appear in the embryonic period. The meconium is formed of glandular secretion, bile and swallowed amniotic fluid. It is a green pasty material excreted by fetuses and neonates.

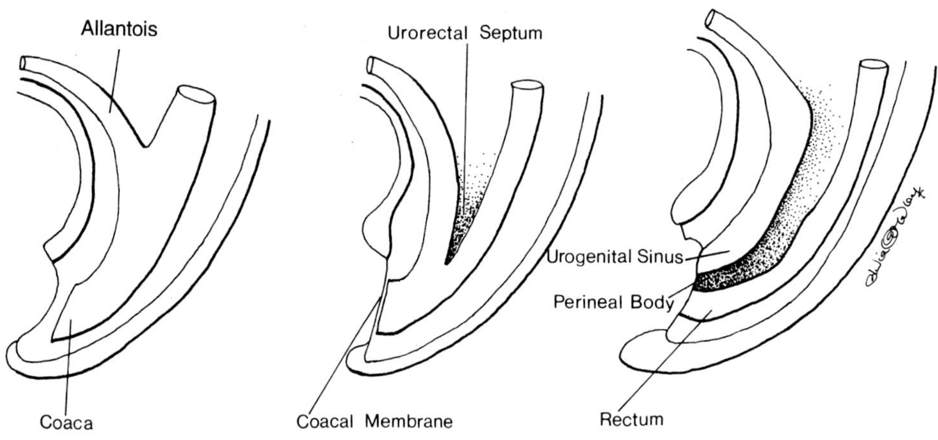

Figure 13.4
Division of the cloacal region.

IV. HINDGUT

Hindgut derivatives include: the distal one-third to two-thirds of the transverse colon, the descending colon, the sigmoid colon, the rectum and the cloaca. The hindgut is supplied by the inferior mesenteric artery. The dilated terminal part of the hindgut, the cloaca, receives the allantois ventrally and the mesonephric ducts laterally.

The opposed layers of the surface ectoderm of the proctodeum and the endoderm of the cloaca form the cloacal membrane, which separates the cavity of the hindgut from the surface (Fig. 13.1). In the angle between the allantois and the hindgut, a wedge of mesenchyme, the urorectal septum, invaginates the endoderm. The urorectal septum proliferates and grows toward the cloacal membrane, dividing the cloaca into two parts: the urogenital sinus ventrally and the anorectal canal dorsally (Fig. 13.4).

The urorectal septum finally reaches the cloacal membrane, fuses with it, and forms the primitive perineum. Thus, the cloacal membrane becomes divided into the dorsal anal membrane and the ventral urogenital membrane. Mesenchymal proliferation around the anal membrane forms an ectodermal depression, the anal pit. Soon thereafter, the anal membrane ruptures and the digestive tract communicates with the amniotic cavity, thus establishing the anal canal. The lower third of the anal canal thus originates from the ectoderm and is supplied by the branches of the internal pudendal artery.

In general, the intestinal epithelium undergoes 1) proliferation, which temporarily occludes the lumen; 2) recanalization, during which programmed cell death reestablishes patency; 3) cellular differentiation; and 4) functional maturation of different types of cells. The mesenchymal layer surrounding the epithelial tube differentiates into smooth muscle layers, connective tissue and blood vessels. During recanalization, secondary small spaces appear just beneath the epithelium. These small spaces (the secondary lumina) coalesce to form larger spaces, which are promptly filled by mesenchymal tissue. The mesenchyme in the spaces proliferate and push the overlying epithelial layer toward the lumen as finger-like intestinal villi. The fetal colon also contains villi that disappear before birth.

V. THE MESENTERIES

The mesentery is a double peritoneal layer that anchors the intestine in the midline of the abdomen. After folding of the embryo, the caudal part of the foregut and the entire midgut and hindgut are suspended in the peritoneal cavity by the dorsal mesentery. The ventral mesentery disappears, except where it is attached to the caudal part of the foregut (Fig. 13.2).

A. Ventral Mesentery

The ventral mesentery is formed from the caudal part of the septum transversum. As a result of the enormous growth of the liver in the septum transversum, the ventral mesentery becomes thin and gives rise to three structures in relation to the diaphragm and the ventral abdominal wall:

1. The falciform ligament extends from the liver to the ventral abdominal wall. The inferior free margin of the falciform ligament contains the umbilical vein, which after birth becomes obliterated and forms the ligamentum teres. The part of the ventral mesentery surrounding the liver becomes the hepatic capsule.
2. The lesser omentum extends from the liver to the ventral border of the stomach and to the duodenum, forming the gastrohepatic and duodenohepatic ligaments. The free margin of the lesser omentum contains the common bile duct, hepatic artery and portal vein.
3. The coronary ligament is part of the ventral mesentery between the liver and diaphragm. The area enclosed within it is the bare area of the liver.

B. Dorsal Mesentery

The dorsal mesentery is formed from fusion of the splanchnopleuric mesoderm of the left and right sides, which extends from the dorsal abdominal wall to the dorsal border of the gut. It is given the following descriptive names at successive levels (the form *meso* combines with an organ):

1. Mesoesophagus
2. Mesogastrium
3. Mesoduodenum
4. Mesocolon
5. Mesorectum
6. In the region of the jejunum and ileum it is called the mesentery proper. The spleen and dorsal pancreatic bud develop between the layers of mesogastrium and mesoduodenum.

C. Formation of the Lesser Peritoneal Sac

Isolated small intercellular clefts appear in the mesogastrium (Fig. 13.5). The clefts coalesce to form a single cavity known as the lesser peritoneal sac or omental bursa. This cavity expands cephalically and transversely in the mesogastrium, becoming situated behind the stomach and to the right side of the esophagus. When the diaphragm completes its development, the cephalic extension of the lesser sac is cut off and forms a small infracardiac bursa, which later disappears. With enlargement of the stomach, the lesser sac expands further between the layers of the elongated mesogastrium. As the stomach rotates, the elongated mesogastrium (the greater omentum) hangs from the greater curvature of the stomach down in front of the developing intestines. Later, the layers of the greater omentum fuse together, obliterating the space between them. As a result of the rotation of the stomach and the duodenum, the right part of the peritoneal cavity becomes incorporated in the lesser sac. The right free border of the ventral mesentery (lesser omentum) forms the anterior boundary of the entrance to the lesser sac (the epiploic foramen) (Fig. 13.5).

The lesser sac now is composed of two recesses:

1. The superior recess lies posterior to the lesser omentum and extends behind the caudate lobe of the liver as far as the diaphragm.
2. The inferior recess lies posterior to the stomach and runs for a variable distance in the greater omentum.

The spleen develops between the layers of the mesogastrium, and as the stomach rotates the left surface of the mesogastrium fuses with the peritoneum over the left kidney. The mesogastrium between the spleen and kidney forms the lienorenal ligament. The mesogastrium between the stomach and spleen forms the gastrolienal ligament.

D. Mesenteries of the Intestines

Rotation of the stomach and duodenum causes the duodenum and pancreas to become pressed against the dorsal body wall. The mesoduodenum fuses with the peritoneum and subsequently disappears; the duodenum and pancreas then become retroperitoneal (Fig. 13.5). During rotation of the midgut loop, the mesentery proper of the jejunum and ileum twists around the origin of the superior mesenteric artery. As the ascending mesocolon disappears, the mesentery assumes a new line of attachment that passes from the duodenojejunal junction to the ileocecal junction. The cecum and appendix do not possess true mesentery, but there is a peritoneal fold associated with the appendix that contains the appendicular artery.

As the intestines assume their final position, the ascending and descending colon become pressed against the posterior body wall, and the mesocolon of these portions fuses with the peritoneum and finally disappears. The ascending and descending colon become retroperitoneal. The transverse mesocolon first becomes attached to the anterior surface of the pancreas and covers the duodenum. Later, it fuses with the posterior layer of the greater omentum. The mesocolon of the sigmoid colon persists, and the mesorectum disappears.

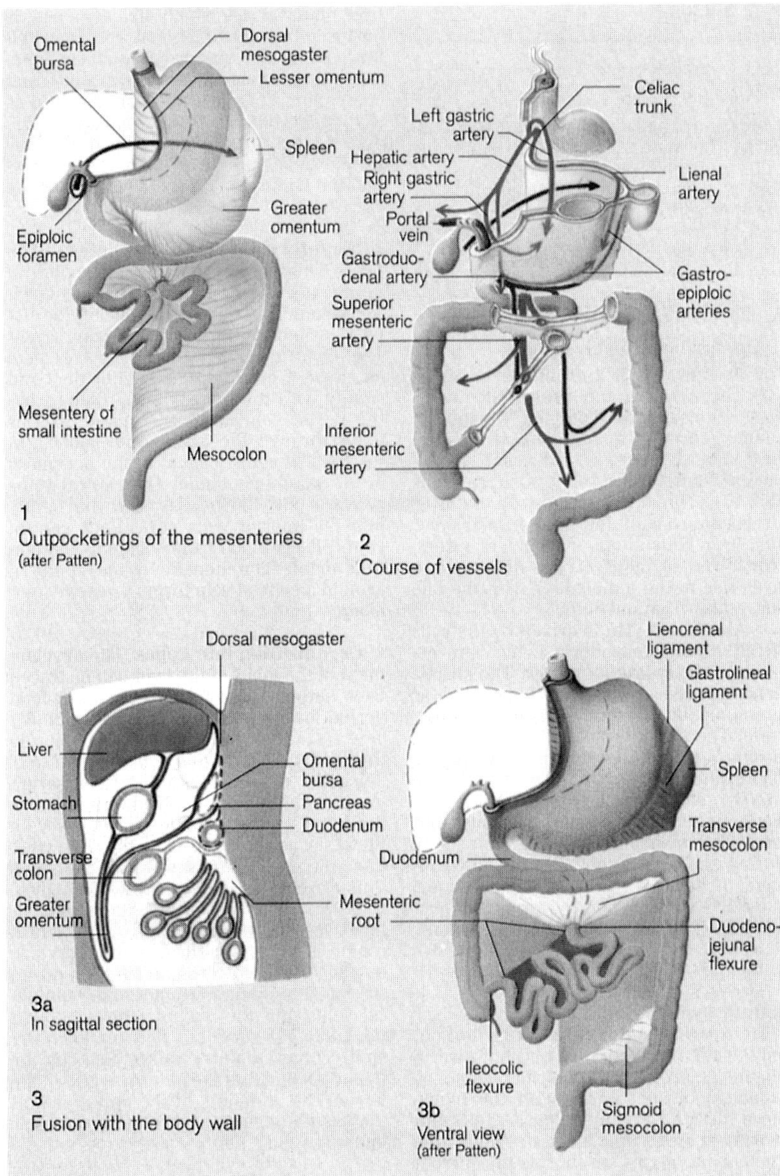

Figure 13.5

Derivation of situs (see Color Plate 13.5). Reproduced with permission from Thieme Medical Publishers Inc., New York, 1995, *Color Atlas of Embryology*, Ulrich Drews, Chapter 7: Gastrointestinal Tract.

VI. CONGENITAL ANOMALIES OF THE DIGESTIVE SYSTEM

A. Atresia and Stenosis

Atresia is an interruption in the continuity of the lumen. It may result from a diaphragm, or fibrous growth, or absence of a segment of the gut. It is generally

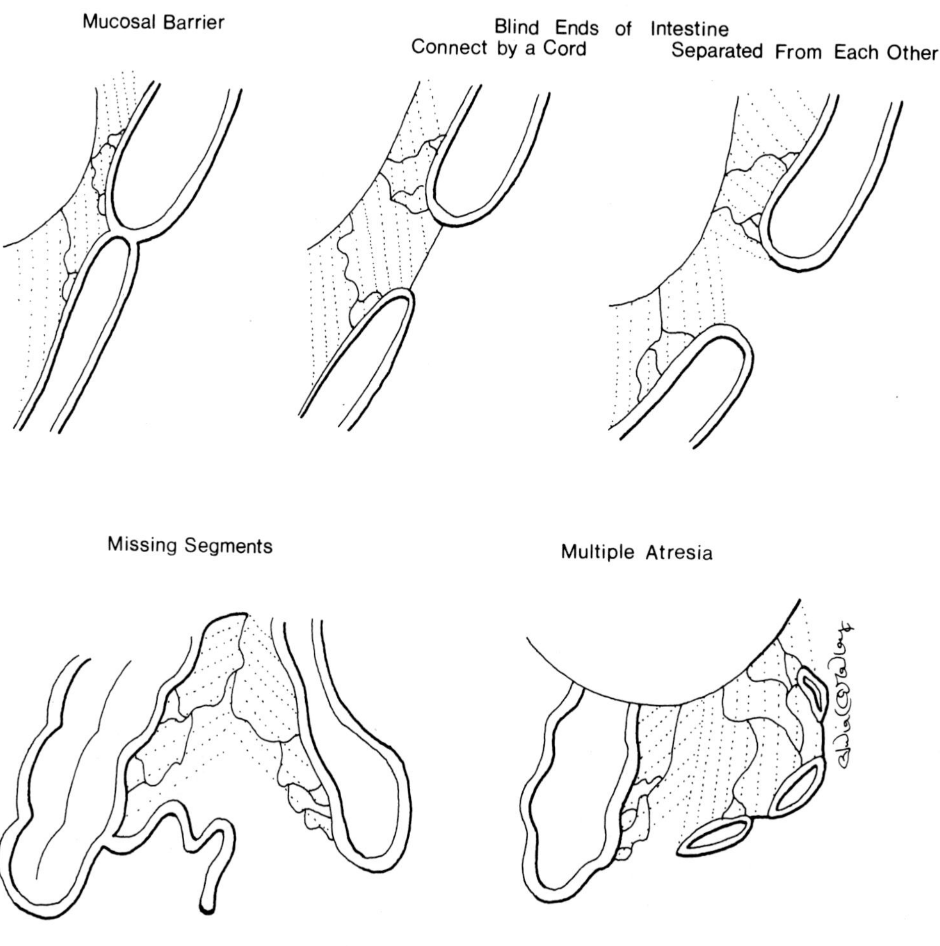

Figure 13.6
Intestinal atresia.

attributed to failure of recanalization, but histopathological findings indicate that many of these lesions arise after organogenesis is complete. Other causes implicated in atresia include internal intestinal prolapse (intussusception), excessive proliferation of epithelium, intrauterine inflammation and local interruption of blood supply (ischemia). Regional ischemia appears to be the major cause. Volvulus (large distention of the intestine) intussusception, entrapment and compression of a segment of intestine and any defect in the anterior body wall lead to regional ischemia. The most commonly affected areas are the proximal esophagus, duodenum and anorectal region (Fig. 13.6).

Stenosis is narrowing of the lumen that may result from incomplete recanalization, hypertrophy of the muscular layer or compression due to ectopic tissue around the gut (Fig. 13.7).

1. Esophageal Atresia

Esophageal atresia is often (90%) associated with tracheoesophageal fistula, which is discussed with development of the respiratory system (Chapter 10, Fig.

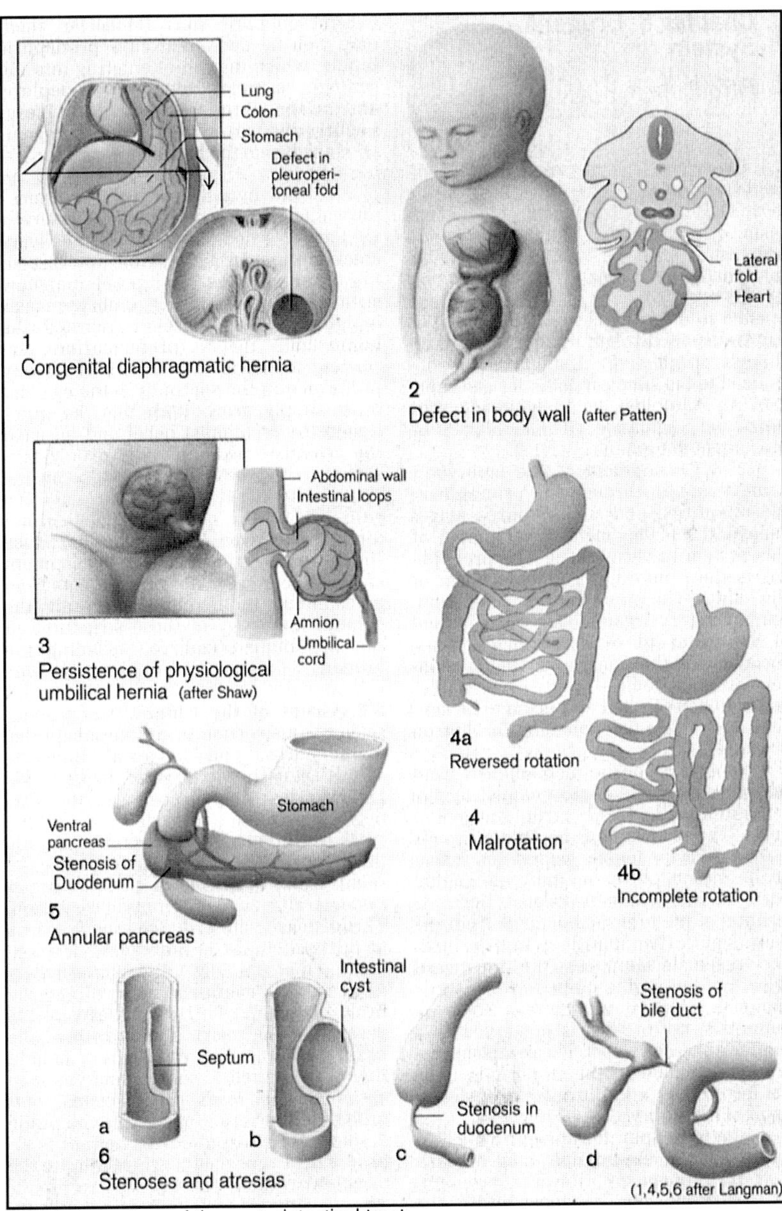

Figure 13.7
Malformations (see Color Plate 13.7). Reproduced with permission from Thieme Medical Publishers Inc., New York, 1995, *Color Atlas of Embryology*, Ulrich Drews, Chapter 7: Gastrointestinal Tract.

10.6). Other anomalies associated with this are of the cardiovascular system and anorectal atresia. Possible morphologic forms are shown in Fig. 10.6. In these conditions, impaired swallowing by the fetus leads to polyhydramnios. After birth, difficulty in swallowing, projectile vomiting (regurgitation) and choking are indications of obstruction in the gastrointestinal tract.

2. Duodenal Atresia and Stenosis

This condition occurs intrinsically due to failure or incomplete recanalization of the duodenum, or extrinsically by compression from malrotation of the gut or annular pancreas. It is often associated with Down's syndrome, tracheoesophageal fistula and anorectal atresia. It most commonly occurs at the entrance of the common bile duct. Regional absence of muscular wall and segmental dilation have been noted with atresia.

3. Pyloric Stenosis

This consists of hypertrophy of the pyloric muscles mainly affecting the circular fibers. In some cases, it appears to be more functional than anatomical. There is considerable narrowing of the pyloric lumen. Although the cause is unknown, the autonomic ganglion cells are fewer than have normally been reported in this region. Several hours after feeding, the newborn undergoes projectile vomiting accompanied by constipation and weight loss. Hereditary factors are important, and it is three to four times more common in males than in females.

4. Extrahepatic Biliary Atresia

Biliary atresia is an absence of an extrahepatic duct, such as the hepatic duct or bile duct, and/or intrahepatic biliary ducts. The bile canaliculi are present. The clinical presentation in these varies. It is more common in females than in males. Although failure of recanalization has been attributed to a pathogenetic mechanism, other causative factors suggested include inflammation due to infection and ischemia. According to these findings, extrahepatic biliary atresia follows an injury to the developed duct system. It is the most frequent cause of persistent jaundice in newborns. Many cases are associated with duodenal atresia and stenosis. The gallbladder may fail to develop.

5. Anorectal Atresia — Anorectal Agenesis

Anorectal atresia is common, having a live birth incidence of 1 in 2500. Male preponderance has been reported. Most of these involve an imperforate anus. About three-fourths have associated anomalies, such as partial sacral agenesis, tracheoesophageal fistula, urogenital defects, and anomalies of the foregut and cardiovascular system. Down's syndrome occurs with low anorectal anomalies. Because of its broad-range involvement, about 30 types have been described. For simplicity, these are divided into low- and high-placed atresia.

a. Low anomalies

   Low anomalies are situated at or below the levators. These are attributed to defective formation of the anal pit.

   i. Membranous atresia results from failure of the anal membrane to perforate (Fig. 13.8).

   ii. Anal stenosis may result from excessive regression of the postanal gut, which reduces the termination of intestine. A slight dorsal deviation of the urorectal septum during its caudal growth is also suggested.

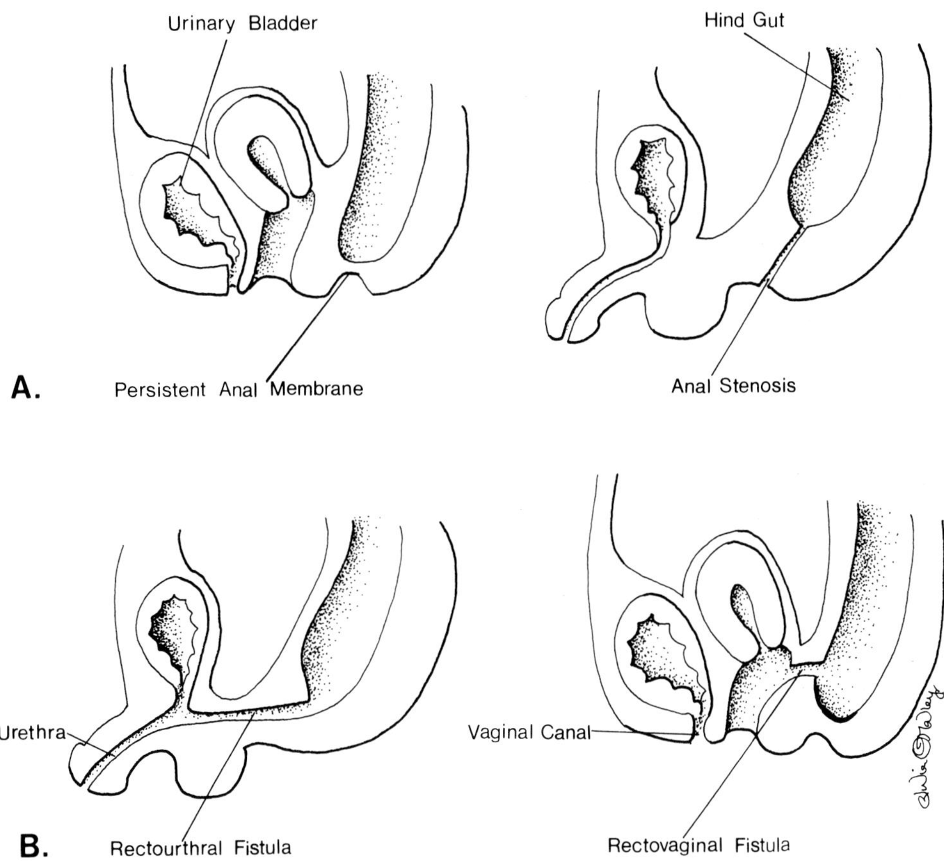

Figure 13.8
Anorectal anomalies: (A) low defects; (B) high defects.

iii. Anal agenesis with (communicating) or without (noncommunicating) fistula may result from incomplete division of the cloaca by the urorectal septum. The anal canal ends blindly but in most cases communicates with the surface by a narrow fistula that opens into the perineum, or into the vulva in females, or into the urethra in males. Defective development of the perineal body is also suggested.

b. High anomalies (situated above the levators)
Development of the cloacal membrane and urorectal septum is important in these anomalies. Excessive regression of the postanal gut beyond the urorectal septum may cause atretic rectum, which no longer communicates with the urogenital sinus. In these conditions, the rectum usually reaches the upper surface of the pelvic floor, but the anal canal is absent. Devascularization of the lower gut, teratogens, short intestines (hypoplasia) and various other possibilities have been suggested as causative factors.

i. Anorectal agenesis, communicating or noncommunicating: This is the most common type of anorectal anomaly. There is usually a fistula, which in males opens into the urinary bladder or urethra and in females opens into the vagina.

ii. Rectal atresia: The anal canal and rectum are separated by an atretic segment of the rectum. This may be a type of general intestinal atresia.

c. Persistent cloaca: This may result from failure of proliferation of mesenchyme in the urorectal septum. It occurs mostly in females and is associated with vaginal obstruction. The urinary bladder and rectum open into a single wide opening.

The low anomalies are relatively simple to correct, but the high anomalies require prolonged surgical care for many years.

In general, the proposed pathogenesis of intestinal atresia includes failed canalization, abnormal mesenchymal and epithelial growth, regional ischemia (which can be caused by volvulus, intussusception and entrapment), fetal hypoperfusion, inflammation, mucosal adhesions and trauma (amniocentesis).

B. Persistent Vitelline Duct

This is the most common anomaly of the intestinal tract, with occurrence in 2% of live births. This family of anomalies has some form of vitelline duct remnants 2 inches in length located 2 feet from the iliocolic valve on the antimesenteric border of the ilium (Fig. 13.9).

1. A vitelline (Meckel's) diverticulum is the most common of these anomalies. It is a blind pouch situated on the ilium, representing a remnant of the proximal portion of the yolk stalk. It is often asymptomatic but contains ectopic pancreatic or gastric tissue. Secretion from ectopic tissue occasionally causes inflammation or ulceration.

2. A vitelline fistula is persistence of a patent vitelline duct that opens on the body surface through the umbilicus. It forms a direct communication between the intestinal lumen and the outside of body.

3. A vitelline cord or ligament is a fibrous cord that keeps the intestine connected to the umbilicus. The intestine occasionally rotates about the cord, causing volvulus, which results in intestinal obstruction.

4. A vitelline cyst results when both ends of a persistent vitelline duct become fibrous, leaving the middle portion patent. The mucosa in this portion continues to secrete, forming a cyst. As it remains connected to the umbilicus and the midgut, it may cause volvulus.

5. A vitelline sinus results when a small portion of the vitelline duct remains and opens onto the skin at the umbilicus. It loses its connection with the gut.

C. Abnormal Rotation of the Gut

This usually involves the midgut and arises as an anomalous rotation and positioning of the gut during its return (reduction) to the abdominal cavity. It is

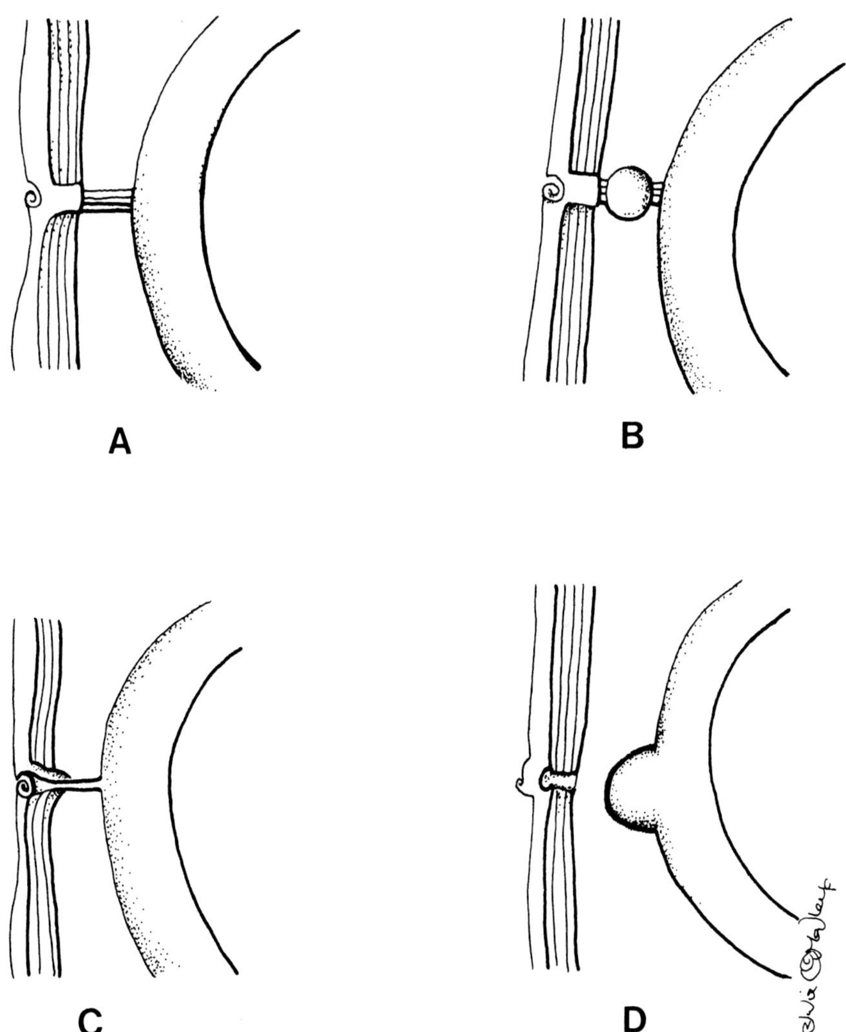

Figure 13.9

Remnants of the vitelline duct: (A) fibrous cord; (B) vitelline cyst; (C) vitelline fistula; (D) Meckel's diverticulum.

prevalent in trisomies 13, 18 and 21. Some abnormalities of rotation are incomplete or reversed processes. In these, the gut may be fixed incompletely within the peritoneal cavity and present as a mobile cecum. Occasionally, the mesocolon persists at the ascending and descending colon. In most cases, they are asymptomatic but may lead to volvulus or strangulation of the gut.

1. Nonrotation is an incomplete rotation in which the midgut loop does not undergo the final 180° rotation. In this condition, the colon and cecum (distal limb of the midgut loop) return first and settle on the left side of the abdominal cavity, and the small intestine (proximal limb) returns later to lie on the right side of the abdominal cavity

2. Malrotation: the midgut loop fails to complete the final 90° rotation (Fig. 13.7). In this condition the duodenum lies in front of the superior mesenteric artery, and the colon moves behind the mesentery of the small intestine, causing duodenal obstruction.

3. Reversed rotation occurs when the intestinal loop rotates clockwise. It may cause an obstruction (strangulation) of the superior mesenteric artery.

D. Duplication of Gastrointestinal Tract

This can occur in any part of the gastrointestinal tract, but most frequently it is in ilium. It develops during recanalization of the gut. In some cases, duplication results in a closed cyst; in other cases, duplication is tubular, communicating with the main lumen. The mucosa in the segments may be different.

E. Pancreatic Anomalies

1. Heterotopic (ectopic) pancreatic tissue is attributed to metaplasia (differentiation of pluripotent endodermal cells), abnormal transplantation of pancreatic cells or differentiation of submucosal glands as pancreatic tissue. Heterotopic pancreatic tissue may proliferate, bulge into the lumen and cause intussusception. About 6 to 10% of Meckel's diverticula contain heterotopic pancreatic tissue.

2. Annular pancreas results from abnormal migration of the ventral pancreatic bud. The two portions of the ventral pancreatic bud migrate in opposite directions and completely surround the duodenum and cause duodenal obstruction. Vomiting begins a few hours after birth.

3. Cystic fibrosis of the pancreas is a disease involving secretory epithelium. This disease also involves secretory epithelium of the lungs, kidneys and liver. This is one of the most frequent fetal genetic disorders and results mainly from a decrease in fluid and salt secretion, leading to thick mucus that blocks the duct system. In lungs, thick mucus accumulation in the airways causes pulmonary distress. Thick mucus from abnormal pancreas results in intestinal obstruction known as meconium ilium. Development of pancreatic tissue is deficient. An increased amount of fibrous connective tissue and acinar and ductular dilatation have been shown. In some cases, remarkably pale and calcified pancreas was described. The primary defect is in chloride transport: chloride channels are blocked and water is retained in the cells, resulting in thick dry mucus. It is inherited as an autosomal recessive trait, and the gene is located on chromosome 7. Insertion of a normal gene using retrovirus is being tried. Although the prevalence is 1:2000, it varies in different parts of the world.

F. Disorders of Motility

1. Small left colon syndrome is a transient suppression of peristalsis associated with complication of maternal diabetes and has also been associated with maternal use of psychotropic drugs.

2. Aganglionic megacolon (Hirschsprung's disease) is a markedly distended pelvic colon with a constricted distal portion. It is produced by functional obstruction caused by the absence of peristalsis in the segment just distal to the distention. The accumulation of intestinal contents causes dilation and hypertrophy of the proximal (normal) colon. The constricted segment lacks parasympathetic ganglion cells. It is attributed to failed and/or defective migration of neural crest cells. The migration theory does not explain the cases in which the aganglionic segment is bounded proximally and distally by the normal segment. The many other explanations include the following: the milieu of the gut wall is unable to attract or sustain the neuroblast, or the milieu of the gut wall inhibits the migration of neuroblasts in affected segment, or the neuroblast may be destroyed en route or in enteric plexi. Accumulation of laminin in the gut wall may serve as a signal to end neural crest cell migration. Cytomegalovirus infection has been reported in some cases. It has been associated with trisomies 18, 21 and 22, intestinal atresia and malrotation, and anomalies of the cardiovascular, urinary and skeletal systems. Familial cases are common, and it is more frequent in males.

G. Abdominal Wall Defects

1. Omphalocele results when intestinal loops do not return into the body cavity. A defect of the umbilical ring and adjacent abdominal wall is reported. The herniated intestinal loops are covered by a sac that is made up of the amnion and peritoneal membrane. The abdominal cavity is probably secondarily too small to receive the herniated intestinal loops. It is associated with cardiac, urogenital, skeletal and nervous system anomalies. Trisomy 18 shows high association with this (Fig. 13.7).
2. Gastroschisis is a herniation of abdominal contents through a congenital defect in the abdominal wall. The herniation sac (found in an omphalocele) is not present, suggesting that the intestinal loops returned normally into the body cavity, but due to the defect in the musculature of the abdominal wall the gut protrudes out again. It is likely that compression of a herniated gut and mesentery produces ischemia, which could lead to intestinal atresia, stenosis and necrosis of the gut wall. On the other hand, gastroschisis may be due to ischemia. Interruption of the vitelline (omphalomesenteric) artery leads to necrosis and loss of abdominal wall lateral to the umbilicus. It is also suggested that an epithelial abnormality at the amnioectodermal junction may produce an abdominal wall defect.
3. Prune-belly syndrome is an abdominal wall distention caused primarily by lower urinary tract obstruction. Due to distention, proliferation of mesoderm in the abdominal wall is retarded, and muscular tissue fails to develop. Intestinal pathology such as malrotation, atresia, stenosis and fistula are common. This syndrome will be discussed in Chapter 14 (Urinary System).

TABLE 13.1. Summary of Development of the Digestive Tube and Its Derivatives

Age (days)	Structure
22	Incorporation of yolk sac into foregut and hindgut
24	Biliary diverticulum appears and oropharyngeal membrane ruptures
28	Lung buds and dorsal pancreas appear
34	Formation of primary and lobar bronchi
35	Ventral pancreas and spleen present
38	Division of cloaca and herniation of gut loop
49	Separation of rectum and bladder, and development of segmental bronchi
56	Recanalization of duodenum
63	Reduction of umbilical hernia

REVIEW QUESTIONS — Chapter 13

Select the most appropriate answer.

1. All of the following are endodermal in origin except the:
   a. pancreatic acinar cells
   b. α-cells of the pancreas
   c. β-cells of the pancreas
   d. pancreatic duct epithelial cells
   e. endothelial cells

2. All of the following are derived from the yolk sac proper except the:
   a. allantois
   b. vitelline duct
   c. pharyngeal clefts
   d. midgut loop
   e. cloaca

3. All of the following are derived from the foregut except the:
   a. pharyngeal pouch
   b. parotid gland
   c. submandibular gland
   d. thyroid gland
   e. trachea
4. The contents of liver are derived from the:
   a. biliary diverticulum
   b. septum transversum
   c. vitelline veins
   d. umbilical veins
   e. all of the above
5. Which of the following is affected by rapid growth of the liver?
   a. herniation of the midgut loop
   b. rotation of the stomach
   c. rotation of the midgut loop
   d. migration of the ventral pancreatic bud
   e. all of the above
6. The hepatic diverticulum grows into the:
   a. septum transversum
   b. pleuropericardial canal
   c. pleuroperitoneal canal
   d. connecting stalk (umbilical cord)
   e. dorsal mesentery
7. Which of the following statements regarding development of the pancreas is not true?
   a. most of the pancreatic mass develops from the dorsal pancreatic bud
   b. part of head is derived from the ventral pancreatic bud
   c. the duct of the ventral pancreatic bud is retained as the main pancreatic duct
   d. the islets of Langerhans are derived from the mesenchyme
   e. the muscular layer in the main duct is derived from the mesenchyme
8. When the duodenum rotates, it carries the ventral pancreatic bud dorsally, where it fuses with the dorsal pancreatic bud, and:
   a. the ventral pancreatic bud forms the major portion of the adult pancreas
   b. the main pancreatic duct forms from the duct of the ventral bud
   c. the dorsal pancreatic bud gives rise to the head of the pancreas
   d. the accessory pancreatic duct, if present, joins the common bile duct to enter the duodenum
   e. all of the above statements are true
9. The splanchnic mesoderm gives rise to the:
   a. lining of the abdominal cavity
   b. lining of the pleural cavity
   c. serosal lining of the gut
   d. epithelial lining of the gut
   e. epithelial lining of the bladder

10. The caudal folding of the embryo causes incorporation of the yolk sac primarily into the:
    a. hindgut
    b. allantois
    c. cloacal membrane
    d. body stalk
    e. umbilical cord
11. After separation from the trachea, the esophagus:
    a. elongates and its lumen becomes obliterated
    b. undergoes 90° clockwise rotation
    c. appears as a dilated part of the foregut
    d. acquires ventral mesentery
    e. all of the above are correct
12. The urorectal septum that divides the cloaca is located:
    a. in the gubernaculum
    b. between the allantois and hindgut
    c. between the mesocolon and ventral mesentery
    d. in the septum transversum
    e. between the mesonephric duct and hindgut
13. All of the following are derived from the septum transversum except the:
    a. falciform ligament
    b. lesser omentum
    c. fibrous pericardium
    d. central tendon of diaphragm
    e. dorsal mesogastrium
14. All of the following are derived from the caudal (distal) limb of the midgut loop except the:
    a. distal ileum
    b. vitelline duct
    c. ascending colon
    d. proximal part of transverse colon
    e. cecum
15. Which part of the developing digestive tract makes three 90° counterclockwise rotations?
    a. the stomach
    b. the duodenum
    c. the midgut loop
    d. the hindgut
    e. none of the above
16. Which of the following marks the distal end of the foregut?
    a. the biliary diverticulum
    b. the superior mesenteric artery
    c. the vitelline duct
    d. the cecal swelling
    e. the celiac trunk

17. The dorsal mesogastrium develops into the:
    a. lesser omentum
    b. coronary ligament
    c. falciform ligament
    d. greater omentum
    e. all of the above

18. The midgut:
    a. is the last part of the digestive system to be enclosed into the body cavity
    b. remains connected to the yolk sac until the late embryonic period
    c. develops mainly in the extraembryonic coelom
    d. grows unequally in its different parts
    e. all of the above are correct

19. All of the following undergo rotation except the:
    a. stomach
    b. duodenum
    c. ventral pancreas
    d. midgut
    e. hindgut

20. The biliary diverticulum:
    a. grows from the foregut
    b. grows into the septum transversum
    c. gives rise to ducts of the liver
    d. gives rise to the hepatocytes
    e. all of the above

21. During early development, the cloaca is formed by the:
    a. incorporation of the allantois into the body of the embryo
    b. incorporation of the yolk sac into the caudal region of the embryo
    c. migration of the mesoderm from the primitive streak
    d. induction of the endoderm by the notochord
    e. all of the above contribute to formation of the cloaca

22. The structure that develops within the mesoduodenum is the:
    a. vitelline duct
    b. gall bladder
    c. spleen
    d. biliary diverticulum
    e. dorsal pancreatic bud

23. The first diverticulum of the yolk sac is the:
    a. allantois
    b. vitelline diverticulum
    c. stomodeum
    d. foregut
    e. hindgut

24. Which of the following is peritoneal during development and becomes retroperitoneal later?
    a. the kidneys
    b. the liver
    c. the second part of the duodenum
    d. the transverse colon
    e. the urinary bladder

25. Which of the following forms an axis for the rotation of the stomach?
    a. the celiac trunk
    b. the superior mesenteric artery
    c. the inferior mesenteric artery
    d. the left gastric artery
    e. none of the above

26. The cecum and vermiform appendix arise from the:
    a. foregut
    b. midgut
    c. hindgut
    d. vitelline duct
    e. none of the above

27. The spleen develops in the:
    a. foregut
    b. midgut
    c. hindgut
    d. somatic pleura
    e. dorsal mesogastrium

28. The vitelline duct detaches from the gut when:
    a. the midgut loop is formed
    b. the midgut loop rotates the initial 90°
    c. the midgut herniates into the connecting stalk
    d. the midgut returns to the abdominal cavity
    e. the cecum descends to its permanent position

29. Rotation of the stomach results in all of the following except:
    a. its dorsal border moves to left
    b. the left vagus nerve moves to the ventral side
    c. the ventral part of the stomach grows rapidly
    d. the caudal (pyloric) end of the stomach moves upward and to the right
    e. the ventral mesogastrium is pulled to the right

30. Which of the following develop(s) in the extraembryonic coelom?
    a. lung buds
    b. the pancreas
    c. the spleen
    d. the ileum
    e. none of the above

31. During development of the gut, all of the following become retroperitoneal except the:
    a. spleen
    b. most of the duodenum
    c. pancreas
    d. ascending colon
    e. descending colon

32. The midgut loop, during its rotation, twists around the origin of the:
    a. celiac trunk
    b. superior mesenteric artery
    c. inferior mesenteric artery
    d. biliary diverticulum
    e. dorsal pancreatic bud

33. A child was born with jaundice, which persisted more than a week. This might be caused by:
    a. a solid bile duct
    b. a noncannulated cystic duct
    c. an absence of gall bladder
    d. duodenal atresia involving the entrance of the bile duct
    e. any of the above could be the cause

34. On one gastrointestinal tract radiograph, most of the small intestinal loops are on the right side of the abdominal cavity, and the colon and cecum are on the left side. This may have been caused by:
    a. failure of the midgut loop to undergo the final 180° rotation
    b. failure of the midgut loop to undergo the final 90° rotation
    c. clockwise rotation of the midgut loop
    d. malrotation of the duodenum
    e. a short hindgut

35. Persistent projectile vomiting in an infant may be caused by any of the following except:
    a. an annular pancreas
    b. esophageal atresia
    c. incomplete rotation of the gut
    d. pyloric stenosis
    e. tracheoesophageal fistula

36. Congenital fibrocystic disease may involve all of the following except the:
    a. heart
    b. pancreas
    c. ileum
    d. kidneys
    e. liver

37. Congenital megacolon (Hirschsprung's disease) is caused by:
    a. failure of sympathetic innervation in portion of colon distal to distention
    b. malrotation of the gut
    c. nonrotation of the gut
    d. an absence of ganglionic cells distal to the distended segment
    e. all of the above could contribute to this condition
38. Atresia is an interruption in the continuity of the lumen. It is generally attributed to failure of recanalization. It could occur in the:
    a. proximal esophagus
    b. pyloric region of the stomach
    c. duodenum
    d. anorectal region
    e. any of the above
39. Volvulus of the gut is:
    a. a large distention of part of a normally developed gut situated proximal to the stenotic part
    b. a large distention of part of the gut located distal to the stenotic part
    c. the aganglionic part of the gut
    d. narrowing of part of the gut
    e. a noncanalized part of the gut
40. All of the following are abnormal hernias (movement of the gut out of abdominal cavity) except:
    a. inguinal hernia
    b. umbilical hernia
    c. diaphragmatic hernia
    d. femoral hernia
    e. strangulating hernia
41. Defective division of the cloaca by the urorectal membrane may result in:
    a. anal stenosis
    b. anal agenesis without fistula
    c. anal agenesis with fistula
    d. an imperforate anus
    e. all of the above
42. All of the following result in abdominal distention except:
    a. Meckel's (vitelline) diverticulum
    b. megacolon
    c. volvulus
    d. prune-belly syndrome
    e. cloacal dysgenesis
43. Abnormal rotation of the gut is associated with the:
    a. esophagus
    b. stomach
    c. duodenum
    d. midgut
    e. hindgut

44. The vitelline duct:
    a. is a remnant of the yolk sac
    b. if it persists after birth, may cause inguinal hernia
    c. is a remnant of the allantois
    d. is a persistent vitelline vein
    e. becomes the ligamentum teres

45. An abnormal rotation of the midgut loop may result in:
    a. congenital umbilical hernia
    b. Meckel's (vitelline) diverticulum
    c. strangulation of the gut
    d. intussusception
    e. all of the above

46. An abnormal migration of the ventral pancreatic bud may result in:
    a. an annular pancreas
    b. an ectopic pancreas
    c. an accessory pancreatic duct
    d. a small pancreas
    e. all of the above

47. Urorectal fistula is caused by:
    a. abnormal partition of the cloaca
    b. failure of fixation of the hindgut
    c. failure of the proctodeum to develop
    d. premature rupture of the anal membrane

ANSWERS TO REVIEW QUESTIONS

1. e	2. c	3. b	4. e	5. a
6. a	7. c	8. b	9. c	10. a
11. a	12. e	13. e	14. b	15. c
16. a	17. d	18. e	19. e	20. e
21. b	22. e	23. a	24. c	25. e
26. b	27. e	28. d	29. e	30. d
31. a	32. b	33. e	34. a	35. c
36. a	37. a	38. e	39. d	40. b
41. e	42. a	43. d	44. a	45. c
46. e	47. a			

CHAPTER

14

URINARY SYSTEM

The urinary and genital systems are closely associated in their development. Both arise from the intermediate mesoderm. In the human, two nephric structures, the mesonephros and metanephros, as temporary and permanent excretory organs respectively, arise during development (the concept of pronephros does not apply in humans). Development of the urinary system depends on directed growth and mutual inductive influence of different structures. Committed mesenchymal cells, under the influence of signals from the extracellular matrix, proliferate and form nephrogenic cords (urogenital ridges) on both sides of the aorta. Mesonephric tubules form in the ridge and gain connection to the mesonephric duct. The mesonephric duct opens into the cloaca.

In the distal (caudal) portion of each mesonephric duct, a ureteric bud grows out and induces the overlying intermediate mesoderm to give rise to a metanephric blastema. The metanephros develops from an epithelial component (ureteric bud) and a mesenchymal component (metanephric blastema). This epitheliomesenchymal interaction is important in development of the kidneys. The ureteric bud gives rise to the ureter, pelvis, calyces and collecting ducts, whereas the metanephric blastema gives rise to glomeruli, the capsule and tubules (the nephrons). The nephrons continue to grow throughout the fetal period.

The cloaca, which receives the allantois and the mesonephric duct, becomes divided into the urogenital sinus and the rectum. The expanded superior portion of the urogenital sinus, which is continuous with the allantois, transforms into the bladder; the inferior portion, in the male, becomes the pelvic urethra and the penile urethra; in the female, it becomes the pelvic urethra and the vestibule.

During the fetal period, the kidneys shift from a pelvic to an abdominal position. The urinary system is the third most commonly affected by abnormalities. The incidence of urinary tract anomalies reported in neonatal autopsies is from 7 to 11%.

I. THE MESONEPHROS

In the human embryo at 22 days gestation, committed mesenchymal cells in the anterior intermediate mesoderm proliferate to form the nephrogenic tissue termed the urogenital ridge (Figs. 14.1 and 14.2). In this ridge, segmentally arranged sets of nephrogenic (epithelial) cords appear in cephalo-caudal sequence. The most cephalic three or four sets of nephrogenic cords disappear by day 24 and are designated as nephrotomes by some authors. Simultaneously, in each urogenital ridge, a solid longitudinal rod, the mesonephric duct, develops close to the surface ectoderm and soon opens into the cloaca (Fig. 14.2). The growth of the mesonephric duct is induced and guided by a zone of reduced pressure (adhesion gradient) within the extracellular matrix between the ectoderm and endoderm. Canalization of the rods transforms these into ducts. Under the influence of the mesonephric ducts, the number of nephrogenic cords increases progressively. The nephrogenic cords develop into the mesonephric tubule. Tufts of capillaries, the glomeruli, appear within nephrogenic tissue and gain connection with branches of the dorsal aorta. With the formation of glomeruli, the mesonephric tubules become S-shaped, and one end of each tubule expands around the glomerulus as a glomerular capsule, while the other end opens into the mesonephric duct. Hence, the tubules establish continuity with a capillary network at one end, and with a collecting system at the other, forming mesonephric excretory units, the nephrons. Urine formation begins with the establishment of excretory units. As more and more mesonephric units differentiate caudally, more cranial units degenerate. By the early second trimester, the mesonephros ceases to be functional. In males, some of the caudal mesonephric tubules persist to become efferent ductules of the testes, and mesonephric ducts become the epididymis and vas deferens.

II. THE METANEPHROS

Early in the fifth week, an epithelial diverticulum called the ureteric bud grows out of each mesonephric duct near its entrance to the cloaca (Fig. 14.1). Precommitted mesenchymal cells of intermediate mesoderm condense around the ureteric bud as the metanephric blastema. As in the mesonephros, epitheliomesenchymal interaction is important. The ureteric bud (epithelial component) and metanephric blastema (mesenchymal component) exert reciprocal inductive influences. The ureteric buds give rise to a collecting system that includes the ureter, pelvis, calyces and collecting ducts, whereas the metanephric blastema gives rise to nephrons.

The ureteric bud divides dichotomously about 15 times (Fig. 14.3). Each branch of the later generation expands to form an ampulla; the surrounding mesenchyme is transformed into epithelial vesicles. The vesicle elongates and bends. Preen-

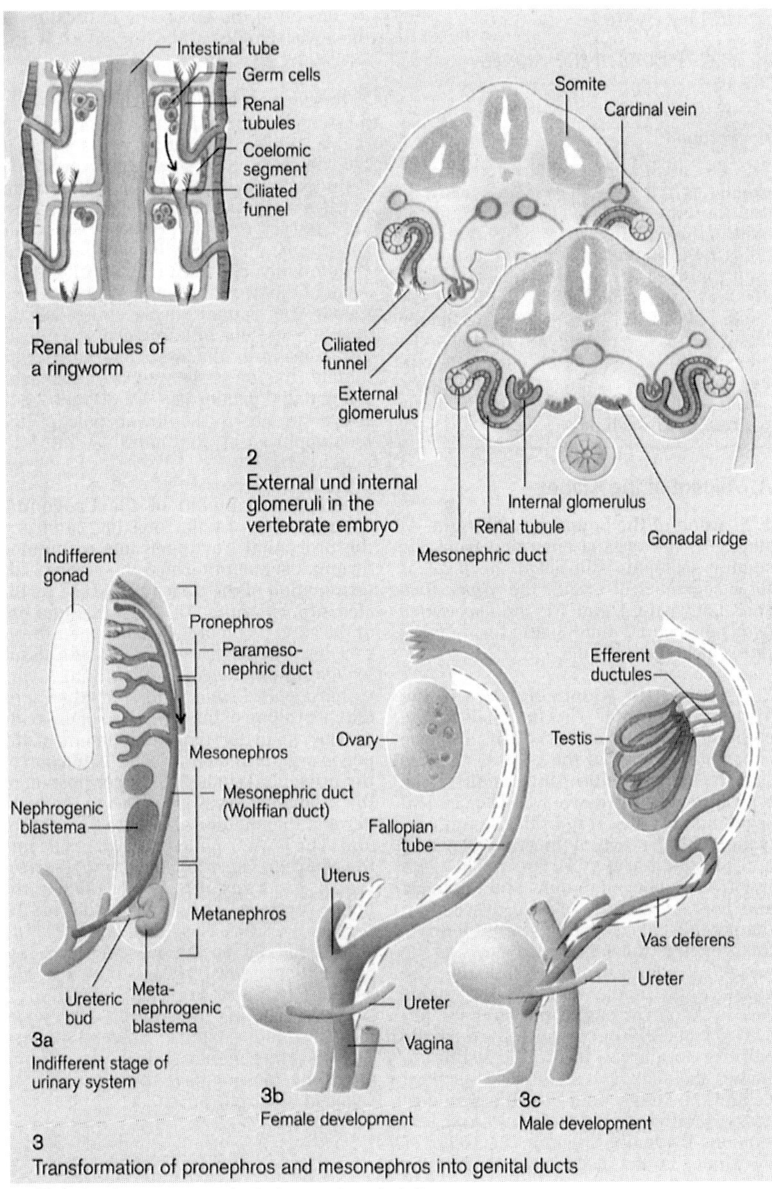

Figure 14.1

Overview: urogenital system (see Color Plate 14.1). Reproduced with permission from Thieme Medical Publishers Inc., New York, 1995, *Color Atlas of Embryology,* Ulrich Drews, Chapter 8: Urogenital System.

dothelial cells migrate into the bend and give rise to a capillary tuft, the glomerulus. Each glomerulus receives an afferent arteriole. An efferent arteriole leaving the glomerulus breaks up into the capillary network around the developing tubules. The vesicle elongates and differentiates into secretory tubules. The differentiation of tubules starts at the portion related to the glomerulus and progresses toward the

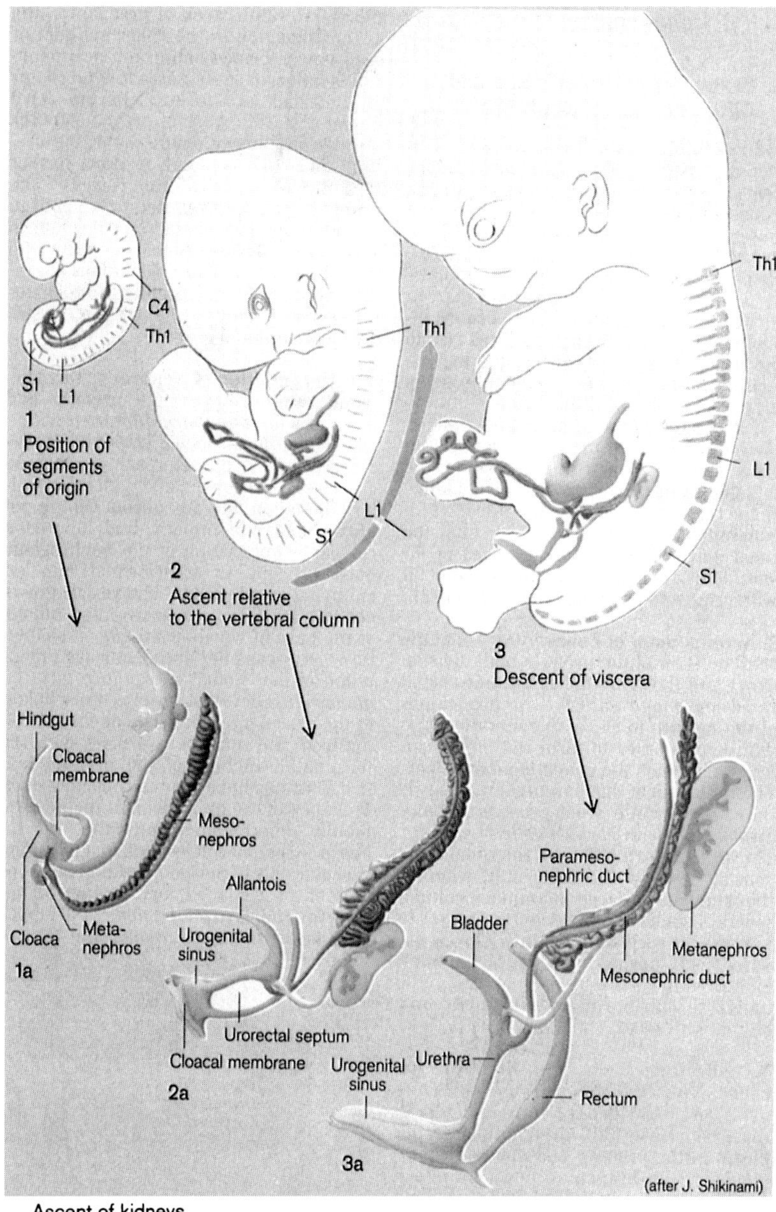

Figure 14.2

Ascent of kidneys (see Color Plate 14.2). Reproduced with permission from Thieme Medical Publishers Inc., New York, 1995, *Color Atlas of Embryology*, Ulrich Drews, Chapter 8: Urogenital System.

distal end. The epithelium at the proximal portion of the tubule becomes thin and surrounds the glomerulus, giving rise to the glomerular capsule. The capsular epithelium then undergoes transformation to form podocytes. The distal end of the tubule then fuses with an ampulla (collecting duct) to establish continuity of the urinary pathways. Hence, formation of the functional units of the metanephros, the

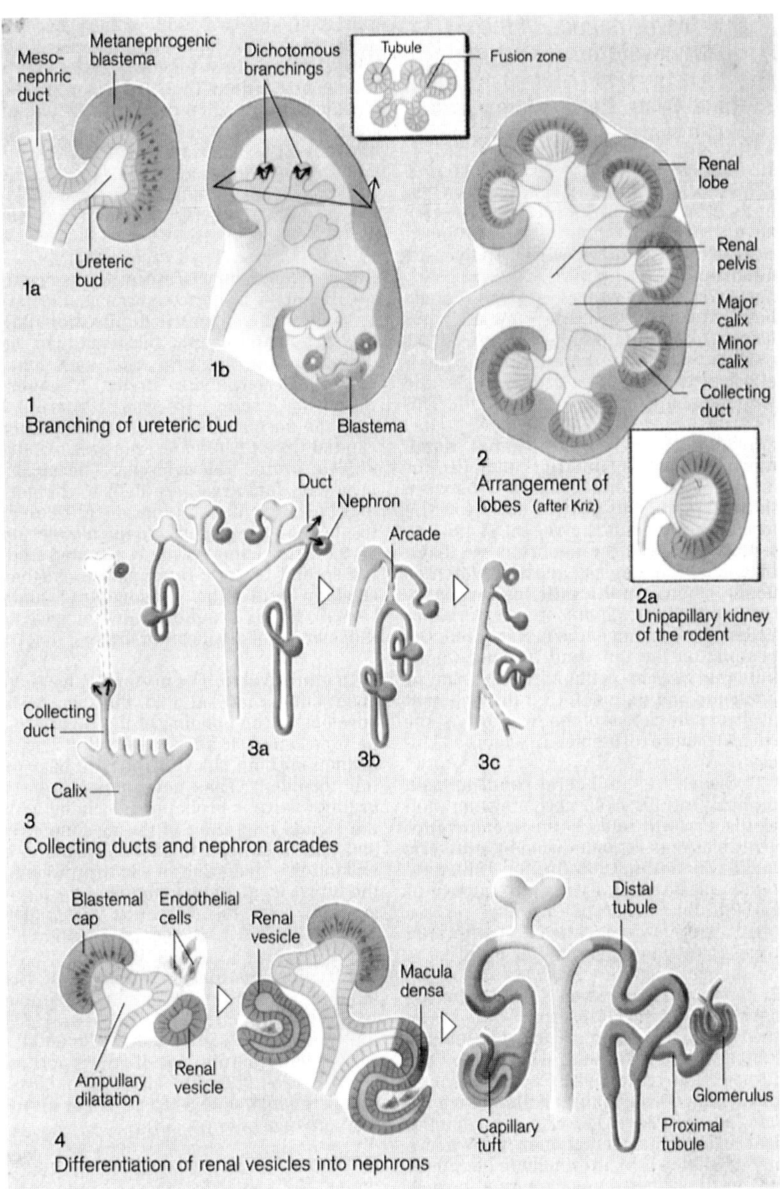

Figure 14.3

Differentiation of the kidney (see Color Plate 14.3). Reproduced with permission from Thieme Medical Publishers Inc., New York, 1995, *Color Atlas of Embryology*, Ulrich Drews, Chapter 8: Urogenital System.

nephrons, involves three types of intermediate mesodermal cells: 1) epithelial cells from the ureteric bud, 2) mesenchymal cells from the metanephric blastema and 3) ingrowing endothelial cells.

During the transformation of mesenchymal cells to epithelial cells, several extracellular proteins such as type I and type III collagen and fibronectin are replaced by type IV collagen, laminin and heparin sulfate proteoglycans. These substances influence the polarization and further differentiation of cells. These preinduced cells then attract the ingrowth of endothelial cells by release of growth factors. Uninduced cells lack this capability. The presence of transferrin and several other factors such as insulin-like growth factor, transforming growth factor-α, epidermal growth factor and nerve growth factor have been shown to affect tubular development.

As mentioned above, differentiation of the ureteric bud and the metanephric blastema depends on inductive signals. If the ureteric bud fails to form or is abnormal, the metanephric blastema does not develop. Conversely, if the metanephric blastema is defective, orderly branching and growth of the ureteric bud does not occur. Several hours of contact with ureteric bud ampulla and metanephric blastema is required to induce formation of the epithelial vesicle and further development of the nephrons.

Each ampulla divides. One branch retains the nephron formed prior to division, and the other branch induces formation of the new nephron. Ampullary division continues until about 20 weeks of gestation, giving rise to about 12 generations of nephrons. Each generation results in a layer of nephrons being laid on the preceding generation. The earliest formed nephrons are those deepest in the kidney. After completion of nephrogenesis, the kidneys increase in size by hypertrophy of the nephrons. Each human kidney contains about 800,000 nephrons.

III. COLLECTING SYSTEM

The distal stalk of the ureteric bud persists as the ureter. The first generation of branching of the ureteric bud coalesces and gives rise to the pelvis. The next three generations coalesce to form major calyces, and the next four generations coalesce and form the minor calyces, each of which receives multiple remaining generations of collecting tubules (Fig. 14.3). Although electrolytes and metabolic waste are handled by the placenta, urine from the metanephros forms a major part of the amniotic fluid that is essential for normal growth of the fetus. Any condition that interferes with production of urine or prevents its draining into amniotic cavity leads to oligohydramnios.

IV. ASCENT OF THE KIDNEYS

The kidneys initially develop in the pelvic region on either side of the aorta. Relative growth and straightening of the body bring the kidneys into the abdomen. During this ascent, the kidneys are displaced laterally and their hila, which were situated ventrally, shift medially. This shift is partly due to overgrowth of the lips of the hila and partly due to actual rotation (Fig. 14.2). In their pelvic position, the kidneys are supplied by the branches from the common iliac arteries. As they ascend at each level, the renal arteries arise from the cephalic segment of the aorta. The inferior branches disappear. The permanent renal arteries are the most cranial branches supplying the kidneys at that level. Ascent of the kidneys stops when they contact the adrenal gland. Because of this process of sequential appearance and disappearance of branches of the aorta supplying the kidneys, 25% of the population may have two or more renal arteries.

V. URINARY BLADDER

During the sixth week, the cloaca is divided into the posterior rectum and anterior urogenital sinus by the urorectal septum (Fig. 14.1). The urogenital sinus is continuous with the allantois, which has a dilated base. The urogenital sinus is divided into a cephalic dilated vesical segment, a middle narrow pelvic segment and a caudal phallic segment. The vesical segment forms the urinary bladder. The middle pelvic segment in the female forms most of the urethra; in the male it forms the membranous urethra and prostatic urethra. The caudal phallic segment (also known as the definitive urogenital sinus) is closed by the urogenital membrane. In females it gives rise to a small portion of the urethra and vestibule and in males it forms the penile urethra.

The most caudal parts of the mesonephric ducts (distal to ureteric buds) are incorporated into the dorsal wall of the bladder, contributing to the connective tissue of the trigone. Consequently, the ureters and mesonephric ducts open separately; the ureters shift cephalolaterally and open into the bladder, whereas the mesonephric ducts migrate caudomedially. In the male, the orifices of the mesonephric ducts are brought close together by relative growth of the bladder, and they open into the prostatic urethra. In females, the caudal ends of the mesonephric ducts degenerate. The trigone extends between the ureteric openings and the internal urethral orifice. The endoderm from the urogenital sinus spreads onto the trigone. On the apex of the urinary bladder, the attenuated part of the allantois solidifies into a fibrous cord, the urachus, which ultimately persists as the median umbilical ligament.

VI. CONGENITAL ANOMALIES OF THE URINARY SYSTEM

A. Abnormalities of the Kidney

1. Renal agenesis results when the ureteric bud fails to induce the metanephric blastema. Possible explanations include failure of the ureteric bud to penetrate and make contact with the metanephric blastema, early degeneration or regression of the ureteric bud, absence of blastema or failure of the blastema to respond to induce substances. Absence of both kidneys occurs in about 1 in 6000 live births. It is more frequent in males (3:1), and in about half of affected fetuses. There are no other malformations except those associated with oligohydramnios. Oligohydramnios, which results from failure of urine production, is the main intrauterine consequence of bilateral renal agenesis. It was first described by E.L. Potter in 1946 as a fetus with a flattened nose, prominent epicanthic fold, and low-set and posteriorly rotated ears. The legs are flexed (bowed), the feet

are inwardly rotated, and the hands are large. The fetuses also have a reduced chest circumference with pulmonary hypoplasia. The majority of female fetuses with bilateral renal agenesis also have abnormalities in the reproductive system, whereas the majority of males have normal reproductive systems. In some cases, anomalies of the GI tract, cardiovascular system and skeletomuscular system are found. As the placenta is the major excretory organ, this condition is compatible with prenatal life but not with postnatal life. Unilateral renal agenesis is an incidental finding. The present kidney undergoes compensatory hyperplasia, and the number of nephrons is increased. Fetal urine production and postnatal renal function are not substantially compromised.

2. Renal hypoplasia. These kidneys have a reduced number of nephrons below that expected for their developmental age. They also have a diminished number of papillae and calyces. This condition causes renal insufficiency in infancy, and development of hypertension is a major complication.
3. Accessory kidney results from early and excessive branching of the ureteric bud. The extra kidney usually is small and completely separated from the normal kidney. Its ureter generally drains into the normal kidney.
4. Renal ectopia. A kidney located outside its normal position is called ectopic. One or both kidneys could be ectopic. Pelvic kidneys are more frequent. They lie below the pelvic brim, and their ascent is abnormal. In crossed renal ectopia, both kidneys are located on one side of the body. They frequently are fused and are thought to arise from a common metanephric blastema. The ureter of the lower kidney crosses to open into the contralateral side of the bladder. Hydronephros, hydroureter or dysplastic changes could develop in these kidneys (Fig. 14.4).
5. Horseshoe kidney results from fusion of the lower poles of kidney before their ascent. The fusion consists of a connective tissue and superficial cortex bridge. The renal mass thus formed is displaced to the pelvic brim. Two separate ureters course ventrally and drain into the bladder. The ascent of this renal mass is hindered by the inferior mesenteric artery. The proposed hypotheses for this condition are medial displacement of the metanephros by the umbilical arteries, convergence of the ureteric buds, and abnormal migration of nephrogenic tissue. Although horseshoe kidneys are asymptomatic, pelvic kidneys in general are subject to increased incidence of infection and obstruction. In any occasion when kidneys do not reach the adrenals (also in agenesis or ectopia), the adrenal glands become diskoid (Fig. 14.4).
6. Renal cystic disease. A variety of developmental anomalies can lead to cyst formation in fetal kidneys. Renal cystic change is the most frequent morphological aberration identified in fetuses and neonates. A few of the conditions are described here (Fig. 14.4).
   a. Renal dysplasia is the most common cause of kidney cysts. Dysplasia is defined as abnormal development of nephrons and ducts, resulting in total or partial renal malformation. In these kidneys, the architecture of the cortex and medulla is disorganized, the glomeruli are small and

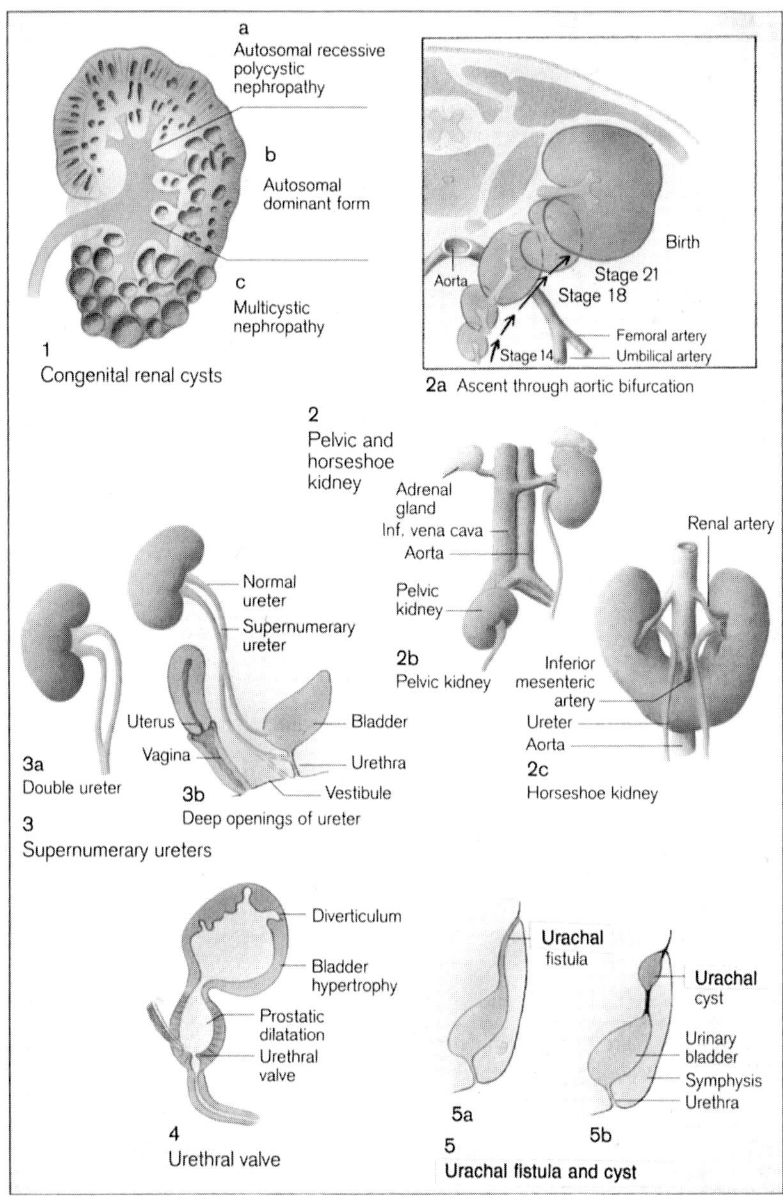

Figure 14.4

Malformations of the urinary system (see Color Plate 14.4). Reproduced with permission from Thieme Medical Publishers Inc., New York, 1995, *Color Atlas of Embryology,* Ulrich Drews, Chapter 8: Urogenital System.

immature, and the tubules are simple and atrophic. Cyst formation is variable and can involve any part of the kidney. Dysplastic kidneys are usually large but insufficient. It is believed that this results from some injury to the ureteric bud that interferes with normal differentiation and metanephric induction of ampullae. It is further believed that urethral obstruction caused damage to the collecting duct ampullae by intraluminal back pressure. Urinary tract obstruction underlies the great majority of cases of cystic dysplastic kidney. Bilateral cystic dysplasia may cause oligohydramnios and pulmonary hypoplasia. Unilateral cystic dysplasia often manifests as an abdominal mass. The foci of nodular renal blastema are thought to be possible precursors of Wilms' tumor.

b. Polycystic renal disease includes two hereditary forms. One is autosomal recessive, or infantile, in type; the other is autosomal dominant, or adult. The kidneys in autosomal recessive type are greatly enlarged and contain numerous elongated radially oriented cysts. The cysts are lined by simple cuboidal epithelium and originate from collecting tubules. This abnormality arises after differentiation. The old hypothesis of failure of nephrons to unite with collecting tubule is not accepted anymore. Abnormal development of the intrahepatic biliary system is a constant finding in these patients. Renal failure and anuria lead to oligohydramnios and pulmonary hypoplasia. Autosomal dominant type renal failure causes symptoms at age 30–40 years. The kidneys in the fetus are enlarged and may be examined ultrasonographically when a parent is known to have this. The cysts in these are small and not radially oriented. Biliary system development is normal. At least two genes, located on the short arm of chromosome 16, have been identified. An abnormal basement membrane structure and aberrant tubular epithelium proliferation have been attributed here. Meckel–Gruber syndrome is a lethal autosomal recessive disorder characterized by polycystic kidneys, intrahepatic biliary proliferation and polydactyly.

7. Renal tubular dysgenesis. The proximal and/or distal tubules stay undifferentiated. The kidneys show an increased density of glomeruli that are separated by immature tubules.

B. Anomalies of the Ureter

1. Duplex ureter results from branching of the ureteric bud before it enters the metanephric blastema. In some cases, an accessory ureteric outgrowth from the mesonephric duct has been found. It may result in supernumerary kidney. Frequently, two separate pelvises and two proximal ureters are present. The ureters usually merge, and a single ureter enters the bladder. If both ureters remain separated, the one from the upper pole is ectopic and enters the bladder inferiorly. In the male, it may open into the prostatic urethra (50%), or into a seminal vesicle (33%) or any other part of the urethra. In females it may open into the urethra (40%) or into the vestibule (35%) or vagina (25%). Hydronephrotic and dysplastic changes can develop in these cases. A single ureter can also have an ectopic connection (Fig. 14.4).

2. Urethral obstruction. Ureterovesical junction obstruction with hydronephrosis is the most frequent anomaly identified by ultrasonographic examination of the fetal urinary tract. The obstruction is caused by aplasia or hypoplasia of the

ureteric lumen at its entry to the bladder. Urethral stenosis, duplex and ectopic ureter, compression of the ureter with an aberrant blood vessel and megaureter can also cause hydronephrosis. In general, any lower urinary tract obstruction is considered to be the primary factor in renal dysplasia. It can also lead to massive distention of the ureter, called megaureter.

C. Anomalies of the Urinary Bladder

1. Absent bladder or small bladder is associated with renal agenesis or dysplasia.
2. Exstrophy of the bladder is a defect in the lower abdominal wall exposing the mucosal surface of the bladder. The mucosa of the bladder is continuous with margins of the abdominal wall. The pubic bones are widely separated. Incomplete closure of the abdominal wall is caused by failure of the mesoderm to invade the central lower abdominal region, where a large wedge of cloacal membrane persists. When the cloacal membrane disintegrates, the epithelia of the urogenital sinus and hindgut are exposed. The extent of exstrophy depends on the deficiency of mesodermal migration and the extent of cloacal membrane rupture. It has been suggested that an overdeveloped cloacal membrane may delay or hinder mesenchymal movement and fusion. It usually includes epispadias.
3. Epispadias is an opening of the urethra on the dorsum of the penis. Urinary incontinence occurs. According to one theory, it is caused by a caudal shift of the lateral primordia of the genital tubercle. Others suggest the lack of median fusion of the mesenchyme of the lower abdominal wall near the genital tubercle, which leads to rupture of the urogenital sinus dorsally. The corpora cavernosa approach each other below the opening.
4. Defective partition of the cloaca results in anomalies of urinary, genital and anorectal structures. If the inferior part of the urorectal septum does not close, a rectourethral fistula develops between the urogenital sinus and rectum. Failure of the cloaca to divide results in rectocloacal canal. The undivided region of the cloaca receives the urethra, the vagina and the rectum. In the female, vesicovaginal or vesicorectal fistula develops. In the male, vesicourethral or vesicorectal fistula develops.

D. Remnants of Urachus

Remnants of urachus result in cysts, umbilical sinuses, vesicocutaneous fistula or apical bladder diverticulum. Normally, the allantois obliterates to form a fibrous cord, the urachus. If there is persistence of lumen between the two fibrosed urachal segments, a urachal cyst develops. Persistence of lumen at the umbilical end of the urachus causes an umbilical (urachal) sinus. Persistence of urachal lumen at bladder end results in an apical diverticulum of the bladder. If the entire urachus stays patent, a vesicocutaneous (urachal) fistula drains urine from the umbilicus (Fig. 14.4).

E. Anomalies of the Urethra

1. Posterior urethral valves are epithelial folds that project from the mucosa of the urethra near the base of the bladder. These are the most common cause of lower urinary tract obstruction. This results in overdistention of the urinary bladder, megaureters and hydronephrotic kidneys. In complete block, urine is not added to amniotic fluid, and this leads to oligohydramnios (Fig. 14.4).
2. Prune-belly syndrome results from lower urinary tract obstruction. Most cases have urethral atresia, stenosis, kinking or posterior urethral valves. The fetus shows a markedly distended smooth abdomen. This marked distention of the abdomen hinders or blocks migration of the somatic mesoderm into the anterior body wall. Therefore, the muscles are absent or hypoplastic. In most cases, the testes are undescended and the prostate is absent or hypoplastic. The condition generally affects males. In late gestation or the neonatal period, the urinary system ruptures and fluid from the peritoneal cavity drains into the amniotic cavity. Prune-belly thus results from deflation. Some investigators attribute primary maldevelopment of the prostate to it. The prostate fails to develop normally because of disrupted mesenchymal–epithelial interaction.

REVIEW QUESTIONS — Chapter 14

Select the most appropriate answer.

1. Development of the urogenital ridges is under the influence of signals from the:
   a. extracellular matrix
   b. surface ectoderm
   c. related endoderm
   d. somatic mesoderm
   e. splanchnic mesoderm
2. The ureteric bud grows out of the:
   a. mesonephric tubules
   b. mesonephric duct
   c. allantois
   d. cloaca
   e. urogenital sinus

3. The cloaca is divided by the:
   a. urorectal septum
   b. metanephric blastema
   c. allantois
   d. mesonephric duct
   e. ureteric bud

4. The structure derived from mesonephric duct is the:
   a. metanephric diverticulum
   b. metanephric blastema
   c. oviduct
   d. ureter
   e. all of the above

5. The penile urethra is derived from the:
   a. ureteric bud
   b. mesonephric duct
   c. paramesonephric duct
   d. urogenital sinus
   e. urachus

6. The mesonephric cords in the embryo enter into the:
   a. urethra
   b. cloaca
   c. urinary bladder
   d. allantois
   e. ureter

7. In males, the mesonephros is retained as all of the following except the:
   a. epididymis
   b. rete testes
   c. ductus deferens
   d. ureter
   e. appendix of the epididymis

8. During development of the kidneys, some cells become lodged in the developing glomerular capsule and give rise to:
   a. podocytes
   b. Bowman's capsule epithelium
   c. macula densa cells
   d. glomerular capillaries
   e. none of the above

9. The cranial part of the urogenital sinus is continuous with the:
   a. allantois
   b. yolk sac
   c. cloaca
   d. urethra
   e. rectum

10. The metanephrogenic diverticulum (ureteric bud) appears as a dorsal bud from the:
    a. metanephric blastema
    b. intermediate mesoderm
    c. urogenital sinus
    d. cloaca
    e. mesonephric duct

11. Embryologically each metanephric tubule consists of two parts that become confluent at the junction of:
    a. Bowman's capsule and the proximal tubule
    b. the proximal tubule and the loop of Henle
    c. the loop of Henle and the distal tubule
    d. the distal tubule and the collecting duct
    e. the collecting duct and the minor calyx

12. The unsegmented caudal portion of the intermediate mesoderm gives rise to the:
    a. pronephros
    b. mesonephros
    c. mesonephric duct
    d. ureteric bud
    e. metanephrogenic mesoderm (blastema)

13. The kidneys develop in the pelvic region. Relative growth of the caudal part of the body causes the kidneys to ascend. This ascent of kidneys stops when
    a. the kidneys receive renal arteries
    b. the kidneys complete their rotation
    c. the upper pole of the kidneys touches the adrenal glands
    d. the right kidney comes in contact with liver
    e. all of the above contribute in this process

14. During their ascent, the kidneys rotate:
    a. medially and their hila become medial
    b. laterally and their hila become medial
    c. ventrally and their hila become medial
    d. dorsally and their cortex becomes lateral
    e. ventrally and their cortex becomes medial

15. In the development of the urinary system, the ureters are formed from the:
    a. pronephric duct
    b. mesonephric duct
    c. allantois
    d. metanephric diverticula
    e. metanephric blastema

16. Differentiation of a nephron starts at the:
    a. segment related to the glomerulus
    b. segment related to the collecting duct
    c. proximal convoluted tubule
    d. distal tubule
    e. lateral division of the ureteric bud

17. The mesonephric tubules:
    a. in the male form the efferent tubules of the testes
    b. in the female degenerate to form the gubernaculum ovary
    c. in the male, some tubules degenerate to form gubernaculum testes
    d. in the female may persist as the epoophoron
    e. all of the above are true

18. The urogenital membrane is made up of the:
    a. embryonic endoderm, mesoderm and ectoderm
    b. embryonic endoderm and ectoderm
    c. endoderm from the allantois
    d. urorectal septum mesoderm
    e. prochordal plate and the overlying embryonic ectoderm

19. The female urethra is derived from the:
    a. mesonephric duct
    b. paramesonephric duct
    c. urogenital sinus
    d. allantois
    e. surface ectoderm

20. The derivatives of the endoderm include all of the following except the:
    a. lining of the ureter
    b. lining of the urinary bladder
    c. prostate gland
    d. lining of the female urethra
    e. lining of the vestibule

21. In a female fetus, the phallic segment of the urogenital sinus forms:
    a. part of the urinary bladder
    b. the membranous urethra
    c. the prostatic urethra
    d. the vestibule
    e. all of the above

22. The most caudal portion of mesonephric ducts:
    a. are incorporated into the dorsal wall of the urinary bladder
    b. in a male fetus, migrate caudally and medially to open into the urethra
    c. in a female fetus, after their migration, they degenerate
    d. contribute to the connective tissue of the trigone
    e. all of the above are correct

23. Which of the following organs does not migrate during development?
    a. the suprarenal glands
    b. the ovaries
    c. the testes
    d. the kidneys

24. After birth, the allantois persists as the:
    a. umbilical cord
    b. urachus
    c. urethra
    d. genital tubercle
    e. vitelline duct

25. After birth, the size of the kidneys is increased by:
    a. hypertrophy of existing nephrons
    b. addition of new nephrons
    c. expansion of the calyces and pelvis
    d. dilatation of the tubular system
    e. all of the above

26. On a routine X-ray, a mass was seen fused to the lower pole of the left kidney. Biopsy of the mass revealed it to be ovarian tissue. During development it could have been caused by
    a. nonseparation of the urogenital ridge into urinary and genital portions
    b. persistent mesonephros
    c. failure in development of the left ovary
    d. fusion of the left kidney with the left ovary during its migration
    e. failure in development of the paramesonephric duct

27. In the child described above:
    a. the left ovary failed to develop
    b. this condition may cause sterility
    c. the left kidney would not function
    d. ectopic pregnancy may occur due to this mass
    e. none of the above is true

28. Horseshoe kidney:
    a. results from nondivision of metanephric blastema
    b. may be caused by early duplication of the ureteric bud
    c. is normally found at level LV2
    d. is persistent mesonephric kidneys
    e. results from fusion of the caudal poles of the metanephric kidneys

29. Any condition that interferes with production or drainage of urine in a fetus my lead to:
    a. oligohydramnios
    b. overdistended urinary bladder
    c. megaureter
    d. hydronephrotic kidney
    e. all of the above

30. Children having prune-belly syndrome have a distended abdomen that is full of amniotic fluid. This may have resulted from:
    a. deflation of the urinary system by rupture and drainage into the peritoneal cavity
    b. obstructive anomalies in the lower urinary tract
    c. urethral atresia or stenosis
    d. enlargement of posterior urethral valves
    e. all of the above may contribute to this condition

31. Failure of the urethral folds to fuse during embryonic life would be expected to result in:
    a. epispadias
    b. exstrophy of urinary bladder
    c. horseshoe kidney
    d. hypospadias
    e. multiple ureters

32. A persistent open allantois may result in:
    a. Meckel's diverticulum
    b. the presence of meconium at the umbilical region
    c. umbilical sinus
    d. posterior urethral valves
    e. none of the above

ANSWERS TO REVIEW QUESTIONS

1. a	2. b	3. a	4. a
5. d	6. d	7. d	8. d
9. a	10. e	11. d	12. e
13. c	14. b	15. d	16. a
17. e	18. b	19. c	20. a
21. d	22. a	23. a	24. c
25. a	26. d	27. e	28. e
29. e	30. e	31. d	32. c

CHAPTER

15

GENITAL SYSTEM

The development of the genital system can be divided into three stages. For the first stage, sex determination involves differentiation of the indifferent gonad into either testes or ovary and is under genetic control. In the second stage, differentiation of ducts or internal genital organs occurs; during the third stage, differentiation of the external genital organs occurs. The development and differentiation of both internal and external organs depends on the presence of male or female gonads and their endocrine products.

In humans, gonadal development begins at about the 5th week. Germ cells migrate into the gonadal ridge during the 6th week. The gonads are identical (indifferent) in both sexes at this stage. After this, further development of the gonads is completely dependent on the sex chromosome complement. The gonads of embryos with a normal Y chromosome develop as testes; gonads without Y chromosomes develop as ovaries (Fig. 15.1).

Both mesonephric and paramesonephric ducts form in all embryos. Testes secrete Mullerian duct inhibitory factor (MDIF), which causes regression of paramesonephric ducts; while testosterone retains the mesonephric ducts in male. In the absence of testes, the mesonephric ducts regress and the paramesonephric ducts are retained.

The external genitalia also begin as identical (indifferent) in both sexes. Under the influence of testosterone, the external genitalia differentiate as male. Although congenital anomalies of the genital system result from many different causes, most of these are caused by chromosomal defects. In other cases, defective gonadal secretion is attributed. Some of these disorders are discussed in this chapter.

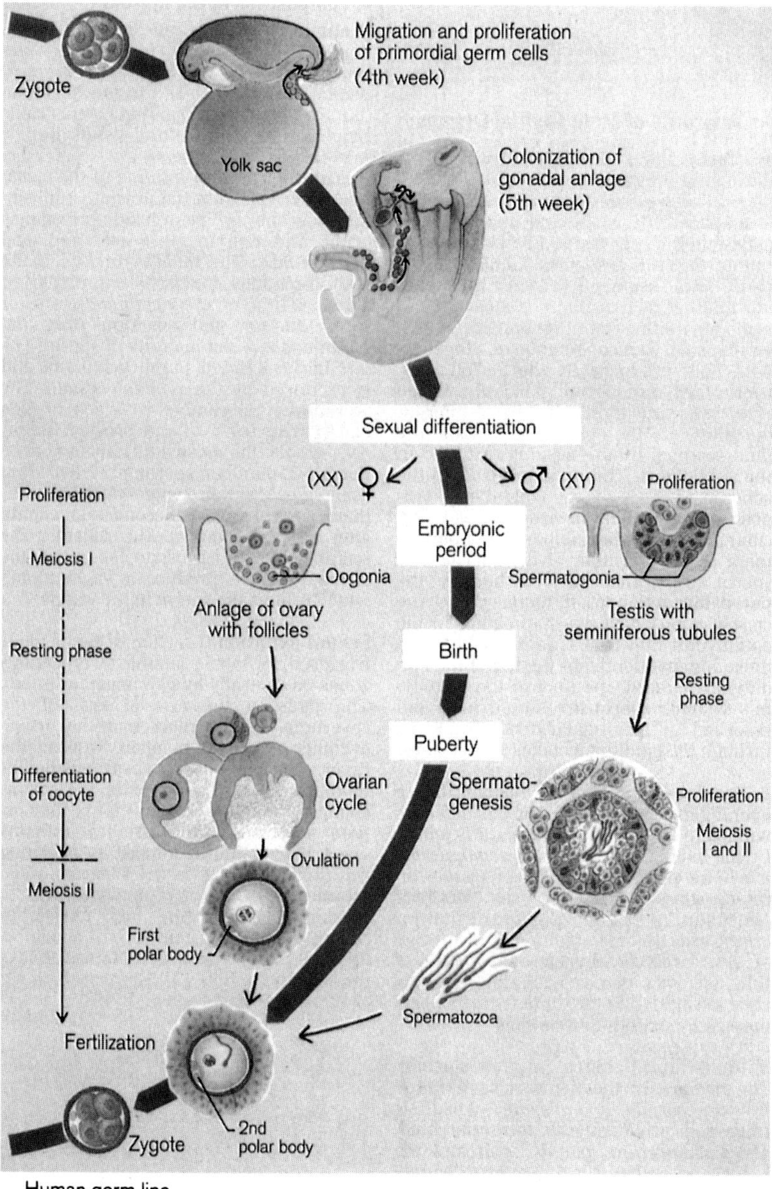

Figure 15.1
Overview: human germ line (see Color Plate 15.1). Reproduced with permission from Thieme Medical Publishers Inc., New York, 1995, *Color Atlas of Embryology*, Ulrich Drews, Chapter 1: Reproduction

I. GONADS

A. Indifferent Stage

In humans, bilateral bulges appear as the urogenital ridge at about the 5th week of development. The mesonephric cells induce the coelomic mesothelial cells to aggregate in the medial aspect of this ridge. Proliferation of the coelomic mesothelium and condensation of the underlying mesenchyme cause a separation between the lateral mesonephros (urinary part) and the medial gonadal (genital) part. The mesonephric cells, the coelomic mesothelial cells and the mesenchymal cells aggregate into cellular plates called the primary sex cords. By the end of the 5th week, large primordial germ cells arise extragonadally from the endoderm (hypoblast) and are clearly detected in the posterior wall of the yolk sac. From here, about a thousand of them start migrating toward the newly formed gonadal ridges. Primordial germ cells migrate through the dorsal mesentery by various active and passive movements in response to a permissive extracellular substance. The germ cells are attracted toward the gonadal ridge by chemotactic substances secreted by the sex cords in the gonadal ridge. As the germ cells penetrate the gonads, they stop their ameboid movement. The germ cells invest the sex cords; their arrival further stimulates development of the sex cords. The differentiated cells of the sex cords (ovarian follicular cells in females, Sertoli cells in male) nourish and regulate the development and maturation of germ cells. Thus, the sex cords are essential for germ cell development. Conversely, if the germ cells do not arrive in the gonadal ridge, neither sex cords nor gonads develop. The primordial germ cells that fail to reach the gonadal ridge usually disappear. In rare instances, they survive in an ectopic site and may give rise to teratomas. In response to mitogenic factors, germ cells and sex cord cells continue to proliferate, the gonads become prominent and separated from the mesonephric ridge and suspend from the gonadal mesentery (Fig. 15.2).

B. Testes

There is a tendency for the human embryo to develop as a female. The transformation of an indifferent gonad to testes is an active process that depends on many factors. The presence of a Y chromosome determines the development sequence toward male. The expression of a gene, the sex-determining region (SRY) located on the short arm of the Y chromosome, results in production of testes-determining factor (TDF), which initiates development of the testes. Embryos where the SRY is not expressed, even if a Y chromosome is present, will develop as female. Under the influence of TDF, the peripheral (cortical) portion of sex cords degenerates, whereas the somatic cells in the deeper (medullary) portion differentiate into Sertoli cells. The germ cells are also tightly packed in this portion of the sex cords. The sex cords become separated from the surface (coelomic) mesothelium (epithelium) by a layer of connective tissue termed the tunica albuginea. The deepest portion of the sex cords that does not contain germ cells becomes the rete testes. The testicular (sex) cords gradually develop lumen and transform into seminiferous and straight tubules. The mesenchyme between the seminiferous tubules gives rise to interstitial cells of Leydig, which start to secrete testosterone at about 9 weeks (Fig. 15.1). The rete testes connect with a few mesonephric tubules that become efferent ductules. This union between mesonephric tubules and seminiferous tubules does not begin until the middle of prenatal life and is not completed until puberty. This reduction in contact with the mesonephros is important, because the mesonephros exerts a feminizing influence on developing gonads. The primordial germ cells in the testes divide slowly by mitosis, but they do not enter meiosis. The cells lining the

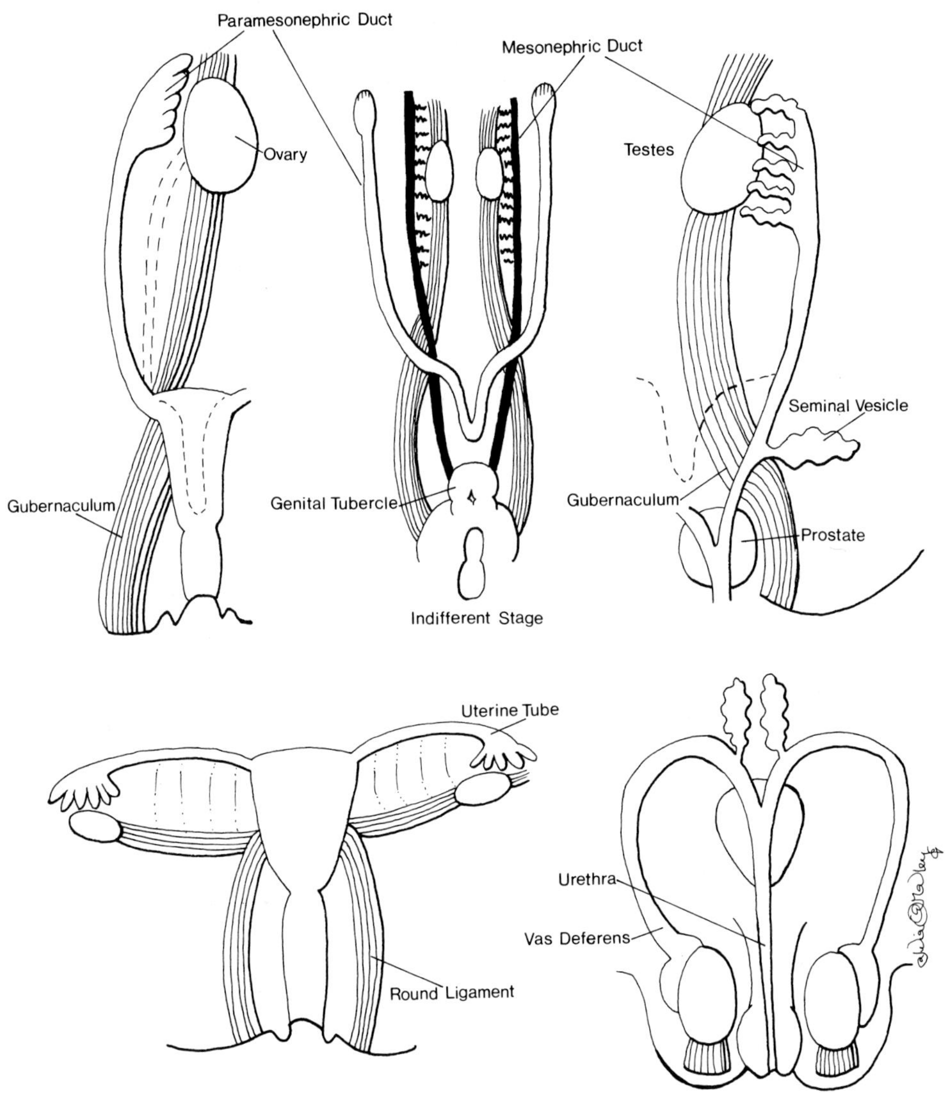

Figure 15.2
Development of the genital system.

seminiferous tubule, now called sustentacular (Sertoli) cells, inhibit germ cells from entering meiosis until puberty (Figs. 15.3 and 15.4; Table 15.1).

C. Ovaries

In embryos having no Y chromosome (or in an embryo where the SRY is not expressed), regardless of the number of X chromosomes, gonads continue their indifferent phase by growing further. The primordial germ cells remain concentrated at the outer (cortical) region because the primary sex cords in the inner (medullary) region are not well developed. In the early fetal period, secondary sex

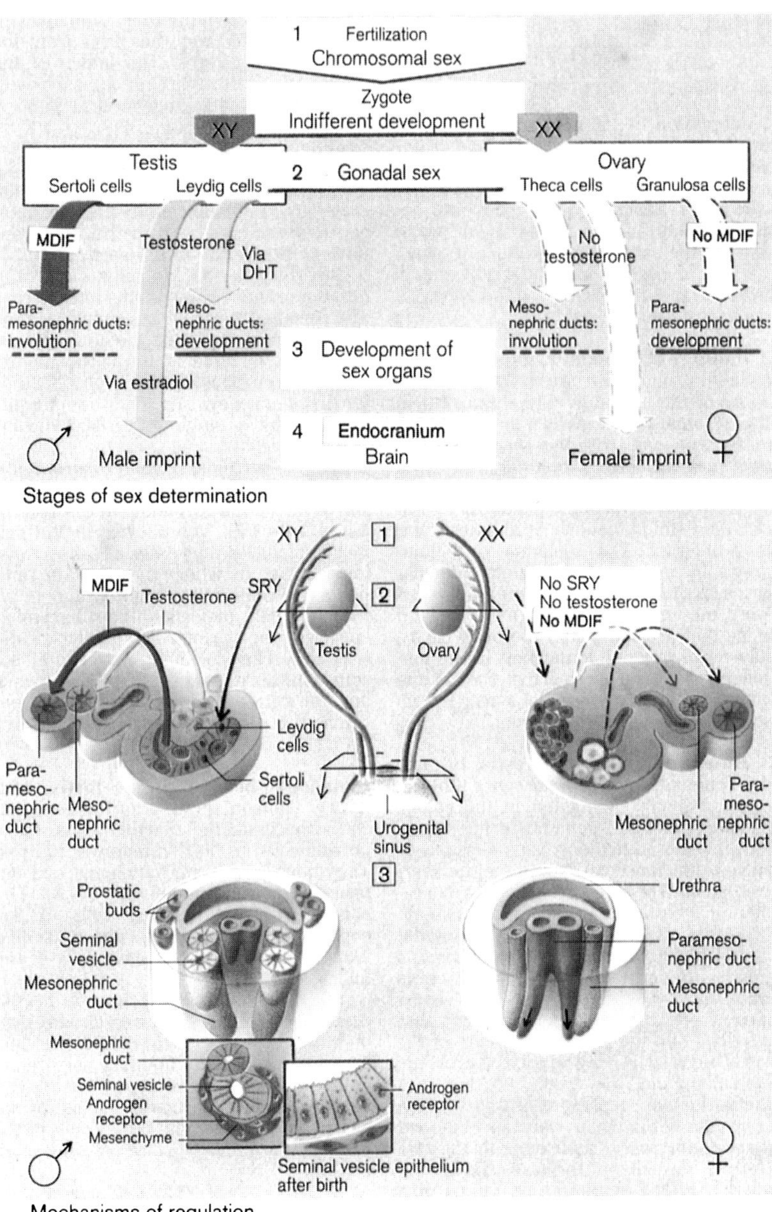

Figure 15.3
Sex determination (see Color Plate 15.3). Reproduced with permission from Thieme Medical Publishers Inc., New York, 1995, *Color Atlas of Embryology*, Ulrich Drews, Chapter 8: Urogenital System.

cords form by the addition of light coelomic epithelial cells and dark mesonephric cells. The germ cells are incorporated into these sex cords, where they proliferate by mitosis and give rise to oogonia. The proliferation of oogonia ceases during the third trimester, and no new oogonia are added. The presence of viable germ cells is very important for differentiation of the ovary. If the germ cells are abnormal or

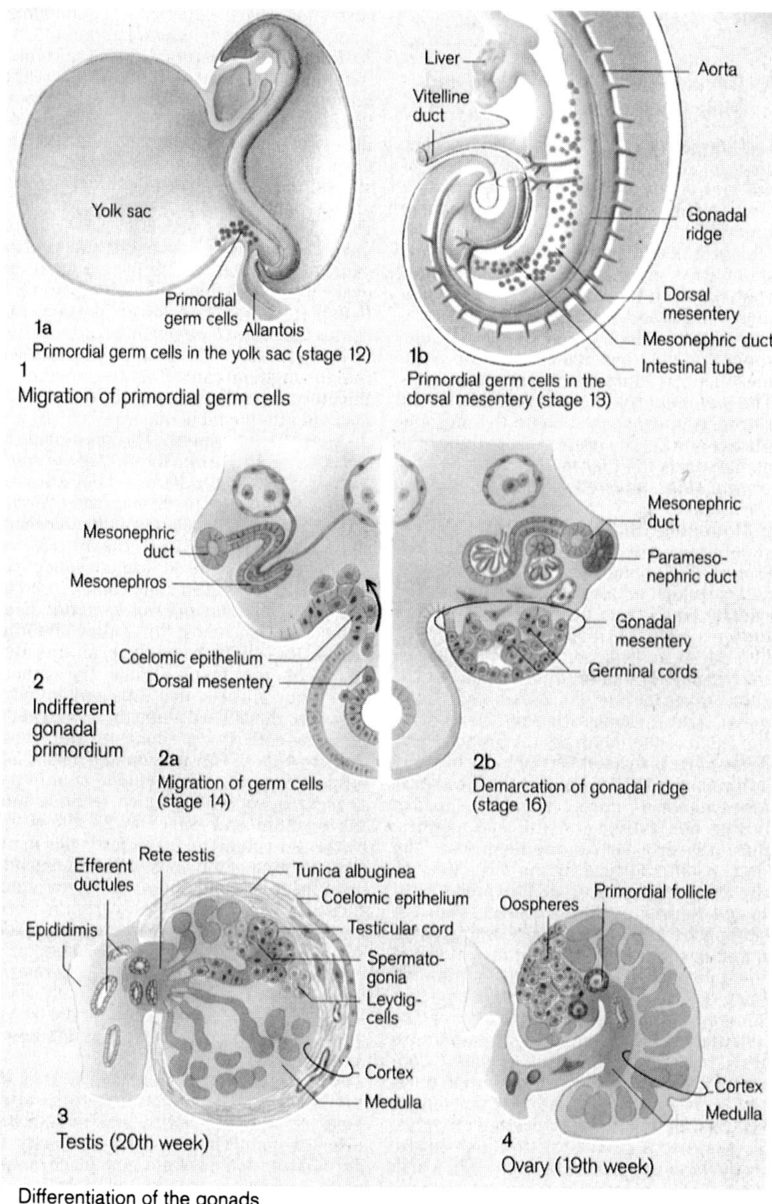

Figure 15.4

Differentiation of the gonads (see Color Plate 15.4). Reproduced with permission from Thieme Medical Publishers Inc., New York, 1995, *Color Atlas of Embryology*, Ulrich Drews, Chapter 8: Urogenital System.

fail to reach the gonad, the growing genital ridge regresses to streak gonads. True differentiation of the ovary starts by triggering the oogonia to enter meiosis. The mitotic oogonia, now termed primary oocytes, are arrested at the diplotene phase of prophase I. At this point, all the chromosomes have paired, crossover has occurred, and all primary oocytes remain in this stage until their ovulation (at puberty) or degeneration (Fig. 15.1). The primary oocytes become surrounded by flattened somatic (sex cord) cells to form primordial follicles. Two types of cells have been recognized in primordial follicles: 1) meiosis-inducing and 2) meiosis-inhibiting cells. Although further development of follicles is delayed until puberty, follicular cells regulate differentiation of oocytes. The medullary region of the developing ovary becomes a dense core of mesonephric cells that give rise to rete ovarii. The primary sex cords in the medullary region are devoid of germ cells and start degenerating. The rete ovarii do not connect to the mesonephric tubules, and no communication forms between the gonads and mesonephric duct (Figs. 15.3 and 15.4).

II. GENITAL DUCTS

A. Indifferent Stage (Fig. 15.2)

All embryos, male or female, initially have both mesonephric and paramesonephric ducts. Development of the mesonephric duct is discussed in the chapter on the urinary system (Chapter 14). The paramesonephric ducts appear in the 7th week as longitudinal invagination of the coelomic epithelium along the urogenital ridge lateral to the mesonephric duct. The invaginated epithelium becomes a cord that grows caudally and medially to terminate on the urogenital sinus. The cords from both sides approach each other and begin to fuse before reaching the urogenital sinus. These cords then acquire lumen in a craniocaudal direction. The cranial end of these paramesonphric (Mullerian) ducts opens into the coelomic cavity, and the caudal end opens into the urogenital sinus.

B. Male Genital Ducts

The sustentacular (Sertoli) cells in developing testes secrete MDIF, which is a glycoprotein whose molecular structure closely resembles that of transforming growth factor β (TGF_{β}). The interstitial (Leydig) cells of testes begin secreting testosterone, which binds to androgen receptors on the mesonephric duct and stimulates it to continue to grow. As mentioned above, some of the mesonephric tubules join the rete testes and become efferent ductules. The part of the mesonephric ducts adjacent to these forms the epididymis. The main part of the mesonephric duct acquires smooth muscle and later becomes the ductus deferens. Near the urogenital sinus, an epithelial outgrowth from the mesonephric duct (ductus deferens) gives rise to seminal vesicles. The ductus deferens continues as the ejaculatory duct that opens into the prostatic urethra. Some mesonephric tubules, the cephalic to efferent ductules, may persist as the paraepididymis, and part of the mesonephric duct cephalic to the epididymis may form the appendix epididymis (Figs. 15.3 and 15.5). The prostate and bulbourethral glands grow out from the urethra. A mesenchymal–epithelial interaction is essential for development of seminal vesicles and these glands. The androgens influence the

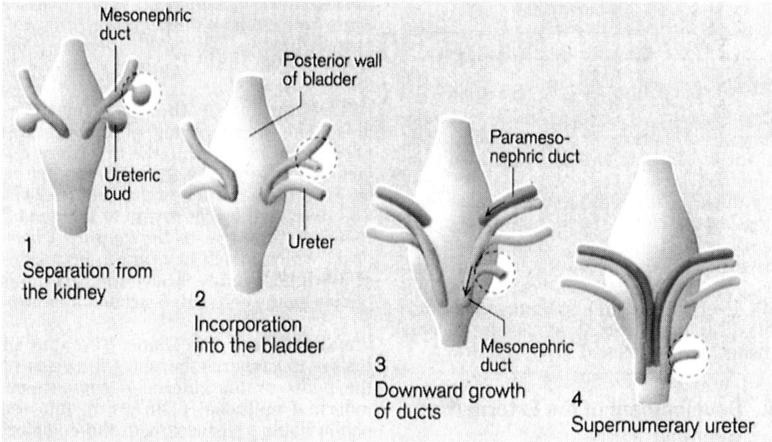

Separation of the ureter from the mesonephric duct

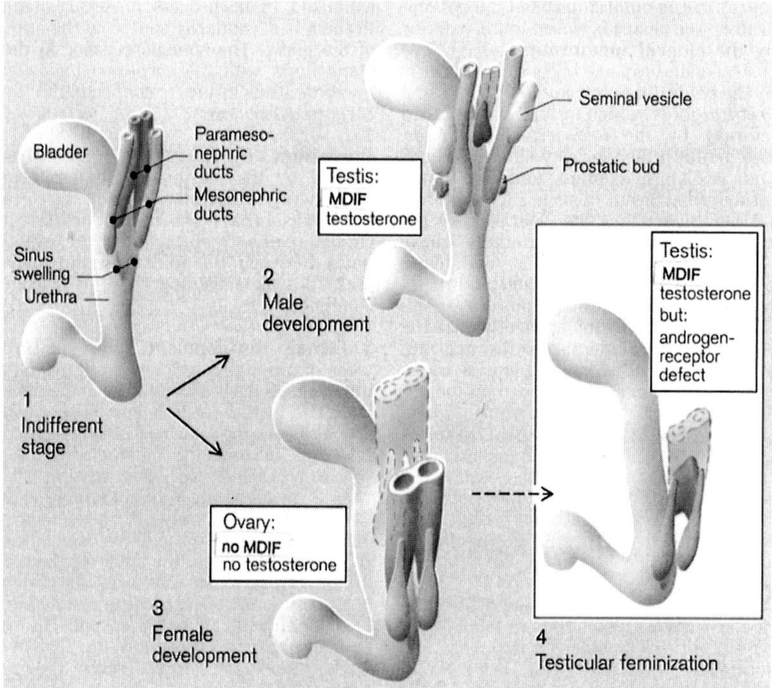

Downward growth of the vaginal anlage

Figure 15.5

Downward growth of the vagina (see Color Plate 15.5). Reproduced with permission from Thieme Medical Publishers Inc., New York, 1995, *Color Atlas of Embryology*, Ulrich Drews, Chapter 8: Urogenital System.

mesenchymal cells, which in turn induce the associated epithelium to differentiate into glands. Under the influence of MDIF, the paramesonephric (Mullerian) duct starts to regress. Only a small portion of it remains as the cephalic part of the prostatic utricle. The cephalic portion of the paramesonephric duct sometimes persists as appendix testes.

C. Female Genital Ducts

In absence of a Y chromosome there is no SRY gene, no TDF, no Sertoli cells and no Leydig cells, but there is testosterone. If no gonads are present, development proceeds toward the female. In these embryos, the mesonephric ducts regress and the paramesonephric ducts develop further and give rise to a female genital tract (Fig. 15.3). The paramesonephric ducts grow caudally, approach each other and begin to fuse. Fusion of the ducts is complete at about the 10th week. The middle wall degenerates, and a single cavity is formed in the fused part, which is now called the uterus. Although the caudal-most portion of the uterus, the cervix, arises from the paramesonephric duct, the epithelium of the urogenital sinus grows over it to form the mucosa of the cervix. The cephalic unfused portion of the paramesonephric duct becomes the uterine tubes, which open into the coelomic cavity. In summary, the paramesonephric duct epithelium lines the uterine tube, the uterus and the endocervix in females, and part of the prostatic utricle and appendix testes in males (Fig. 15.5).

The mesonephric and paramesonephric ducts are enclosed in peritoneal folds. As the paramesonephric ducts swing medially, these peritoneal folds become the broad ligament of the uterus that divides the pelvic peritoneal compartment into the uterorectal pouch and the uterovesicle pouch. Formation of the vagina remains in dispute. The most accepted hypothesis is adopted herein. At the entrance of the fused paramesonephric duct, bilateral thickenings appear and unite with the tip of the paramesonephric duct (Fig. 15.6). These sinovaginal bulbs are commonly said to be derived from the urogenital sinus. The tissue in the sinovaginal bulbs continues to thicken, and it becomes the vaginal plate. It is also suggested that the vaginal plate may contain paramesonephric and mesonephric components. During the second trimester the vaginal plate cells desquamate, forming vaginal lumen. Simply, the upper part of the vagina is derived from the paramesonephric ducts, and the lower part is derived from the vaginal plate. A partition between the vaginal plate and the urogenital sinus persists as the hymen. It usually breaks shortly before or shortly after birth. The lower end of the vagina continues down along the urethra and opens separately into the vestibule. Development of the female reproductive system is influenced by estrogens secreted by the fetal ovaries. The mesonephric ducts generally disappear in the female. The caudal part of these may persist as ducts of epoophoron, and remnants of the mesonephric tubule may appear as epoophoron and paroophoron in the broad ligament. The urethral and paraurethral glands correspond to the prostate glands. The lesser and greater vestibular glands arise from the vestibule.

III. EXTERNAL GENITALIA

A. Indifferent Stage

The mesenchyme surrounding the cloacal membrane proliferates and forms the cloacal folds (Fig. 15.7). During division of the cloaca (demarcated by the urorectal septum), the cloacal folds are also divided into the ventral urethral folds and the dorsal anal pit. The urethral folds unite cranially to give rise to the genital tubercle.

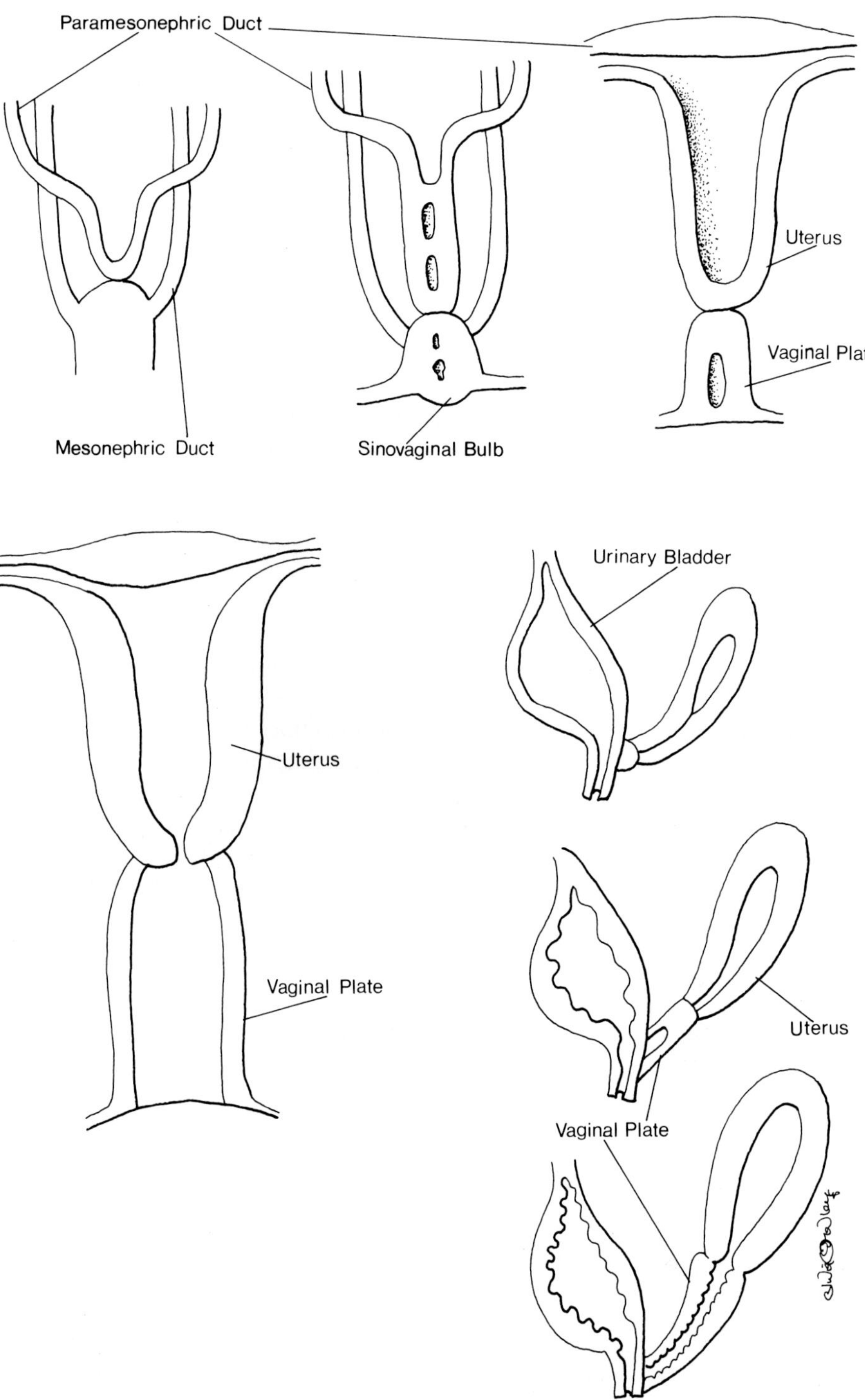

Figure 15.6
States of development of the female genital tract.

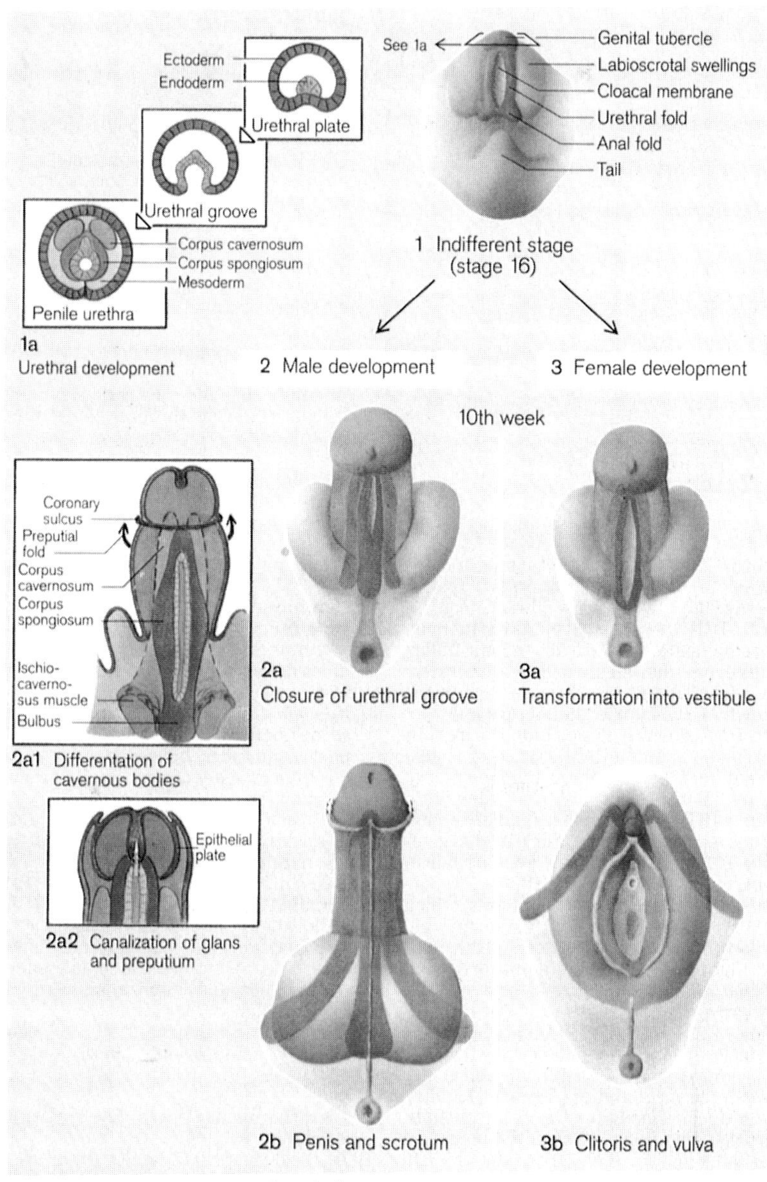

Figure 15.7
External genitalia (see Color Plate 15.7). Reproduced with permission from Thieme Medical Publishers Inc., New York, 1995, *Color Atlas of Embryology*, Ulrich Drews, Chapter 8: Urogenital System.

The cavity of the urogenital sinus now bounded by urethral folds is known as the urethral groove. This urethral groove extends into the genital tubercle. A pair of new swellings, the genital (labioscrotal) swellings, appear lateral to the urethral folds. The genital tubercle enlarges to form the phallus. When the urogenital (cloacal) membrane disintegrates, the urethral groove loses its floor, making it open to the exterior. A median septum from the endoderm develops and grows deeper as the urethral plate. The lower margin of the urethral plate disintegrates to give rise to a new definitive urethral groove, which continues with the phallic part. Hence, the definitive urethra is lined partly by endoderm and partly by ectoderm.

B. Male External Genitalia

In the male, the tissues around the urogenital sinus secrete an enzyme, 5α-reductase, which converts testosterone to dihydrotestosterone. Dihydrotestosterone influences the growth, maintenance and fusion of the male external genitalia. As the phallus (genital tubercle) elongates rapidly and forms the penis, it pulls the urethral folds forward and closer to each other. Soon thereafter, the urethral folds fuse, enclosing the definitive urethral groove forming the penile urethra (Fig. 15.7). A groove around the distal end of the penis defines the glans. A skinfold, the prepuce, grows over the glans. The penile urethra does not extend into the glans. At the tip of the glans, an ectodermal cell cord grows inward, extending to the lumen of the penile urethra. This cord becomes canalized and establishes a continuous channel from the tip of the penis to the base of the urinary bladder. Hence, the vesicourethral canal forms the proximal half of the prostatic urethra, the definitive urethral groove forms the distal half of prostatic urethra and penile urethra, and the ectodermal ingrowth gives rise to the urethra in the glans. Again, the prostatic buds grow out of the prostatic urethra, and the bulbourethral and urethral glands develop from the penile urethra. The genital (labioscrotal) swellings migrate caudally, grow toward each other, and fuse to form scrotal sacs.

C. Female External Genitalia

Differentiation of the external genitalia toward female occurs when there is either no dihydrotestosterone (DHT) or there is no tissue responsiveness to DHT. DHT may be lacking because there is no testosterone, or because there was no 5α-reductase to convert testosterone to DHT. Defective responsiveness of this tissue to DHT may be due to receptor abnormalities (Fig. 15.7). In the absence of DHT, the genital tubercle stays smaller, bends caudally and forms the clitoris. The urethral plate does not extend to it. The urethral folds on both sides of the phallic portion of the urogenital sinus do not fuse and persist as labia minora. The open urogenital sinus remains as a cleft, the vestibule, in which the urethra and vagina open. The genital (labioscrotal) swellings become the labia majora.

IV. DESCENT OF THE GONADS

A. Descent of the Testes

The gonads and kidneys develop retroperitoneally, and testicular descent occurs behind the peritoneal layer, transferring them from the abdomen to the scrotal sacs.

The descent is commonly divided into two phases: a) transabdominal descent and b) transinguinal descent.

1. Transabdominal descent brings the testes down to the level of the inguinal ring. A gelatinous cylindrical mass of undifferentiated mesenchyme appears on either side of the vertebral column during the embryonic period. The cephalic end of this tissue becomes attached to the lower poles of the gonads, and the caudal end extends through the region of the inguinal canal and becomes firmly attached to the genital swellings. This tissue (ligament) is now known as the gubernaculum. This also receives a contribution from the regressing mesonephros, and is sometimes referred to as the inguinal ligament of the mesonephros. The gubernaculum swells in its caudal portion and enlarges the inguinal canal. With enlargement of the inguinal canal, an extension of the peritoneal sac, called the processus vaginalis, develops adjacent to the gubernaculum. The processus vaginalis extends to the scrotal sac. As the gubernaculum swells, it becomes shorter and displaces the testes down into the inguinal region. Some maintain that no active descent occurs in the abdominal cavity. Straightening and lengthening of the body may cause the descent.

2. Transinguinal descent of the testes is influenced by testosterone and other androgens. The testes rapidly slide behind the processus vaginalis through the inguinal canal. The second shortening of the gubernaculum is caused by loss of its mucoid matrix, leading to its reduction and regression. This movement of the testes through the inguinal canal is also aided by increased abdominal pressure built by the growing abdominal viscera (Fig. 15.8A). During transinguinal descent, the ductus deferens, which accompanies the testes, acquires coverings from abdominal wall muscle. When the testes reach the scrotum, the part of the processus above the testes obliterates, whereas the part of it associated with the testes persists as tunica vaginalis testes. Testicular descent is caused by a combination of processes. In addition to the well formed inguinal canal, the intact infraabdominal wall, fetal androgens and a normal hypothalamic–pituitary gonadal axis are important. Early testosterone secretion is believed to be regulated by human chorionic gonadotropin (HCG) from the placenta, later by gonadotropic hormones from the pituitary.

B. Descent of the Ovaries

The ovaries also undergo caudal and lateral shift and become suspended in the broad ligament of the uterus. In the absence of testosterone and other androgens, the gubernaculum does not shorten, reduce or regress; rather, it becomes attached to the paramesonephric ducts, near the fused part (uterotubal junction). This attachment changes the direction of pull of the gubernaculum, causing the ovary to be directed medially into the pelvis. The segment of the gubernaculum from the medial pole of the ovary to the uterus (fused paramesonephric ducts) becomes the ovarian ligament, and its segment from the uterus to the labia majora (genital swellings) becomes the round ligament of the uterus. Cephalic to the ovary, regressed mesonephros becomes the suspensory ligament of the ovary.

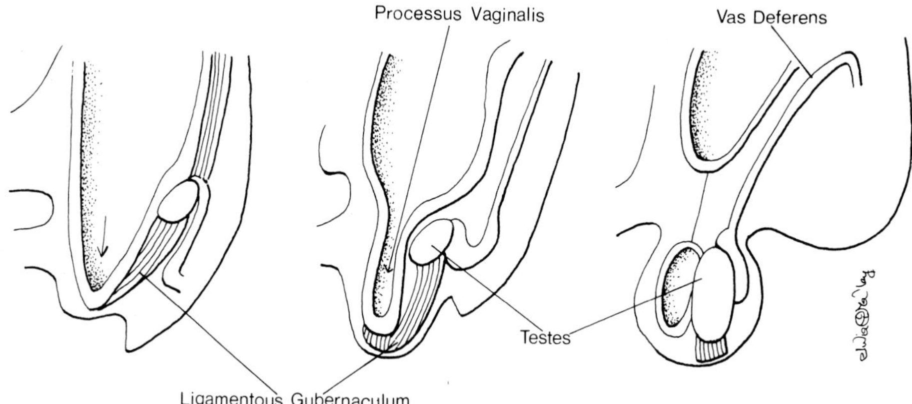

Figure 15.8A
Descent of the testes.

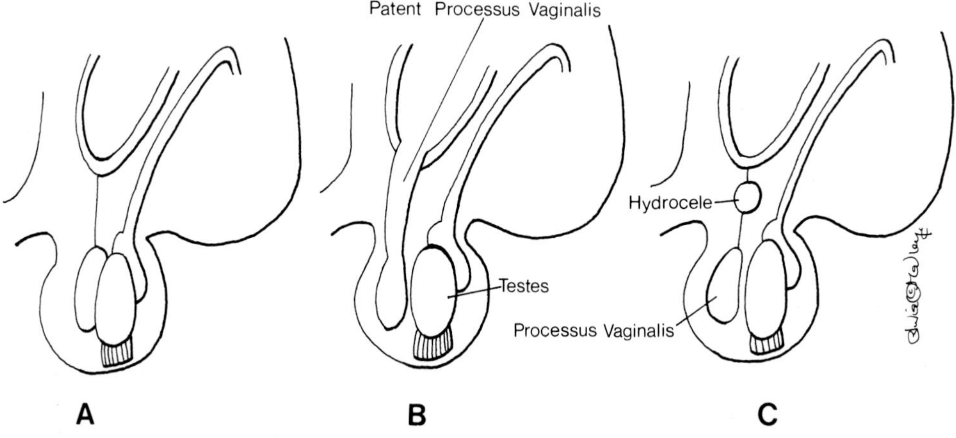

Figure 15.8B
Malformations associated with descent of the testes: (A) normal, for comparison; (B) congenital inguinal hernia; (C) hydrocele.

V. CONGENITAL ANOMALIES OF THE GENITAL SYSTEM

A. Chromosomal Causes of Abnormal Development

1. Monosomy XO (Turner's Syndrome)

In general, embryos with monosomy of an autosomal or sex chromosome are not viable, and about 99% of them are spontaneously aborted. The fetal phenotype consists of subcutaneous edema that is pronounced on the dorsa of the hands and feet, horseshoe kidney, hypoplastic left heart and nuchal cystic hygroma. Those who survive fail to sexually mature at puberty. With absence of the Y chromosome, initially the gonads, the duct system and external genitalia develop as female. The primary follicles fail to develop and the primordial germ cells start to degenerate; in addition, the initial normal germ cell complement is gradually reduced, so that in childhood streak gonads without any germ cells are left. It is also suggested that germ cells become arrested at early prophase I of meiosis, possibly due to abnormal pairings of chromosomes that results in their degeneration. In the absence of normal germ cells, the follicles fail to develop. In the absence of ovaries, the development of ducts and external genitalia remains infantile.

Aside from ovarian dysgenesis, individuals with Turner's syndrome have short stature, a broad chest with wide-set nipples, puffiness in the hands and feet, a short and webbed neck, cervical lymphatic cysts and coarctation of the aorta (in some cases). The most common karyotype is 45,X. Another frequent finding is mosaicism (45,X/46,XX). It is postulated that an active gene on an otherwise inactive X chromosome may account for clinical manifestation.

2. Klinefelter's Syndrome (47,XXY)

The children with this phenotype begin normal male development. Testicular development is normal until late childhood, when testicular basement membrane fibrosis becomes evident. The number of germ cells then decreases progressively and relative prominence of interstitial (Leydig) cells is seen. These males then grow disproportionately tall, have gynecomastia, and show some mental dullness and behavioral problems. With increasing seminiferous tubule hyalinization they become infertile. Although only 55% of these conceptuses survive to term, its incidence is still 1:800 in newborn males. Other congenital anomalies associated with this syndrome have been reported. Some also show mosaicism (XXY/XY, XXXY). Increase in the number of X chromosomes is accompanied by increasing severity of the syndrome.

3. 46,XX Males

These individuals generally develop as male but become infertile in adult life. In adults, the testes show thickening of the seminiferous tubule basement membrane and absence of germ cells. In most of these cases, a portion of the Y chromosome containing the SRY gene is located on one X chromosome short arm. It may have resulted from abnormal crossing over that occurred during meiosis in male germ cells. The presence of an SRY leads to production of TDF, which causes male differentiation of gonads and other parts of the genital system.

B. Hermaphroditism

1. True Hermaphrodites

These individuals have both ovarian and testicular tissues in the gonads. True hermaphroditism may be present in the same gonads as an ovotestes, or one gonad may be an ovary and the other a testes. The external genitalia are usually ambiguous. The condition also includes abnormal phallus size, partially fused genital folds, with or without a vaginal opening, and hypoplastic uterus and uterine tubes. The karyotype in the majority of cases is 46,XX, with no evidence of Y DNA or an SRY gene. The other karyotype found is 46,XX/46,XY. Hypoplastic gonads are devoid of germ cells. Most will also have hypospadias or a female-type urethra. The condition is extremely rare in humans.

2. Male Pseudohermaphrodites

Pseudohermaphrodites generally have gonads of one sex and the external genitalia of the opposite sex. A male pseudohermaphrodite is an individual whose karyotype is 46,XY, whose gonads are testes, and whose external genitalia is female or ambiguous in appearance. This condition may have various causes, some of which include:

a. Androgen receptor disorders (testicular feminization)
In these disorders, the gonads are normal testes, but the androgen receptors on target tissue are either qualitatively or quantitatively abnormal. Testosterone and DHT cannot influence an indifferent genital system. Hence, this differentiation of the mesonephric duct and male external genitalia does not occur normally. Sertoli cells secrete MDIF; thus, regression of the paramesonephric duct occurs normally. The fetus will then have female-type external genitalia without a uterus, uterine tubes or an upper vagina. The testes remain at an abdominal location or in the inguinal canal or in the labia majora. Fusion of the genital folds varies depending on the degree of insensitivity of androgens. The individuals are infertile males having a small phallus (Fig. 15.9).

b. 5α-reductase deficiency
This condition is characterized by a pseudovagina and perineoscrotal hypospadias. The normal development of the male genital duct system is dependent on testosterone, whereas development of male external genitalia is dependent on DHT, which is formed by conversion of testosterone in the presence of 5α-reductase. Deficiency of 5α results in absence of DHT, and the external genitalia remain unfused and appear to be female at birth. The development of testes and vas deferens is normal. Under the influence of MDIF, the paramesonephric ducts regress. In some cases, a sudden surge in testosterone at puberty may cause a dramatic change of the external genitalia and accessory glands into male structures. The presence of testosterone during intrauterine life and puberty influences brain differentiation toward the sense of male gender identity. Nowadays, when the condition is suspected at birth, testosterone and DHT levels are measured. These children can be treated with specific amounts of DHT and surgical correction of hypospadias. The response to testosterone alone is poor. The treated individuals are fertile and are able to reproduce (Table 15.1).

c. MDIF deficiency will allow normal androgen production and normal development of male genitalia, but female ducts (uterus and uterine tubes) also develop. These males characteristically present with undescended (cryptorchidism) or ectopic testes. It is inherited as an X-linked or autosomal recessive trait. Children having this condition are at 20 times greater than normal risk for gonadal neoplasia.

d. Testosterone biosynthesis defects are caused by various enzyme deficiencies. Enzymes such as 17α-hydroxylase and 17,20-lyase are needed for

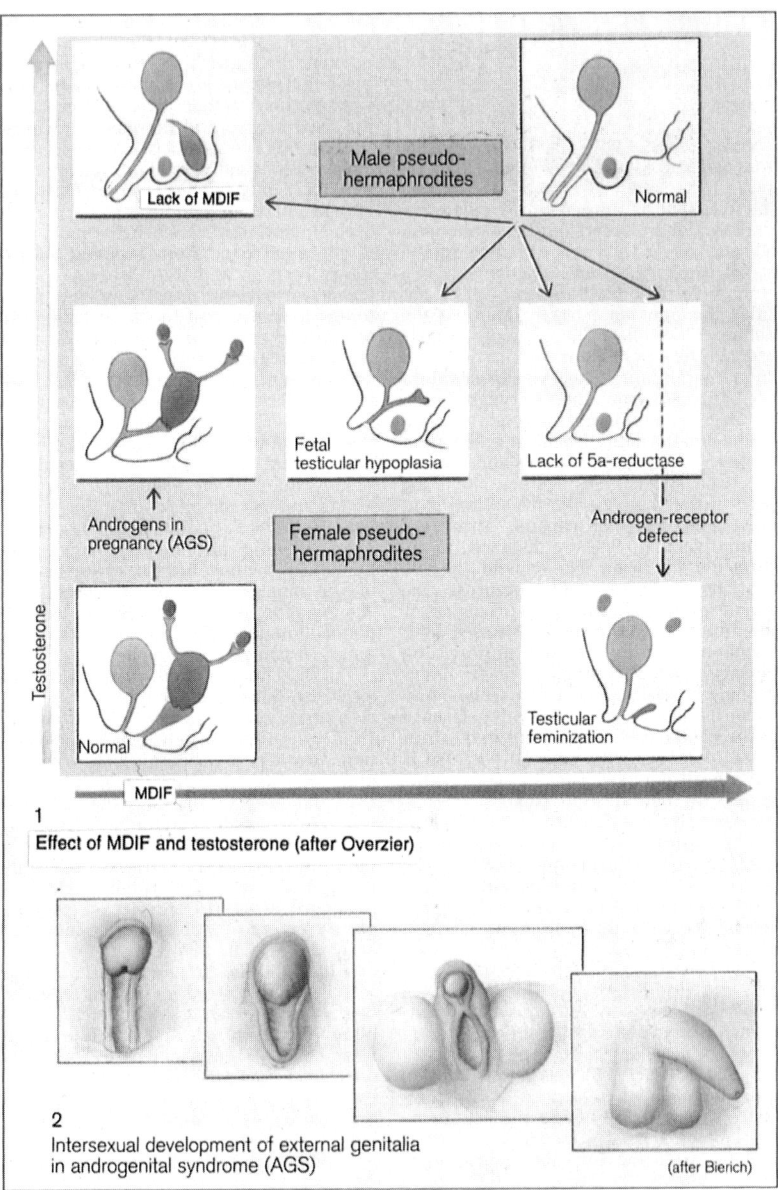

Figure 15.9
Intersexuality (see Color Plate 15.9). Reproduced with permission from Thieme Medical Publishers Inc., New York, 1995, *Color Atlas of Embryology*, Ulrich Drews, Chapter 8: Urogenital System.

synthesis of steroid hormones from cholesterol. Deficiencies of these enzymes are grouped under androgenital syndromes. The block in synthesis of testosterone will thus affect all testosterone-derived androgens, which leads to malformation of many structures, including the mesonephric duct and brain. Since the MDIF secreted by Sertoli cells is normal, paramesonephric duct derivatives do not develop. This is a very rare condition.

3. Female Pseudohermaphrodites

A female pseudohermaphrodite has 46,XX karyotype. The gonads are ovaries, and the external genitalia are ambiguous or male in appearance. In most cases, the virilizing hormones are androgens, which are secreted by hyperplastic adrenals. In some cases, virilization can be caused by administration of synthetic progestin compounds, or by maternal virilizing hormones or maternal virilizing tumor, or cloacal dysgenesis (Fig. 15.9).

a. Adrenogenital syndromes are conditions that are caused by insufficiency or absence of various enzymes required for synthesis of steroid hormones, especially androgens. As mentioned above, some of these cause pseudohermaphroditism in males, but in female fetuses the deficiency commonly results in excess production of androgens, leading to virilization. Deficiency of 21α-hydroxylase is the most common of all. This enzyme is required for conversion of progesterone to aldosterone. When this conversion is blocked, the precursors form excess androgens. Classically, this syndrome is associated with low aldosterone, and the newborn has problems with electrolyte disturbances and blood pressure. As adrenal production of excess androgens occurs after differentiation of the internal genital organs, only the development of external genitalia is affected. These children exhibit clitoral enlargement and fusion of the urethral and genital folds. Aldosterone abnormalities are potentially life-threatening. 21α-hydroxylase deficiency is an autosomal recessive trait, and recurrence risk in the affected families is high (25%). Prenatal diagnosis in the first trimester is by chorionic villus sampling, and in the second trimester by typing of amniocytes and measurement of 17-hydroxyprogesterone in the amniotic fluid. Prenatal treatment with dexamethasone to prevent masculinization of the affected fetus has been successful.

b. Virilization by other causes
Any exogenous androgens, either by maternal drug ingestion (progestins or androgens) or excess androgen production due to maternal neoplasia, may cause virilization of otherwise normal female fetus. Although this condition is rare, careful history of the mother is important.

c. Cloacal dysgenesis is often referred to as female pseudohermaphroditism, because it is characterized by ambiguous genitalia, an enlarged clitoris with a urethra at its tip, posteriorly fused labia minora and a persistent urogenital sinus. These fetuses usually have no anal, urethral or vaginal openings. The perineal skin is smooth. This condition is always associated with urinary tract obstruction, which leads to megaureter, cystic kidneys, marked ascites (urine passes through uterine tube to abdomen) and oligohydramnios. It is postulated that these anomalies result from maldevelopment of the cloaca. Defect in cloacal septation, failure of fusion of the urorectal septum with the cloacal membrane, and a defective cloacal membrane have been attributed as causes.

C. Gonadal Dysgenesis

This is characterized as gonads that have not developed properly to either testes or ovaries. These gonads usually exhibit severe internal disorganization and absence of germ cells.

1. Pure Gonadal Dysgenesis (Swyer's Syndrome)

These individuals have 46,XY but appear as females of normal stature. Although gonadal development starts as male, only streak gonads with primitive tubules are present at birth. Some cases are inherited as X-linked recessive condition, but multiple causes have been suggested for pure gonadal dysgenesis.

2. Mixed Gonadal Dysgenesis

These individuals have 45,X/46,XY chromosomal mosaicism, suggested to result from a loss of the Y chromosome in some of the early embryonic cells. These fetuses may develop as normal males, or males with hypospadias, or males with undescended testes, or females with the characteristics of Turner's syndrome. They may have half a uterus on one side and vas deferens on the contralateral side, or they may have an infantile uterus. The gonads may develop as unilateral testes with a contralateral streak or dysgenetic bilateral testes. In general, development of the genital duct system is according to the gonad present on that side of the body.

3. Primary Hypogonadism

These individuals develop normally until puberty or early adulthood and then fail to undergo the changes associated with puberty. The hypothalamus and pituitary are normal, and high levels of gonadotropins are produced, but the gonads do not respond to the gonadotropins. In affected females (46,XX), the gonads are streaks similar to those seen in Turner's syndrome. There may be occasional primary follicles in the streaks. In males, the primary defect is failure of the Leydig cells to produce sufficient amounts of testosterone, which leads to small testes, usually without any germ cells.

4. Secondary Hypogonadism

This is caused by defective production of gonadotropin by the anterior pituitary. The defect may be caused by insufficient secretion of gonadotropin-releasing hormone by the hypothalamus. The females (46,XX) are characterized as Turner's syndrome and males as Klinefelter's syndrome.

D. Anomalies of the Genital Tract

1. Uterine anomalies
   Most of the uterine anomalies are found in conjunction with other abnormalities. Isolated uterine anomalies are rare.
   a. Complete or partial atresia of both paramesonephric ducts results in absence of uterus and uterine tubes (Mullerian aplasia). More than half of these cases have some renal anomalies. Vaginal development may also be affected.
   b. Uterine hypoplasia is underdevelopment of the caudal part of the paramesonephric ducts, resulting in a rudimentary uterus.
   c. Duplication of the uterus results from nonfusion of the caudal part of the paramesonephric ducts (Fig. 15.10).
      i. In the extreme form, the uterus is doubled (uterus didelphis).
      ii. A uterus with two cervices and two openings into the vaginal canal, which may also have a median septum.
   d. A uterus with a partial septum, causing a depressed fundus (uterus arcuate).
   e. The bicornate uterus has two horns entering a common vaginal canal.
   f. The unicornate uterus is a hemiuterus with one uterine tube. One paramesonephric duct fails to develop.
   g. A unilateral partial atresia of the cervix is a hemiuterus with a rudimentary uterus attached to its fundus. Only one side is normal.
   h. Cervical atresia: a hypoplastic cervix without any opening into the vaginal canal.
2. Agenesis of the vagina caused by failure of induction.
3. Vaginal atresia caused by failure of canalization.
4. Postpubertal adenosis of the glandular epithelium of the vaginal wall. This is associated with maternal ingestion of diethylstilbestrol (DES) for prevention of spontaneous abortion or premature delivery. This synthetic hormone is consid-

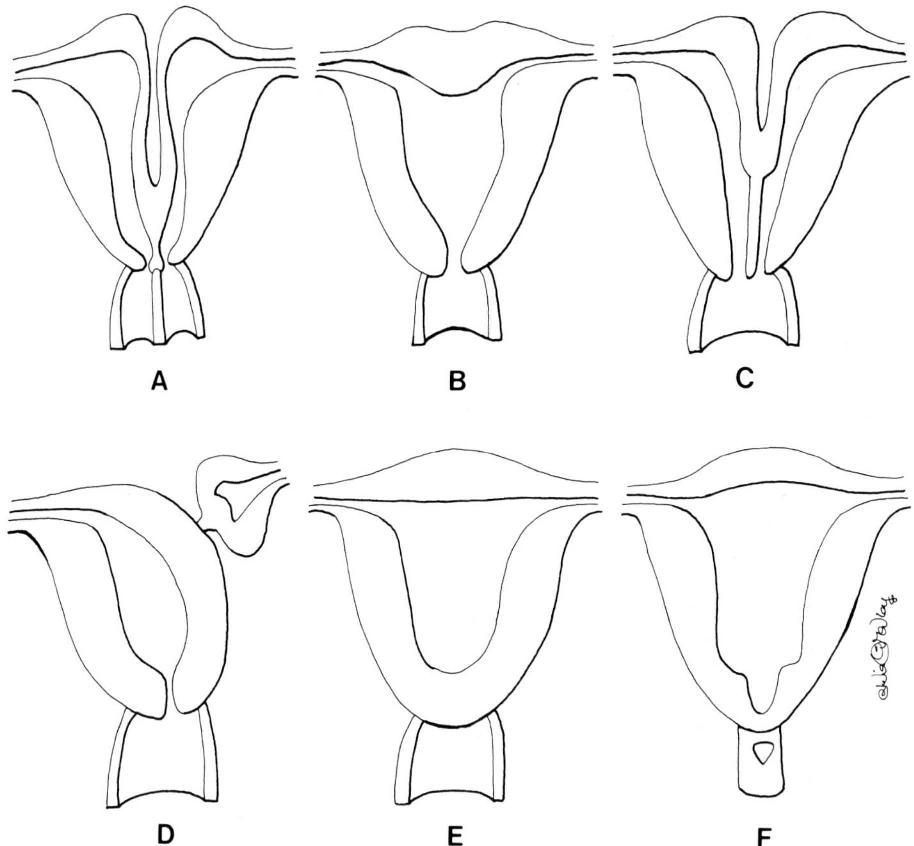

Figure 15.10
Uterovaginal malformation: (A) uterus didelphy; (B) uterus arcuate; (C) bicornuate uterus; (D) unilateral partial atresia of the cervix; (E) cervical atresia; (F) vaginal atresia.

ered to be a transplacental teratogen for affected female fetuses and causes increased chances of cervical and vaginal carcinoma after puberty.

5. Unilateral or bilateral absence of the ductus deferens results from a failure of either one or both mesonephric ducts to differentiate. It may be due to absence or insufficiency of testosterone or nonresponsiveness of the mesonephric tissue to testosterone. The absence of ductus deferens is usually associated with kidney agenesis. If only the proximal part of the mesonephric duct is affected, kidney development is normal but the individual is sterile.

E. Anomalies of External Genitalia

Abnormal development of external genitalia may be caused by chromosomal defect, single-gene mutation or environmental factors. As discussed above, most of these appear as developmental arrest leading to ambiguous development.

1. Agenesis and hypoplasia of the genital tubercle result in an absent or very small penis. It is attributed to insufficiency of fetal testicular hormones and is usually associated with hypopituitarism.

2. Hypospadias is defective development of the ectodermal urethra, usually failure of canalization of the ectodermal urethra. The urethra opens on the ventral surface of the glans penis or penile shaft. Although its incidence is 1 in 300 live births, it is less often seen as an isolated anomaly. It is most associated with androgen deficiency or nonresponsiveness of the urethral folds (partially) or ectodermal urethra to androgens.

F. Anomalies of Gonadal Descent

1. Cryptorchidism is failure of one or both testes to descend normally into scrotum. Its incidence is 1:30 live male births. Spermatogenesis is impaired in undescended testes and may eventually lead to atrophy. There is an increased likelihood of malignancy in undescended testes. It may also involve dysgenesis or agenesis of the testes, absence of, or short, testicular blood vessels, and agenesis of the ductus deferens. Damage to the CNS and premature birth may cause delay in descent. Cryptorchidism is usually associated with other anomalies, including male pseudohermaphrodism, mixed gonadal dysgenesis, MDIF deficiency, osteogenic imperfecta, prune-belly syndrome, trisomies 13 and 18, and Klinefelter's syndrome. Undescended testes may be located in the abdomen just above the deep inguinal ring or in the inguinal canal in the superficial inguinal ring.

2. Ectopic testes result when they deviate from the usual path of descent and may come to lie in the lower abdominal wall, the medial aspect, the thigh, the base of the penis, the perineum or another unusual place. Local fibrosis in the normal path may block descent.

3. Congenital indirect inguinal hernia results from failure of the processus vaginalis to close normally. The persistent peritoneal sac permits an intestinal loop to descend into the scrotum (Fig. 15.8B).

4. Congenital hydrocele results from accumulation of fluid in an unobliterated portion of the processus vaginalis. The abdominal end of the processus occasionally remains open, and, although it is too small to permit herniation, peritoneal fluid passes into the passageway.

5. An encysted hydrocele is associated with the spermatic cord and is caused by closure of the processus vaginalis on both ends, leaving a middle portion patent. Fluid accumulation forms a cyst in the patent portion of the processus vaginalis.

TABLE 15.1. Sex Determination and Differentiation in Males and Females

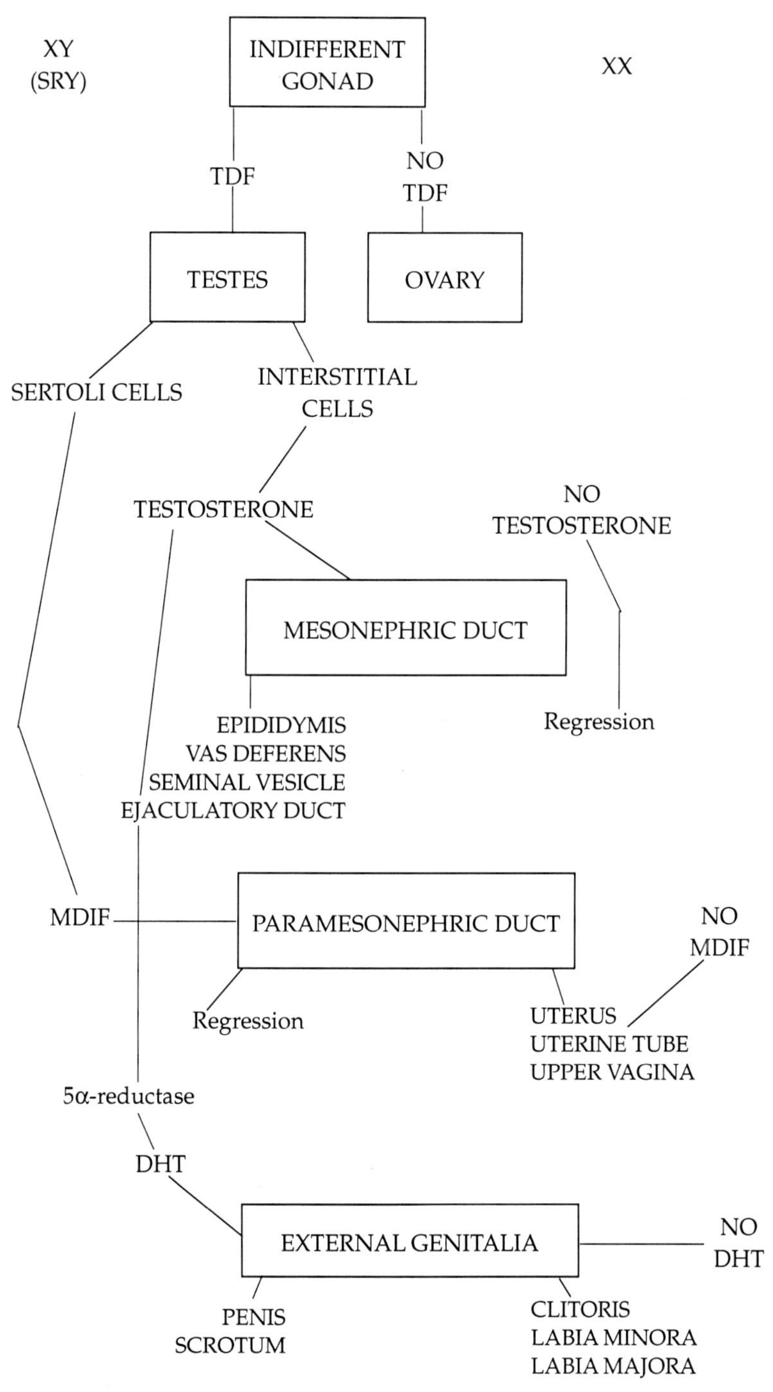

SRY = sex-determining region
TDF = testes-determining factor
MDIF = Mullerian duct inhibitory factor
DHT = dihydrotestosterone

TABLE 15.2. Homologous Structures in Males and Females Derived from Embryonic Urogenital Structures

Embryonic structure	Male	Female
Indifferent gonad	Testis Seminiferous tubules Rete testes Interstitial cells	Ovary Ovarian follicles Medulla Rete ovarii
Mesonephric tubules	Ductuli efferentes Paradidymis	Epoophoron Paroophoron
Mesonephric duct	Appendix of epididymis Ductus epididymis Ductus deferens Seminal vesicle Ejaculatory duct	Appendix vesiculosa Duct of epoophoron Duct of Gartner Part of urethra Part of vagina
Paramesonephric duct	Appendix of testis Prostatic utricle	Hydatid (of Morgagni) Uterine tube Uterus Part of vagina
Urogenital sinus	Urinary bladder Urethra Prostate gland (not middle lobe) Bulbourethral glands Urethral glands	Urinary bladder Urethra Urethral and paraurethral glands Vestibular glands Part of vagina
Genital tubercle	Penis	Clitoris
Urethral folds	Floor of penile (spongy) urethra	Labia minora
Labioscrotal swellings	Scrotum	Labia majora
Gubernaculum	Gubernaculum testis	Ovarian ligament Round ligament of the uterus

TABLE 15.3. Summary of the Development of the Urogenital System

Age (days)	Embryonic structure
24	Pronephros primordium
28	Mesonephros and mesonephric duct
35	Ureteric bud, metanephros and bulging of urogenital ridge
42	Division of cloaca, gonadal primordium (indifferent sex)
49	Paramesonephric duct and differentiated gonads
56	Fusion of paramesonephric ducts in the female
100	Primary follicles in ovary

REVIEW QUESTIONS — Chapter 15

Select the most appropriate answer.

1. The primary sex cords in the medullary region of the gonads become devoid of germ cells and give rise to the:
   a. epididymis
   b. ductus (vas) deferens
   c. ejaculatory tube
   d. rete testes
   e. paraepididymis

2. The cranial (cephalic) end of the paramesonephric ducts opens into the:
   a. urogenital sinus
   b. vaginal canal
   c. allantois
   d. urethra
   e. peritoneal cavity

3. Mullerian duct inhibitory factor (MDIF) is secreted by:
   a. primordial germ cells
   b. interstitial cells of Leydig
   c. sustentacular cells of Sertoli
   d. granulosa cells of follicles
   e. gonadotrophs of the pituitary

4. An epithelial outgrowth from the mesonephric duct gives rise to the:
   a. epididymis
   b. prostate gland
   c. bulbourethral glands
   d. seminal vesicle
   e. all of the above

5. Genetic sex is established at:
   a. fertilization of the ovum
   b. appearance of the genital tubercle
   c. development of Sertoli cells in the fetal testes
   d. elaboration of Mullerian inhibitory factor
   e. formation of the blastomere

6. The primary germ cells arise from the:
   a. intermediate mesoderm
   b. yolk sac
   c. ectoblast
   d. splanchnomesoderm
   e. paraxial mesoderm
7. Differentiation of sex in the embryo occurs first in the:
   a. duct system
   b. gonads
   c. internal genital organs
   d. external genital organs
   e. all of the above differentiate simultaneously
8. Differentiation of the gonads from the indifferent stage is under:
   a. genetic control (karyotype of the embryo)
   b. endocrine control (hormones secreted by the gonads present)
   c. the influence of placental hormones
   d. the influence of maternal hormones
   e. all of the above
9. Differentiation of the internal and external genital organs depends on the:
   a. genetic makeup of the embryo
   b. presence of gonads and their endocrine secretion
   c. placental hormones
   d. maternal hormones
   e. all of the above
10. The genes for testes-determining factors are located on the:
    a. short arm of the Y chromosome
    b. long arm of the Y chromosome
    c. short arm of the X chromosome
    d. long arm of the X chromosome
    e. both the X and Y chromosomes contain these genes
11. The sex cords give rise to all of the following except:
    a. Sertoli cells of seminiferous tubules
    b. ovarian follicular cells
    c. rete testes
    d. straight tubules of testes
    e. primordial germ cells
12. An embryo contains the XX sex chromosomal complex. In the developing gonads:
    a. its medullary region differentiates and its cortex regresses
    b. its cortical region differentiates and its medulla regresses
    c. the presence of an X influences medullary differentiation
    d. neither the presence of the X chromosome nor absence of the Y chromosome influences development of the ovaries
    e. the SRY is expressed

13. Testes-determining factor (TDF) is secreted by the:
    a. early gonads
    b. interstitial cells of Leydig
    c. Sertoli cells
    d. mesonephric duct
    e. paramesonephric duct

14. The mesenchyme of the genital ridge separating the seminiferous tubules gives rise to cells that secrete testosterone that binds to the receptors on the:
    a. yolk sac
    b. paramesonephric
    c. mesonephric duct
    d. metanephric blastema
    e. ureteric bud

15. During the early development of the gonads, which of the following arise from the splanchnic mesoderm?
    a. the mesonephric cells
    b. the coelomic mesothelial cells
    c. the mesenchymal aggregate cells
    d. the germ cells

16. Attraction of primordial germ cells toward developing gonads is influenced by chemotactic substances secreted by the:
    a. yolk sac
    b. mesonephric cells
    c. primary sex cords
    d. mesonephric duct
    e. all of the above

17. In a female fetus, the uterus develops from the:
    a. mesonephric duct
    b. paramesonephric duct
    c. somatic pleura
    d. urogenital sinus
    e. endoderm

18. The homologous structure to the labia majora is the:
    a. scrotum
    b. vestibule
    c. prostate
    d. seminal vesicle
    e. penis

19. Descent of the testes is caused by all of the following except:
    a. increased abdominal pressure
    b. presence of gut in the processus vaginalis
    c. enlargement of the inguinal canal by swelling of the gubernaculum
    d. reduction and regression of the gubernaculum
    e. straightening and lengthening of the body

20. The efferent ductules of the testes develop:
    a. by transformation of some of the mesonephric tubules
    b. from the mesonephric ducts
    c. by transformation of the metanephric tubules
    d. from the primary sex cords of the gonads
    e. from the secondary sex cords

21. The urogenital sinus contributes to all of the following structures except the:
    a. urinary bladder
    b. male urethra
    c. female urethra
    d. labia majora
    e. vestibule

22. In females, the homologous structure to the gubernaculum testis is the:
    a. labia minora
    b. broad ligament
    c. clitoris
    d. suspensory ligament of the ovary
    e. round ligament

23. All of the following are derived from the mesoderm except the:
    a. ureteric bud
    b. metanephric blastema
    c. urethra
    d. ductus deferens
    e. seminiferous tubules

24. The process vaginalis:
    a. is a peritoneal fold that gives rise to the broad ligament of the uterus in females
    b. is a sac of peritoneum that extends through the abdominal wall into the scrotum in the male
    c. is a ligamentous remnant of the mesonephros that causes the testes to descend in the scrotum in males
    d. becomes the round ligament in the female
    e. all of the above are correct

25. In a genetic male (44 x XY) who lacks androgen receptors, each of the following will be increased except:
    a. estradiol
    b. FSH
    c. LH
    d. progesterone
    e. testosterone

26. A child born with ambiguous external genitalia could be:
    a. a female pseudohermaphrodite
    b. a male pseudohermaphrodite
    c. a true hermaphrodite
    d. someone with Klinefelter's syndrome
    e. any of the above

27. The gender identity of the child in question 26 is established by:
    a. gender assignment by the medical personnel on its birth certificate
    b. chromosomal analysis
    c. concentration of maternal sex hormones during pregnancy
    d. appearance of a dominant type of external genitalia
    e. all of the above

28. In monosomy (XO, Turner's syndrome):
    a. the children start normal male development
    b. the number of germ cells in the gonads starts decreasing after puberty
    c. the children grow disproportionately tall
    d. an active gene on a dormant X chromosome may account for the clinical manifestations
    e. all of the above might occur

29. If male pseudohermaphroditism is due to androgen receptor disorders, then:
    a. the testes are normal for the most part
    b. the testosterone and DHT levels are normal
    c. MDIF is secreted by the Sertoli cells
    d. the fetus will have female external genitalia without a female genital tract
    e. all of the above is correct

30. Failure of formation and migration of the primordial germ cells results in:
    a. Turner's syndrome
    b. Klinefelter's syndrome
    c. pseudohermaphroditism
    d. gonadal agenesis
    e. none of the above

31. Failure of the urethral folds to fuse during embryonic life would be expected to result in:
    a. epispadias
    b. exstrophy of urinary bladder
    c. horseshoe kidney
    d. hypospadias
    e. multiple ureters

32. Cryptorchidism (undescended testes):
    a. may result in atrophy of the testes
    b. may be associated with MDIF deficiency
    c. may also involve agenesis of the ductus deferens
    d. is a characteristic of male pseudohermaphroditism
    e. all of the above are correct

33. Incomplete fusion of the paramesonephric ducts during development can result in:
    a. a bicornuate uterus
    b. a double uterus
    c. duplication of the vagina
    d. pelvic kidneys
    e. bladder diverticulum

34. In a newborn child, a large opening was found in the anal region. It was diagnosed as persistent cloaca. Further examination revealed that the mesonephric duct (ductus deferens) and ureter opened into it together. The karyotype of the child was male and the phallus was underdeveloped. In this case, which of the following would also be missing?
    a. the urinary bladder
    b. the urethra
    c. the prostate
    d. the perineal body
    e. all of the above would be missing

35. If the child described above had been female:
    a. the diagnosis would be different
    b. disregarding the prostate, other missing structures would be different
    c. this anomaly would not have occurred
    d. the sex of the child would not make any difference
    e. the external genitalia would have developed normally

36. In the case mentioned above, the basic structure that failed to develop is the:
    a. mesonephric duct
    b. urorectal septum
    c. allantois
    d. metanephric duct
    e. urachus

37. A child was born with ambiguous external genitalia. If the karyotype of this child is male, this condition could have resulted from:
    a. androgen receptor disorder
    b. 5α-reductase deficiency
    c. testosterone biosynthesis defect
    d. MDIF deficiency
    e. any of the above

38. In the above mentioned child, the testes developed normally but there was no vas deferens. Other finds in this child would include:
    a. abdominal testes
    b. absence of a uterus
    c. a small phallus
    d. a blind short vagina
    e. all of the above

39. In this child there was:
    a. no expression of the SRY
    b. no secretion of MDIF
    c. no DTF
    d. no influence of testosterone and DHT
    e. all of the above

40. If the karyotype of the above-mentioned child had been female, the cause of ambiguity of the external genitalia would be:
    a. an absence of the SRY
    b. failure of mesonephric duct regression
    c. virilizing hormones from hyperplastic adrenal glands
    d. nonresponsiveness of tissue to 5α-reductase
    e. any of the above

ANSWERS TO REVIEW QUESTIONS

1. d 2. e 3. c 4. d 5. a
6. b 7. b 8. a 9. b 10. a
11. e 12. b 13. a 14. c 15. c
16. c 17. b 18. a 19. b 20. a
21. d 22. e 23. c 24. b 25. d
26. e 27. b 28. b 29. e 30. d
31. d 32. e 33. b 34. e 35. d
36. b 37. e 38. a 39. e 40. c

CHAPTER

16

ENDOCRINE SYSTEM

The endocrine system includes all the body organs that synthesize and secrete hormones. The chief endocrine organs (glands) are the hypothalamus, hypophysis (pituitary gland), epiphysis, thyroid gland, parathyroid glands, pancreas, suprarenal glands, testes, ovaries and placenta. Other organs that have some endocrine function are the liver, kidneys, stomach, and intestines. Development of some of the above-mentioned organs has been described with other systems. Only the endocrine gland has not been discussed before. The hypothalamus, pituitary, suprarenals, pineal (epiphysis) and islets of Langerhans will be covered here. Although early development of the thyroid gland, parathyroids, is discussed along with the pharyngeal apparatus, their endocrine development will also be included.

The endocrine glands develop from epithelia as a diverticulum that later loses its connection with the surface and becomes ductless glands. Their secretions, hormones, are released into the blood, and for this they acquire a very rich blood supply. Most hormones appear in the blood at 12 to 20 weeks gestation.

Normal development of the fetal endocrine system is for the most part independent of maternal endocrine status, because most hormones do not cross the placental barrier. However, hormones secreted by the placenta reach the fetal circulation. In addition, the immunoglobulins produced during maternal endocrine disease can cross the placenta and may affect development of the fetal endocrine system.

Endocrine function during fetal life is qualitatively and quantitatively different from that during postnatal life.

I. HYPOTHALAMUS

Early development of the brain is described with the central nervous system. In a 32-day old embryo, hypothalamic sulcus appears in the diencephalic region, which indicates formation of the hypothalamus. Differentiation of distinct nuclei and fiber tracts occurs between 6 and 12 weeks of gestation. At 12 weeks, the portal vascular connection between the hypothalamus and pituitary establishes the hypothalamic–pituitary axis. Growth hormone-releasing hormone (GHRH) can be detected at 18 weeks. Somatostatin (SRIF), an inhibitor of growth hormone, is detected at 11 weeks and increases with gestational age until 22 weeks. Thyrotropin-releasing hormone (TRH) can be detected at 8 weeks. The concentration of dopamine, a major prolactin inhibitor, is much higher (than in the adult) between weeks 10 and 15 in hypothalamic tissue. Other releasing hormones from the hypothalamus also appear in the fetal circulation during this period.

II. PITUITARY GLAND (HYPOPHYSIS)

The pituitary gland consists of two main components: the adenohypophysis (anterior pituitary) and the neurohypophysis (posterior pituitary), which develop from two different sources (Fig. 16.1).

A. The Adenohypophysis

The ectoderm situated immediately external to the oropharyngeal membrane is induced by the closely related floor of the diencephalon. A diverticulum from this ectoderm arises in the roof of the stomodeum as the adenohypophyseal primordium (Rathke's pouch). Soon thereafter, the floor of the diencephalon adjacent to this primordium evaginates to form the neurohypophysis diverticulum. The adenohypophyseal primordium becomes vesicular, forming a pouch. The stalk of this pouch becomes narrow and then loses its connection with the roof of the mouth (stomodeum). Under the influence of an epitheliomesenchymal interaction, the anterior wall of the pouch proliferates and sends cell cords into the mesenchyme, forming the pars distalis, its main subdivision. The posterior wall of the pouch, which is in contact with the neurohypophyseal primordium, remains thin and constitutes the pars intermedia. The vesicular lumen decreases and becomes a residual slit (Fig. 16.1). A portion of the adenohypophysis grows upward and surrounds the stalk of the neurohypophysis and is known as the pars tuberalis. Remnants of the adenohypophyseal primordium stalk may persist as the pharyngeal hypophysis. In the pars distalis, the basophilic staining cells appear at 8 weeks and acidophilic cells by weeks 9 to 12. The adenohypophysis is fully differentiated by 16 weeks. The growth hormone levels in the serum can be detected as early as 10 weeks, suggesting that early growth hormone release by the pituitary is independent of hypothalamic control. Growth hormone levels peak at weeks 20 to 24 and steadily decline afterward. This may indicate establishment of the hypothalamic–hypophysial axis and interaction between releasing hormones (GHRH and

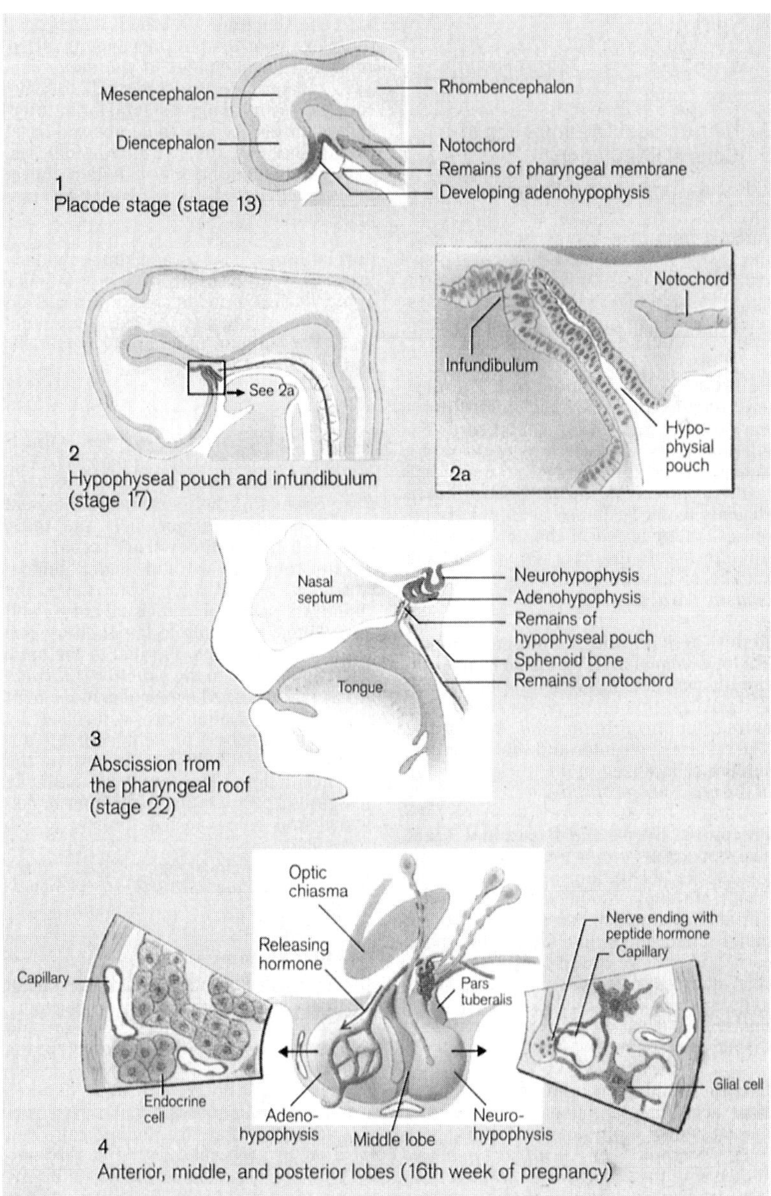

Development of pituitary gland

Figure 16.1
Pituitary gland (see Color Plate 16.1). Reproduced with permission from Thieme Medical Publishers Inc., New York, 1995, *Color Atlas of Embryology*, Ulrich Drews, Chapter 4: Nervous System.

SRIF) and the pituitary. Growth hormone levels are higher in premature infants than in full-term newborns. Adrenocorticotropic hormone (ACTH) is detectable at the 10th week of gestation. Although corticotropin and α-lipotropin are produced in the pars distalis, their further cleavage to form (α) melanotropin and (β) endorphins occurs in the pars intermedia. A marked increase in thyroid-stimulating hormone (TSH) occurs at 16 weeks and peaks at 22 weeks. Another surge in TSH and TRH occurs at birth, reaches a maximum at 30 minutes postnatally and gradually decreases over the next 2 to 3 days.

B. The Neurohypophysis

As mentioned above, neurohypophyseal evagination grows toward the adenohypophyseal pouch, extending from the median eminence of the hypothalamus, forming the infundibular process (pars nervosa). The neuroepithelial cells in the pars nervosa proliferate and differentiate into pituicytes. Neurosecretory cells located in the hypothalamus send their processes to the pars nervosa. The hormones secreted by the neurosecretory cells are transported through these processes and are stored and released from the pars nervosa (Fig. 16.1). The pituitary becomes functionally active in the middle of intrauterine life, and the fetus thereby regulates its own endocrine system. Arginine vasopressin (AVP, ADH) can be detected as early as 10 weeks. Maternal placental electrolytes influence fetal AVP secretion and in turn may alter placental water transfer. An increase in fetal AVP levels may affect fetal amniotic fluid volume, fetal lung fluid production and fetal renal function. Cord blood AVP levels are higher in newborns delivered vaginally than in those delivered by cesarean section. The plasma AVP levels decrease rapidly after birth but remain higher in premature infants with respiratory distress syndrome. There is now reason to believe that the newborn's inability to concentrate urine is probably due to reduced generation of cyclic AMP in response to AVP stimulation.

III. EPIPHYSIS (PINEAL) GLAND

As the neurohypophysis grows down from the floor of the diencephalon, the epiphysis grows upward from the most caudal part of the roof of the diencephalon. The cells in the walls of this diverticulum proliferate, filling the space. Some of the cells differentiate into pinealocytes (neuroglia), and nerve fibers from the epithalamus soon grow into the pineal gland. The function of the pineal gland in humans is not clear. Under the influence of the light–dark cycle, it secretes the hormone melatonin, which inhibits the function of the hypophyseal–gonadal axis.

IV. SUPRARENAL (ADRENAL) GLANDS (Fig. 16.2)

Suprarenal glands consist of two parts, the cortex and the medulla, which develop from two different sources.

A. The Cortex

Cells from the coelomic mesothelium of the posterior body wall separate and migrate between the root of the mesentery and mesonephros to form bilateral cell masses. Although these cells arise from an epithelium, they have the capability to

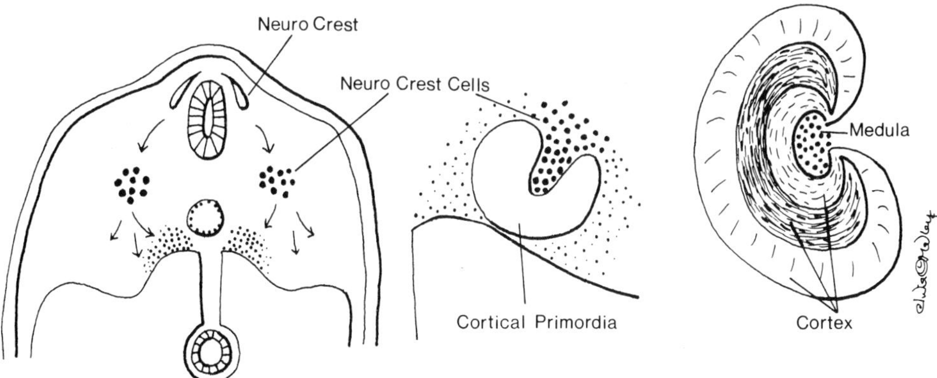

Figure 16.2
Development of the suprarenal glands.

change to mesenchyme. Some cells from the mesonephros may also contribute to these suprarenal cortical aggregations. These cortical cells proliferate and differentiate into acidophilic cells of the fetal zone of the cortex. A second wave of migratory cells surrounds the fetal zone, establishing a definitive or permanent zone. By 8 weeks, the inner fetal zone and outer definitive zones of the suprarenal cortex are apparent. The development of the suprarenal cortex in the early fetal period (up to 20 weeks) is stimulated and maintained primarily by placental human chorionic gonadotropin (HCG). After this period, ACTH is the most important stimulus for its normal growth and maintenance. The suprarenal glands are largest in size between weeks 16 and 24. At birth they are one-third the size of the kidneys. The fetal zone constitutes about 80% of its mass. During the second trimester, steroid synthesis pathways are established between the fetus and placenta. Precursors are formed by the placenta and are converted by the fetal zone of the cortex. During the third trimester, the fetal suprarenal glands become more sensitive to ACTH. In response to ACTH, the definitive (permanent) zone becomes more active and synthesizes cortisol from low-density-lipoprotein (LDL) cholesterol of maternal or fetal origin without requiring placental progesterone. Cortisol and cortisone levels increase during this period. Higher cortisol levels are essential for maturation of the lungs and other tissues. The cortisol levels may decrease at birth, but they peak during the first 12 hours in response to the rise in ACTH. During the first week of life, ACTH decreases to very low levels, and this may account for some postnatal involution of the fetal zone. The zona glomerulosa and zona fasciculata are present at birth, but the zona reticularis is not recognizable.

B. The Medulla

The neural crest cells from the adjacent sympathetic ganglia migrate and reach the mediodorsal aspect of the primitive suprarenal glands at about 8 weeks gestation. These cells slowly invade the developing cortex, giving rise to the suprarenal medulla. Under the inductive influence of the cortex, the medullary cells differentiate into a cluster of sympathetic neurons and glandular (chromaffin) cells. Most of these cells are located in the center of the gland. Epinephrine, norepinephrine and enkephalins are found in the suprarenal gland by 10 weeks. The concentration of medullary hormones increases with gestational age. Glucocorticoids and ACTH are required for conversion of norepinephrine to epinephrine.

V. ISLETS OF LANGERHANS (ENDOCRINE PANCREAS)

The dorsal and ventral pancreatic buds grow out from the primitive foregut and fuse to each other (see Chapter 13, Digestive System). The glandular epithelium forms by budding of the cords of endodermal cells. Many of these outgrowths differentiate into acini, and some groups of cells separate from ducts and move into the parenchyma as small cell nests or clusters. These isolated cells (islets) differentiate into neuroendocrine cells (Chapter 13, Fig. 13.3). Glucagon-containing α-cells appear first and peak at 17 to 19 weeks.

Insulin-containing β-cells appear next and continue to increase during infancy; they later become the most abundant cells in the islets. Somatostatin-containing δ-cells appear next and peak at about 5 months after birth. The cells staining for pancreatic polypeptide are found mainly in tissue derived from the ventral pancreatic bud, and their number decreases with gestational age. Gastric inhibitory polypeptide, β-endorphin and somatomedin C have also been found in the fetal pancreas.

The pancreatic hormones increase from weeks 7 to 20. Maternal insulin crosses the placenta in small amounts and has no effect on carbohydrate metabolism in the fetus. The stimuli for insulin secretion during the early fetal period are cyclic AMP and the agents that increase intracellular cAMP. Glucagon, somatomedin and growth hormones may affect the development and function of β-cells during this period. Although the β-cells become more responsive to amino acids after 16 weeks, insulin secretion in response to glucose remains low until after birth. However, chronic hyperglycemia in maternal diabetes causes early maturation of β-cells. The response of α-cells to amino acids and prolonged glucose changes also increase with gestational age. Aside from promoting metabolism of glucose, lactate and amino acids, insulin is a major stimulator of fetal growth. At birth, the nutrients supplied from the mother stop, and the glucose and insulin levels drop in the blood of the newborn, while the glucagon and catecholamine levels increase. Within a few days, the liver becomes more responsive to low insulin and high glucagon, and blood glucose levels stabilize.

VI. THE THYROID GLAND

The thyroid gland appears as a median endodermal thickening in the floor of the pharynx between pharyngeal arches 1 and 2 (Chapter 11, Fig. 11.1, Pharyngeal Apparatus). The epithelial cells arrange into a bilobate structure that migrates down the front of the neck. It remains connected to the tongue by a narrow stalk known as the thyroglossal duct, marking its origin as the foramen cecum of the tongue. The thyroglossal duct becomes fragmented and disappears, and the thyroid gland loses its connection to the pharynx. During migration, the thyroid primordium becomes associated with ultimopharyngeal bodies (from pharyngeal pouch V), which detach from the pharyngeal wall and implant into the dorsal aspect of the thyroid. It is suggested that ultimopharyngeal bodies contain neural crest cells. The cells from these bodies later differentiate into "C" or parafollicular cells that will secrete thyrocalcitonin. The thyroid primordium is invaded by the surrounding mesenchyme, which breaks its solid mass into cords and plates. The cords and plates finally become small clusters of cells. The colloid begins to appear in the center of the clusters, and follicles are formed. By 11 weeks, the characteristic morphology is apparent. At this stage, the thyroid is capable of concentrating iodides to synthesize thyroid hormones. Severe iodine deficiency during this period may result in neurological damage that is not responsive to postnatal therapy with iodine or thyroid hormones. Conversion of thyroxine (T_4) to triiodothyronine (T_3) does not mature until late in gestation. At birth, TSH levels increase rapidly, and thyroxin and T_3 levels stimulated by TSH also increase rapidly by 24 hours and then gradually decline to normal levels over days 5 to 7 postnatally (Fig. 16.3).

VII. The Parathyroids

The inferior parathyroids develop from the third pharyngeal pouch (Chapter 11, Fig. 11.1), and the superior parathyroids arise from fourth pharyngeal pouch. Usually four in number, they come to lie on the posterior aspect of the thyroid gland. Although they arise from the pouches, it is proposed that their cells are neuroectodermal in origin. Therefore, parathyroids probably contain cells from all three sources: the endoderm, the neural crest and the ectoderm. Fetal parathyroids are responsive to calcium levels, and parathormone has been detected in fetal blood. Severe maternal hypercalcemia may result in marked and prolonged suppression of parathyroid function in newborns. Fetal calcium levels are consistently higher than maternal levels. Only chief cells are present at first. Other cells, like oxyphils, appear during postnatal life (Fig. 16.3).

VIII. CONGENITAL ANOMALIES OF THE ENDOCRINE SYSTEM

A. Hypothalamic–Pituitary Unit

1. Anencephaly

Midline malformations like anencephaly, septooptic dysplasia, and cleft lip and palate are associated to some degree with hypothalamic–pituitary unit dysfunction. Anencephaly provides valuable information about the importance of hypothalamic–pituitary hormones during fetal development. Although in these fetuses diencephalon is missing and pituitary is reduced, TSH and PRL are detectable. GH, FSH, LH and ACTH are not detectable. Therefore, the fetal zone of the adrenals is regressed, but the thyroid is relatively normal. Plasma PRL levels indicate that the

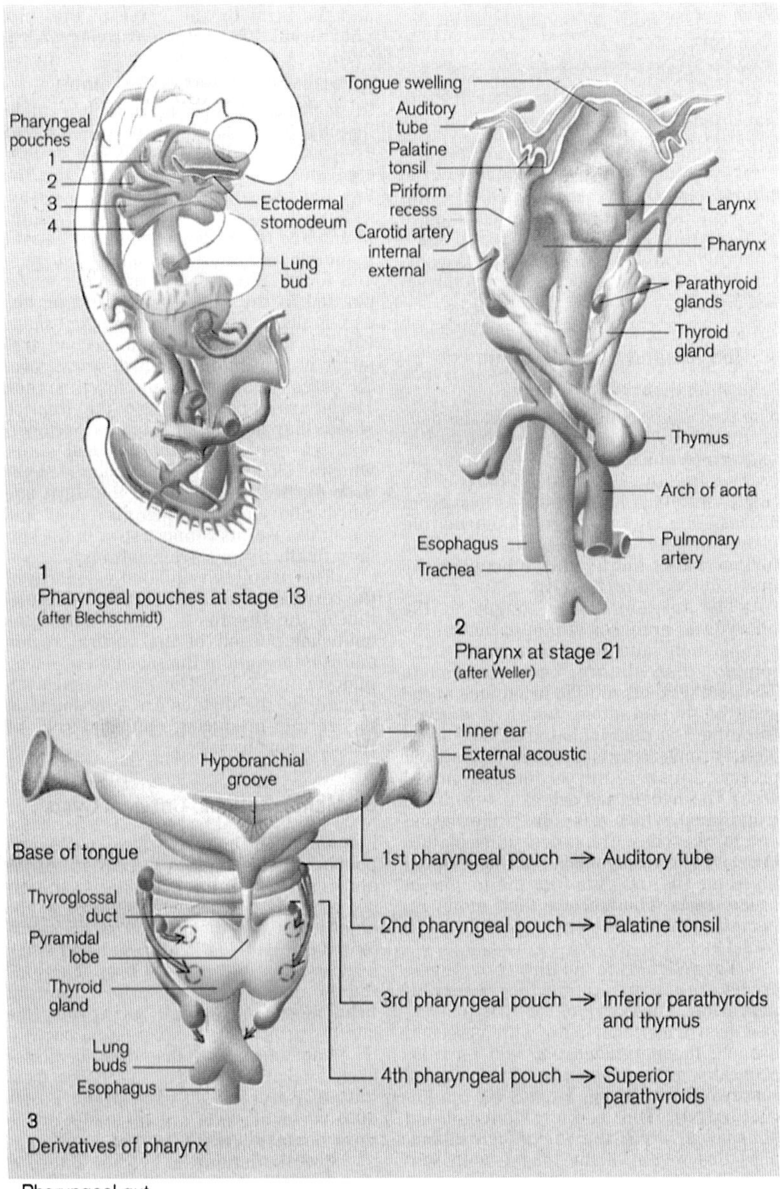

Figure 16.3

Derivatives of pharyngeal gut (see Color Plate 16.3). Reproduced with permission from Thieme Medical Publishers Inc., New York, 1995, *Color Atlas of Embryology*, Ulrich Drews, Chapter 7: Gastrointestinal Tract.

hypothalamus is not needed for PRL secretion. It also indicates that the pituitary–thyroid axis does not require TRH for development. In the male anencephalic, the gonads and genitalia are abnormal and the testes undescended. This indicates that sexual differentiation is not dependent on fetal gonadotropins but that gonadotropins are required for late differentiation of gonads and genitalia. Due to a lack of FSH, the testes have a reduced number of Leydig cells and the epididymis is underdeveloped. Female anencephalics show less marked effects. As mentioned above, the hypothalamic–pituitary–adrenal axis is needed for development of corticotrophs.

2. Congenital Pituitary Aplasia or Hypoplasia

Hypopituitarism may result from an abnormal hypothalamus. In pituitary aplasia, no adenohypophyseal tissue is present. GH, PRL, TSH, ACTH, LH and FSH are not detectable. Hypoplasia of the thyroid, gonads and genitalia is marked, and the fetal zone of the adrenal gland is absent. Despite the GH deficiency, body growth is normal at birth, indicating that intrauterine growth is independent of GH. However, GH is essential for postnatal growth. If not treated, growth impairment occurs. Hypoplastic male genitalia or neonatal hypoglycemia is indicative of this condition. Hypoglycemia results in ACTH or GH deficiency.

B. Suprarenal Anomalies

1. Congenital Adrenal Hypoplasia

Congenital adrenal hypoplasia may be primary or secondary to adrenal unresponsiveness to ACTH or to ACTH deficiency. Small adrenals have also been reported in maternal pregnancy-induced hypertension and in trisomy 13 and 18. The most common cause of adrenal hypoplasia is ACTH deficiency. Although in these fetuses adrenal development is normal until 15 weeks, as it is under the influence of HCG, it starts involution thereafter because of ACTH deficiency. Electrolyte and aldosterone levels are normal. In case of adrenal unresponsiveness to ACTH, the clinical presentation is similar to that above, but the skin is hyperpigmented, indicating increased levels of ACTH.

Primary congenital adrenal hypoplasia is rare. It may be X-linked or autosomal recessive. In this type of adrenal insufficiency, children usually die within 72 hours of birth. Severe steroid deficiencies cause vomiting, hypoglycemia, hyperkalemia and hyperpigmentation. Although cortisol and aldosterone levels are decreased, renin and ACTH levels are increased (compensatory hypersecretions of ACTH). Adrenal hypoplasia caused by pregnancy-induced hypertension is mild in nature. In trisomies 13 and 18, the adrenals are very hypoplastic.

2. Congenital Adrenal Hyperplasia

Adrenal hyperplasia results from overstimulation of the adrenals. Basically, it is an enzymatic defect in steroid synthesis that leads to decreased negative feedback and consequently to hypersecretion of ACTH that overstimulates the adrenals. The precursors either accumulate or are used in production of androgens or other substances. The effect on a fetus depends on the deficiency of particular enzymes. For instance, 11-hydroxylase or 21-hydroxylase deficiencies in a female fetus may result in virilization and ambiguous external genitalia; while in the male they may cause penis enlargement. In some cases, 21-hydroxylase deficiency leads to decreased synthesis of aldosterone and cortisol, which results in salt wasting. On the other hand, 11-hydroxylase deficiency results in accumulation of 11-deoxycorticosterone, which leads to salt retention and hypertension. The adrenal glands of congenital hyperplasia are enlarged at birth.

3. Accessory Suprarenal Cortical Tissue

Accessory suprarenal cortical tissue may be found in the posterior body wall near the kidneys. It arises from the coelomic epithelium. Its effect on suprarenal development depends on its size and degree of activity. The cortical tissue may be accompanied by medullary tissue.

4. Infections

Fetal infections with herpes virus, toxoplasmosis, cytomegalovirus, echovirus and Listeria involve adrenal glands, leading to necrosis, inflammation and dystrophic calcification, which may lead to variable degrees of adrenal insufficiency.

C. Islet Cell Disorder

1. Hypoglycemia

In most cases, hypoglycemia in newborns is transient. It is common in low-birth-weight neonates (small for gestational age), the smaller member of twins, infants of diabetic or toxemic mothers, infants with erythroblastosis and infants who undergo perinatal stress. Hypoglycemia in low-birth-weight neonates is multifactorial. In these infants, glucose production is inadequate, whereas glucose utilization is increased. Decreased glucose production results in low glycogen storage in the liver, and maturation of enzymes needed for gluconeogenesis is delayed. Increase in glucose utilization is due to a relatively large brain, stresses or hyperinsulinism (secondary to intrauterine asphyxia, maternal drugs or disease). Hypoglycemia is also seen in larger infants with fetal distress (e.g., congenital heart disorder, asphyxia, hypothermia and maternal toxemia). In diabetic mothers, glucose and other nutrients cross the placenta, resulting in β-cell hyperplasia, which leads to hyperinsulinism and hypoglycemia. At birth, an abrupt decrease in nutrients and increased utilization of glucose, coupled with inadequate response to glucagon, causes further hypoglycemia. Glucose levels normalize in the first week of life. Erythroblastosis also leads to β-cell hyperplasia. The release of amino acids during hemolysis and excess production of steroids by the adrenals in response to stress and hypoxia influence islet cell hyperplasia. Neonates with erythroblastosis are not large, indicating that hyperinsulinism in these cases does not have a substantial effect on fetal growth.

Persistent hypoglycemia results from enzymatic defects in gluconeogenesis, glycogenolysis, fatty acid oxidation or glucose metabolism-associated hormone deficiencies. All may lead to hyperinsulinism. Persistent hypoglycemia begins after the first week of life. It has been proposed that these disorders result from a basic defect in control of endocrine pancreatic development. Postnatal monitoring of blood glucose in at-risk infants is essential. Delayed diagnosis may lead to neurologic or ophthalmologic disorders. Hypoglycemia convulsions occur in these infants. Any defect in function of α-cells results in glucagon deficiency, which in turn leads to persistent hypoglycemia.

2. Hyperglycemia

Three types of hyperglycemia in newborns have been described: a) transient hyperglycemia in premature infants, b) transient diabetes of newborns and c) permanent diabetes of newborns.

a. Transient hyperglycemia in premature infants is probably due to insulin resistance. Elevated glucose in these infants may lead to hyperosmolarity, cerebral hemorrhage and osmotic diuresis. Hyperglycemia usually responds to treatment and slow infusion of nutrients.

b. Transient diabetes mellitus is seen in infants who are small for their gestational age. It is caused by impaired insulin secretion, which may be due to delayed adenylate cyclase–cAMP system maturation. There is often a family history of diabetes. Intrauterine malnutrition may be another factor. It is characterized by polyurea, severe dehydration and failure to thrive. Fever, infection and gangrene are complications. Although insulin is required temporarily, glucose tolerance usually normalizes over weeks to months.

c. Permanent diabetes of newborns is caused by insulin deficiency. There is intrauterine growth retardation. Like the transient condition, polyurea and dehydration are present, but permanent diabetes requires continuous insulin administration. There may be pancreatic agenesis or hypoplasia with few islets without β-cells. This may have been caused by either β-cell agenesis or postnatal involution, or intrauterine islet destruction.

D. Thyroid Anomalies

1. Hyperthyroidism

These infants are born to thyrotoxic (hyperthyroid) mothers. Increased fetal movement, fetal tachycardia and intrauterine congestive heart failure may occur. The infant may be stillborn or die soon after birth. Characteristic findings in the cases include: severe cardiac hypertrophy, enlarged thyroid, pulmonary hypertension and intrauterine growth retardation. Symptoms in liveborns include irritability, exophthalmos, enlarged thyroid, tachycardia, diarrhea, vomiting and failure to thrive. Microcephaly, premature closure of sutures and growth retardation further complicate this condition.

2. Congenital Hypothyroidism

Congenital hypothyroidism is the most common cause of mental retardation that could be prevented. Its incidence is 1:4000 live births. Most cases are due to primary agenesis of the thyroid gland; other cases are inherited defects in thyroid hormone synthesis or are due to a defect in the hypothalamic–pituitary–thyroid axis. About 10% of these cases of transient hypothyroidism result from intrauterine exposure to iodine-containing drugs or antithyroid drugs. Transient hypothyroidism is not associated with elevated TSH or thyroglobin abnormalities. Clinical signs develop shortly after birth in iodine-deficiency-induced endemic cretinism. It is characterized by dry skin, a protruding tongue, and mental and physical growth retardation. Early therapy with iodine or thyroid hormone can normalize thyroid function. In some cases, although severe myxedema can be prevented, mental retardation and deafness may not be prevented by postnatal therapy.

3. Familial Goitrous Cretinism

This results from an enzymatic defect in the synthesis of thyroid hormones. The low levels of thyroid hormones cause increased release of TSH by the pituitary, which in turn causes enlargement of the thyroid gland. The characteristic features of endemic cretinism do not appear at birth. If not treated, these features appear early in the neonatal period. Early diagnosis and treatment lead to normalized thyroid function.

4. Ectopic Thyroid Tissue

Ectopic thyroid tissue may be present just beneath the tongue or anywhere in the path of its descent. Thyroid tissue fails to separate from its origin and may compromise the trachea or esophagus.

5. Remnants of Thyroglossal Duct

Remnants of the thyroglossal duct are very common findings in children. A persistent thyroglossal duct may rarely remain patent and form a thyroglossal fistula that may open in the lower neck in the median plane.

6. Accessory Thyroid

An accessory thyroid may be found in the neck or in the mediastinum. It may have resulted from early attachment of thyroid cells to adjacent organs.

E. Parathyroid Anomalies

Hyperparathyroidism causes severe abnormalities during intrauterine life, whereas hypoparathyroidism and hypocalcemia, although more common, rarely cause any problems during intrauterine life.

1. Hyperparathyroidism and Hypercalcemia

Primary hyperparathyroidism is usually inherited as an autosomal recessive trait. An inherited paternal gene causes reduced sensitivity to normal levels of ionized calcium, which leads to parathyroid hyperplasia. It manifests in utero as fractures or limb deformities. At birth, neonates rapidly develop hypotonia, lethargy and respiratory distress. Severe vomiting, poor feeding and constipation are also present. If untreated, renal insufficiency and severe hypertension lead to death within months. Parathyroid glands show chief cell hyperplasia.

2. Hypoparathyroidism and Hypocalcemia

a. Transient hypoparathyroidism and hypocalcemia are seen in many premature or stressed neonates. Serum calcium levels start to decrease shortly after birth, a decrease that is less marked in breast-fed infants than in infants fed on formula with a lower calcium/phosphate ratio. A transient condition is also seen associated with maternal diabetes or maternal anticonvulsant drug use. In these cases, increased calcitonin levels, a reduced ability to excrete a phosphate load and decreased responsiveness to PTH may be factors.

b. Persistent hypoparathyroidism is much less common. It results from maternal hyperparathyroidism where high maternal calcium levels result in fetal hypercalcemia that suppresses fetal parathyroid function. It is manifested as neuromuscular irritability or tetany. The symptoms may be precipitated by formula feeding. Persistent hypoparathyroidism may also result from parathyroid aplasia or hypoplasia, or from end-organ resistance to parathormone. DiGeorge's syndrome is a severe condition of persistent hypocalcemia resulting from aplasia or hypoplasia of the parathyroids and is associated with thymic hypoplasia, immunodeficiency, cardiac anomalies and facial dysmorphism. This is due to defective development of the third and fourth pharyngeal pouch derivatives. It is associated with deletion on chromosome 22.

REVIEW QUESTIONS — Chapter 16

Select the most appropriate answer.

1. Which of the following originates from surface ectoderm?
   a. neurohypophysis
   b. adenohypophysis
   c. adrenal cortex
   d. thyroid
   e. parathyroid

2. Cells from the coelomic mesothelium give rise to the:
   a. suprarenal cortex
   b. granulosa cells of the ovary
   c. Sertoli cells of the testes
   d. all of the above
   e. none of the above

3. Which of the following arise from the endoderm?
   a. the pancreatic acinous cells
   b. β-cells
   c. α-cells
   d. pancreatic duct cells
   e. all of the above

4. Maternal placental electrolytes influence the fetal:
   a. pars nervosa
   b. pars distalis
   c. pars intermedia
   d. epiphysis
   e. all of the above

5. Which of the following is derived from two different primary germ layers?
   a. epiphysis
   b. adrenal glands
   c. pancreas
   d. parathyroids
   e. thyroids

6. The neural crest cells from the sympathetic ganglia migrate to form the:
   a. pars nervosa of the pituitary gland
   b. medulla of the adrenal gland
   c. follicular cells of the thyroid gland
   d. inferior parathyroids
   e. pars intermedia of the pituitary gland

7. During fetal life the parathyroid glands:
   a. contain oxyphil cells that degenerate before birth
   b. contain parafollicular cells that secrete hormones to regulate fetal calcium metabolism
   c. develop in situ from cervical somites
   d. migrate cephalically from the thymus to reach the cervical region

8. The parafollicular cells of the thyroid originate from the:
   a. ectoderm
   b. mesoderm
   c. endoderm
   d. neuroepithelium
   e. neural crest

9. The suprarenal glands develop with the:
   a. ventral mesentery
   b. intermediate mesoderm
   c. paraxial mesoderm
   d. neurocrest
   e. dorsal mesentery

10. The pituicytes originate from the:
    a. ectoderm
    b. mesoderm
    c. endoderm
    d. neuroepithelium
    e. neural crest

11. Which of the following statements about familial goitrous cretinism are true?
    a. it results from low levels of thyroid hormones
    b. it may cause increased levels of TSH
    c. thyroid hypertrophy may not be present at birth
    d. it can be corrected by early treatment
    e. all of the above are correct

12. Congenital suprarenal hyperplasia:
    a. causes an increase in adrenal cortical hormones
    b. inhibits release of ACTH by the pituitary
    c. causes masculinization of external genitalia in female fetuses
    d. may lead to adrenal cortical insufficiency, although there is an overproduction of adrenal cortical hormones

13. Congenital hyperparathyroidism manifests as:
    a. in utero limb fractures
    b. limb deformities
    c. rapidly developing postnatal hypotonia
    d. respiratory distress
    e. all of the above may be present at birth

14. Pituitary hypoplasia may result in:
    a. achondroplasia
    b. maldevelopment of the adrenal cortex
    c. maldevelopment of the thyroid gland
    d. hypoplastic gonads
    e. all of the above

15. In congenital adrenal hypoplasia due to unresponsiveness to ACTH:
    a. the electrolyte levels are normal
    b. the aldosterone levels are normal
    c. the skin is hyperpigmented
    d. the ACTH level is increased
    e. all of the above are correct

ANSWERS TO REVIEW QUESTIONS

1. b	2. a	3. e	4. a
5. b	6. b	7. b	8. c
9. b	10. d	11. e	12. c
13. e	14. e	15. e	

CHAPTER

17

EYE AND EAR

I. THE EYE

Components of the eye originate from the neuroectoderm, the surface ectoderm, neural crest cells and the mesoderm. The optic primordium (vesicle) arises as an outpocketing from the wall of the diencephalon; the lens arises from the surface ectoderm.

The mesoderm surrounding the optic primordia contains neural crest cells and gives rise to the choroid and sclera. In general, the development of the eye depends upon: 1) a series of inductive signals, 2) precise alignment of the essential visual components and visual pathway, and 3) coordinated differentiation and position of the lens and cornea. In response to neural ectoderm induction, the optic primordia transform to an optic vesicle. This vesicle induces formation of the lens, and the lens in turn induces formation of the cornea.

A. Early Development

On the 22nd day, a groove, the optic sulcus, appears on both sides in the lateral walls of the diencephalon. These sulci deepen to form the optic primordia, which contact the surface ectoderm. The mesenchyme containing neural crest cells soon migrates between the optic primordia and surface ectoderm. The optic primordia enlarge to form the optic vesicles (Fig. 17.1). During the fourth week, the distal part of the vesicle enlarges and thickens into the optic disk. The narrow proximal part is now known as the optic stalk. The surface ectoderm adjacent to the optic (retinal) disk thickens to form the lens disk (placode). This opposition between optic disk and lens disk is essential. Any interference in this opposition will lead to failure of

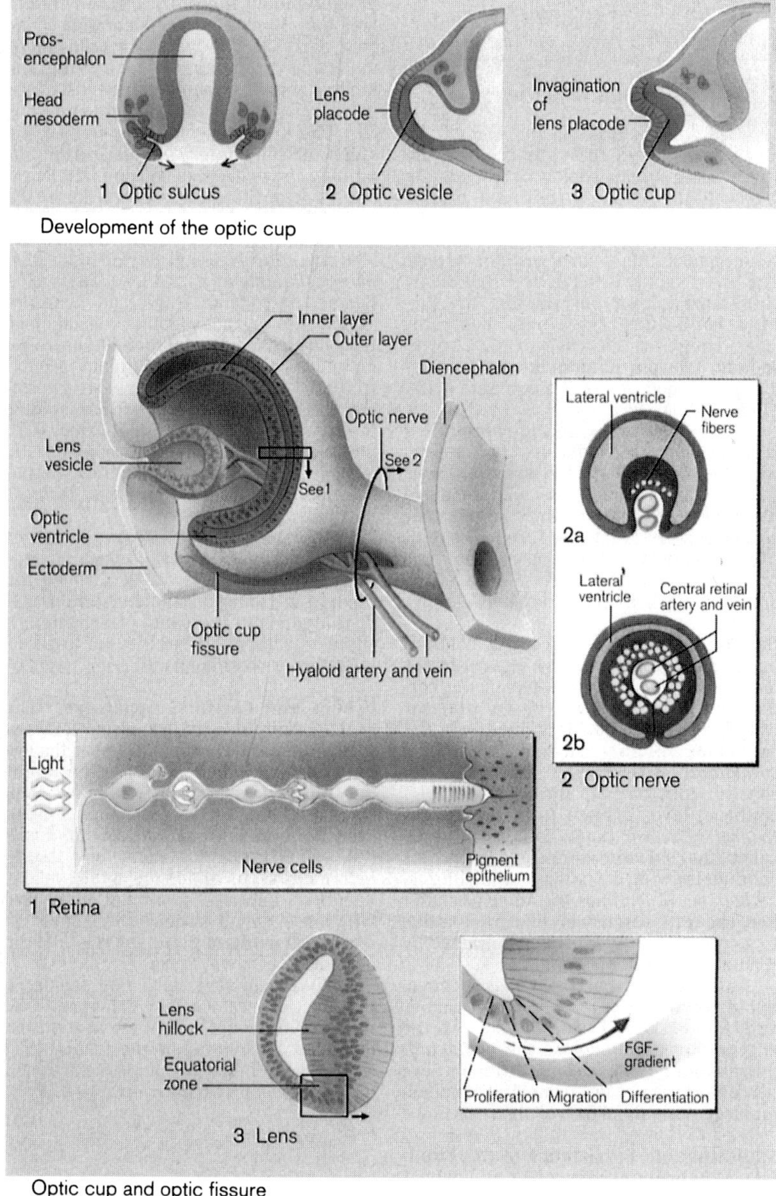

Figure 17.1
Optic cup (see Color Plate 17.1). Reproduced with permission from Thieme Medical Publishers Inc., New York, 1995, *Color Atlas of Embryology*, Ulrich Drews, Chapter 5: Sense Organs.

further development of the eye. The lens disk induces invagination of the optic vesicle to the double-layered optic cup. Formation of the optic cup induces invagination of the lens disk to become the lens vesicle, which detaches itself from the surface ectoderm. Some authors believe that lens formation can occur in the absence of an optic cup because lens induction begins with a series of complex interactions with other tissues before formation of the optic cup. But most authors still agree that the optic vesicle influences growth differentiation and maintenance of the developing lens. Invagination of the optic vesicle is asymmetrical. It occurs at the infranasal quadrant and extends into the optic stalk, leaving a gap that constitutes the retinal (choroid) fissure. The hyaloid artery, a terminal branch of the ophthalmic artery, passes through the retinal fissure to supply the posterior chamber (the space between the lens and retina). Formation of the retinal fissure is essential to direct the nerve fibers from the retina to the brain. Later in development, the retinal fissure closes. The distal part of the hyaloid artery disappears, and its proximal part persists as a central artery of the retina.

B. The Retina (further development of the optic cup) (Fig. 17.1)

Invagination in the optic disk creates the optic cup with an inverted inner layer and an outer layer. The outer layer of the optic cup remains thin, and cells in this layer accumulate pigments to give rise to retinal pigmented epithelium. The inner layer epithelium differentiates into the neural retina. The space between the pigmented (outer) and neural (inner) layers of the optic cup begins to decrease, and persists only as potential cleft. During the seventh and eighth weeks, the neural retinal layer differentiates into neuroblastic layers. The cell layer of the retina that is closer to the pigmented layer proliferates to send waves of cells inward (toward the vitreous body). As the cell number increases, the differentiation of these cells into cell types begins. The first to be recognized are ganglion cells close to the vitreous body. Then the inner and outer nuclear layers take shape. Simultaneously, supporting cells differentiate. This retinal lamination is completed by midterm. The cells of the outer nuclear layer differentiate into photoreceptors. The cells of the nuclear layer send in their processes to establish plexiform layers. All layers of definitive retina are apparent by the eight month. The immature cells remain at the edge of the retina, forming a peripheral concentric growth ring. These cells divide and add new cells to the retina for further growth and development. As new cells migrate inside the ring, cellular differentiation occurs. This way the progeny of a single cell can give rise to different types of retinal cells that are distributed precisely to the different layers.

The axons from ganglion layer cells grow along the innermost layer toward the optic stalk. Positional cues direct these fibers to make very precise connection to the brain centers. Growth of these fibers transforms the optic stalk to optic nerve. Just before entering the brain, two optic nerves join to form the optic chiasma, where half the fibers in each nerve cross over, resulting in combination of the ipsilateral and contralateral fibers on each side. This bundle of mixed fibers grows back to lateral geniculate bodies of the thalamus. Here each retinal axon courses to the correct spot and synapses with the correct target neuron, reproducing a rough map. The rough map is fine-tuned by pruning and secondary axonal reconnections. Both local factors and diffusing molecules appear to guide the growth of these fibers. The visual cortex secretes chemoattractants that direct the growth of the lateral geniculate fibers. The local factor, melanin, present in the peripheral pigmented layer appears to influence the arrangement of axons in the optic stalk. The neuronal cell adhesive molecule (NCAM) may affect the migrating growth cones in their pathfinding. The process of pathfinding is complex, and many other factors may have a role in it. Developmental regulation such as feedback by neural impulses, programmed death of neurons, and rearrangement and rewiring of axons is essential for pathfinding. The two layers (pigmented and neural) of the retina never fuse. The photoreceptors are fully developed at term. The macula is established by 4 months postpartum. Myelination of the optic nerve starts during the 7th month at the optic chiasma and is completed by 1 month postpartum.

C. The Lens

As mentioned above, the surface ectoderm adjacent to the optic disk thickens to form the lens disk (placode). Formation of the lens disk induces invagination of the optic disk, and invagination of the optic disk induces the lens disk to invaginate and become the lens vesicle. The lens vesicle soon detaches itself from the surface ectoderm. The cells situated at the inner (deeper) layer of the lens vesicle elongate to transform into lens fibers. The lens fibers fill the cavity of the vesicle, and the outer (superficial) layer stays thin and forms the lens epithelium. New cells are added from the outer cuboidal epithelium. These newly formed cells migrate toward the periphery. The cells at the peripheral ring divide further and transform into new lens fiber cells. The new lens fiber cells push the older fiber cells centrally. Addition of new lens fiber continues throughout lens growth, forming a very compact center, the nucleus of the lens. The organization and shape of the lens are adapted for transmission of light to specific parts of the retina. The mesenchymal cells surrounding the lens vesicle form the capsule of the lens. All components contributing to the lens become transparent (Figs. 17.1 and 17.2).

In animal studies, if the lens is rotated, the lens fiber cells now in the outer layer become shorter and transform into cuboidal epithelium and resume proliferation, whereas the epithelial cells transform into lens fiber cells. This shows the influence of the retina in lens formation. Chemical signals pass from the retina through the developing vitreous body to stimulate the lens vesicle layer closer to the retina to transform into lens fiber cells. The influence of the retina on the lens remains throughout life. The cells of the lens adopt chemical characteristics according to their location. DNA synthesis, for the most part, occurs at germinating cells at the peripheral ring. RNA synthesis is concentrated in the outer epithelial cells and in newly formed fibers. Protein synthesis is more pronounced in nuclear fiber cells.

D. The Cornea, Sclera and Choroid

The lens vesicle, as it detaches from the surface ectoderm, induces the adjacent ectoderm to become a multilayered transparent structure, the cornea. The mesenchyme surrounding the optic cup gives rise to an outer fibrous thick layer (the sclera) and an inner thin pigmented vascular layer (the choroid). This mesenchyme also migrates between the lens and cornea and splits into two layers. The new cavity formed between these layers is known as the anterior chamber. The outer wall of the anterior chamber continues with the sclera, and the inner layer continues with

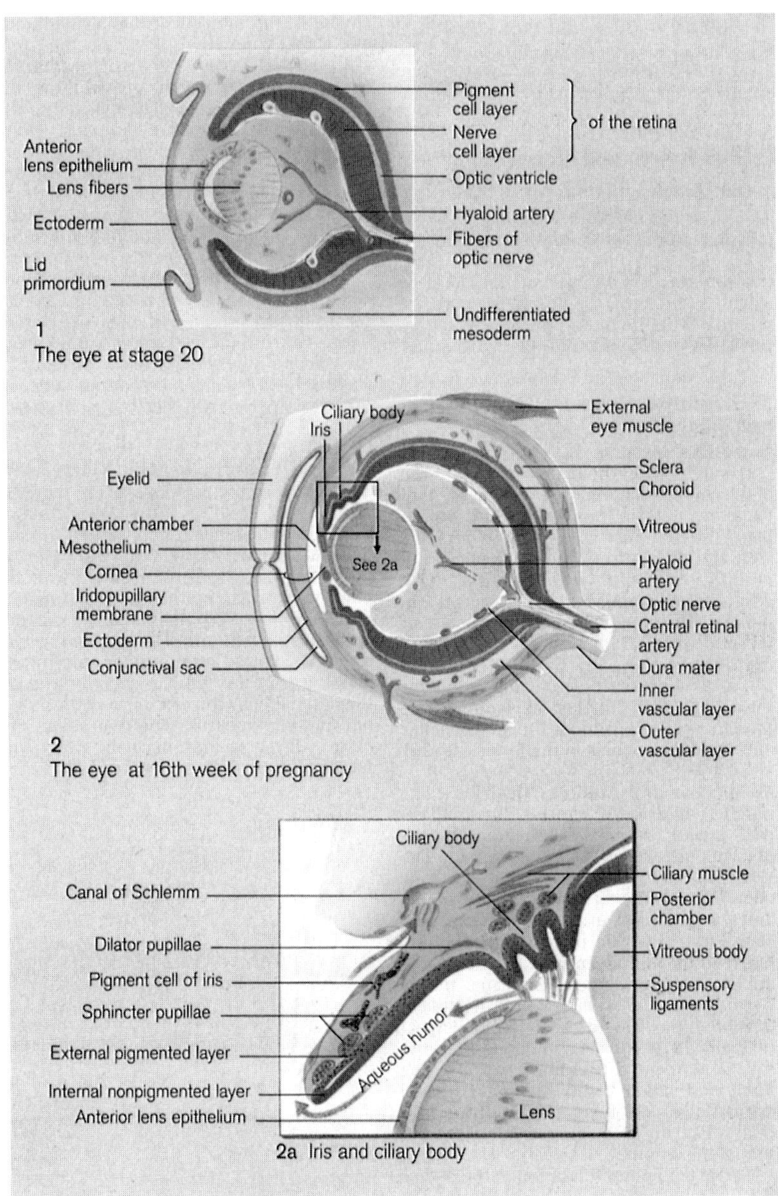

Figure 17.2
Accessory organs of the eye (see Color Plate 17.2). Reproduced with permission from Thieme Medical Publishers Inc., New York, 1995, *Color Atlas of Embryology*, Ulrich Drews, Chapter 5: Sense Organs.

the choroid. The cornea is therefore made up of an anterior (outer) epithelium that is derived from the surface ectoderm, the middle stroma, and the inner endothelium, both of which are derived from mesenchymal neural crest cells. The inner lining of the anterior chamber, which is closely associated with the lens, gives rise to the pupillary membrane. The pupillary membrane soon becomes vacuolated, begins to disappear, and vanishes at birth. Early in its development, the epithelial cells of the cornea secrete epithelial-derived collagen to form the primary stroma. The migrating neural crest cells then convert to fibroblast to give rise to permanent (secondary) stroma. Both outer (epithelial) and inner (endothelial) cells continue to add to the stromal component. Transparency is achieved by removal of most of the water from the stroma. This is mediated by thyroxin. The cornea then curves, conforming to the shape of the eye (Fig. 17.2).

E. The Iris and Ciliary Body

The rim of the optic cup forms the common primordium of the iris and the ciliary body. It undergoes rapid growth, which extends in front of the lens and can be divided into three parts: 1) the optic part, which will designate the future ora serrata (terminal part of the retina); 2) the ciliary part, which will differentiate into the ciliary body; and 3) the iridial part, which will differentiate into the iris. All of these parts contain pigmented and neural layers of retina. Analogous to the rest of the optic cup, these are surrounded by the mesenchyme, which contains neural crest cells. The neural crest cells induce the epithelium of the iridial part to give rise to circumferentially arranged sphincter pupillae and dilator pupillae muscles. Thus, these muscles are unique in their neuroectodermal origin. The stroma of the iris is formed from mesenchymal tissue. The amount and distribution of pigmentation in the iris give it color. The greater the density of pigment cells, the browner will be the color of the eye. The permanent color of the eye gradually develops 6 to 10 months postnatally. The epithelium of the ciliary part proliferates to form a secretory body. Secretion from this ciliary body fills the anterior chamber. Appropriate intraocular fluid pressure is essential for normal development of these parts. Ectomesenchymal cells associated with the ciliary body give rise to ciliary muscle (Fig. 17.2).

F. The Vitreous Body

As the optic cup is forming, mesenchyme fills the cavity of the optic cup and forms a fibrillar mesh and gelatinous substance. This primary vitreous body is supplied by the hyaloid artery. The portion of the hyaloid artery in the vitreous body later regresses, and the primary (vascular) vitreous body becomes the avascular secondary vitreous body.

G. The Eyelids

During the sixth week, two ectodermal folds appear, one on the cephalic and the other on the caudal edge of the eye. These folds of skin rapidly grow toward each other, cover the cornea, meet and fuse by the end of ninth week. The eyelids remain fused until the 28th week. During their fusion, eyelashes and small glands form along their margins. At the lateral edges of the eye, multiple epithelial buds grow inward from the surface ectoderm and differentiate into lacrimal glands. The lacrimal glands are not fully developed at birth. Newborns typically do not shed tears when they cry.

H. Congenital Malformation of the Eye

1. The following congenital malformations are related to the first four weeks of development. Malformations of the eyes during this period are related to maldevelopment of the neural tube. These include anophthalmia, synophthalmia (cyclopia), cystic eye, nonattachment of the retina and aphakia (absence of lens).

   a. Anophthalmia (absence of the eye)
   Anophthalmia may be primary when the optic vesicles (an evagination of the neural tube) fail to develop. 75% of primary anophthalmia cases are bilateral and are not associated with a family history. Anophthalmia may be secondary, so that the entire cephalic end of the neural tube is missing. In degenerative anophthalmia, the optic vesicles begin to form but soon degenerate. Although the eyelids, lacrimal glands, choroid, sclera and extraocular muscles are present, the structures normally induced by the optic cup are missing. Furthermore, the structures that are present (as mentioned above) may be hypoplastic and disorganized. Optic nerves, chiasma and tracts are absent.

   b. Synophthalmia — cyclopia
   This results from fusion of two globes and is associated with severe malformation of the forebrain and nose. The forebrain remains as a single sphere, and there is a proboscis nose. Fused eyes share a joint sclera and two retinas, which continue into fused optic nerves. A proboscis nose is situated above the eye. True cyclopia, where only one eye develops in the center of the forehead, is extremely rare. There is a high incidence of synophthalmia in trisomy 13.

   c. Congenital cystic eye
   Congenital cystic eye results from failure of invagination of the optic vesicle to form the optic cup. The optic vesicle becomes filled with fluid, forming a cyst, so that the cyst is lined by neuroepithelium and is covered by sclera.

   d. Congenital nonattachment of the retina
   In this case, the optic vesicle invaginates to form the optic cup, but the space between the neural retina and pigment retina becomes filled with fluid and keeps the two retinal layers separated from each other. The eyes remain small (microphthalmic) and are usually associated with other eye anomalies. It may be associated with hydrocephalus.

   e. Congenital aphakia (absence of lens)
   This results from failure of the optic vesicle to contact (or induce) the

surface ectoderm. No lens plate forms. Subsequently, no formation of anterior chambers and cornea is induced, and other severe ocular malformations result. In some cases, the lens forms but is reabsorbed due to intrauterine infection or trauma. This condition is known as secondary congenital aphakia.

2. Congenital malformations of the eye related to four to six weeks of development.

    a. Cryptophthalmos
    Cryptophthalmos is complete absence of eyelids. This results when skinfolds do not form. In most cases, there is also microphthalmia. It is usually associated with malformations of the nose, throat and urogenital system.

    b. Coloboma
    A coloboma is a notch, fissure or gap in any eye structure, occurring usually in the iris, retina or choroid. It is attributed to a disturbance in closure of the retinal (choroid) fissure. Closure of the fissure begins in its middle and extends anteriorly and posteriorly. Thus, a typical coloboma is situated inferomedially in the line of the fissure. This is why coloboma of the iris and retina are more common. Failure of complete closure of the retinal fissure is accompanied by optic cup abnormalities and disturbance in the mesoderm. Persistence of blood vessels at the rim of the optic cup and localized defect in the neuroectoderm are also suggested as possible causes. Pigmented epithelium and choroid are missing in the defect. The sclera usually forms properly, but in some cases it becomes thin at the defect and may lead to herniation of the retina, forming a large cyst. There is a high incidence of coloboma in trisomy 13; and it may also be associated with other systemic abnormalities.

3. Congenital eye malformation related to the fetal period. During the embryonic period, the central region of the eye is occupied by the hyaloid artery, the vessels surrounding the lens (tunica vasculosa lentis) and the pupillary membrane. Normally, all of these structures regress before birth as the permanent vitreous body is formed. In some cases, some portion of these structures persists.

    a. Persistent hyaloid artery
    Any portion of the vascular system may persist. The persistent hyaloid artery may continue with the vascular plexus around the lens (persistent tunica vasculosa lentis). More serious is the persistent hyperplastic primary vitreous body. This condition results from proliferation of the transient vasculature of the primary vitreous. It leads to formation of a mass behind the lens, and to loss of vision. The mass is attached laterally to a highly vascular ciliary body and leads to retinal detachment.

    b. Persistent pupillary membrane
    Remnants of pupillary membranes attached to the anterior surface of the iris may result from nonclosure of the pupillary vessels. This is regarded as retarded development of the eye.

c. Congenital glaucoma
During the period from eight months to term, the anterior chamber angle undergoes final differentiation to form the filtration apparatus. Any disturbance in this process leads to congenital glaucoma. Increased intraocular pressure results from obstruction to drainage of intraocular fluid. It may be present at birth or develop later. Congenital glaucoma may be inherited (primary group) or associated with nongenetic diseases (secondary group). The secondary group is usually associated with congenital rubella, abnormal development of the anterior chamber (anterior chamber cleavage syndrome), aniridia (absence of iris), and persistent hyperplastic primary vitreous body.

d. Microphthalmia — abnormally small eye
In microphthalmia, although the eye is small, it contains normal intraocular structures. It may be accompanied by a cyst or coloboma. It is generally associated with persistent hyperplastic primary vitreous, congenital infection, maternal drug use and chromosomal abnormalities.

4. Congenital anomalies of the eye in chromosomal aberration. Abnormalities of the eye have been observed in almost all children born with chromosomal aberrations. Some characteristics of specific chromosomal aberrations are described here.

a. Trisomy 13
Trisomy 13 causes insult during weeks 5 and 6 of gestation. Characteristic findings in most cases are colobomas with metaplastic tissue involving the ciliary body and iris. Optic nerve hypoplasia, cataracts, persistent hyperplastic primary vitreous, retinal dysplasia, corneal opacities and ectopic neural retinal tissue also occur frequently. Severe cases involving malformation of the brain may be associated with cyclopia.

b. Trisomy 18
In trisomy 18, the cornea, lens and retina are commonly affected. A hypoplastic ciliary process, hypoplasia of the iris, retinal fold, hypopigmentation of the posterior retinal pigmented epithelium, abnormalities of all corneal layers, posterior subcapsular cataracts, congenital glaucoma and persistent hyaloid artery have been described.

c. Trisomy 21
Abnormalities of eyes are less severe in trisomy 21. In some cases, of Down's syndrome, cataract, peripheral iris hypoplasia and Brushfield's spots have been observed. Retinal hypoplasia and optic nerve hypoplasia have also been reported. Facial features include mongoloid slant, almond-shaped palpebral fissure, epicanthus, external hypertelorism, convergent strabismus, narrowed interpupillary distance and myopia.

d. Cat-eye syndrome
Cat-eye syndrome is a vertically oriented iris coloboma. This may be due to an extra chromosome consisting of an isochromosome causing a tetrasomy for a portion of chromosome 22.

e. Triploidy
Generalized underdevelopment of the eye (microphthalmus) is a feature of triploidy. Colobomas and dislocated lens may also be present. In some cases, the eyes are normal.

f. Deletion 11p
Aniridia (iris hypoplasia) is the most consistent finding in this condition. Deletion involving the $11p_{13}$ band is well documented. Histologically, a small portion of iris tissue is always present. The condition may be associated with corneal opacification, glaucoma, cataract or optic nerve hypoplasia that may result in visual loss. Aniridia has also been reported in other (nonvisual) abnormalities.

g. Deletion 13q (13q14)
Retinoblastoma may be associated with deletion of the long arm of chromosome 13. The karyotype may appear normal, but recent genetic analysis has shown that deletion of band 14 on this chromosome (13) produces deficiency of a 4- to 7-kilobase segment of mRNA coded by the Rb gene, and the tumor is always associated with deletion 13 (band 14).

5. Congenital retinoblastoma
Congenital retinoblastoma is a malignant tumor of the eye that represents 1% of all malignancies in childhood. Its incidence is 1 in 18,000 to 1 in 23,000 live births. About 9% are recognizable at birth. It may be inherited as an autosomal dominant trait. Inherited tumors are usually bilateral. Intravitreous or intrachoroidal mass is identified grossly and is associated with retinal detachment. Necrosis and calcification are usually present. Invasion to the optic nerve, the choroid and the orbit indicates a poor prognosis. Nonhereditary tumors are unilateral and may be associated with other chromosomal abnormalities.

6. Congenital anomalies of the eye associated with prenatal infections.

a. Rubella
Cataract with absence of significant changes in the lens capsule or epithelium is the major anomaly caused by rubella infection during pregnancy. Other findings often seen are congenital glaucoma, iris atrophy with hypoplasia of the dilator pupillary muscle, corneal clouding, and salt-and-pepper fundus resulting from focal hypo- and hyperpigmentation of the posterior retinal pigmented epithelium. Virus may persist in the lens longer than in other sites. Abnormalities of the eye are more common following maternal infection during the first trimester.

b. Herpes simplex virus
Microphthalmos, cataract and optic nerve atrophy have been reported with intrauterine herpes simplex virus infection.

c. Cytomegalovirus (herpes virus)
Congenital cytomegalovirus produces necrosis in the choroid and retina. Other clinically described anomalies include cataract, coloboma and microphthalmos. Typical intranuclear inclusions are seen in necrotic retinal cells.

d. Toxoplasmosis
Intrauterine infection by *Toxoplasma gondii* produces a necrotizing inflammation of the retina and choroid, especially at the macular region. Free or encysted organisms are present around necrotic areas. Inflammation and necrosis may lead to detachment of the retina and loss of retinal neurons. Severe infection may produce total retinal detachment, cataract and necrosis of the iris and ciliary body.

e. Syphilis
Intrauterine infection by treponemas may produce chorioretinal atrophy and breaks in Bruch's membrane. This also produces a salt-and-pepper fundus. Involvement of the cornea leads to its cloudiness.

7. Teratogens causing congenital eye abnormalities.

a. Alcohol
Fetal alcohol syndrome has been associated with abnormalities of the face and eye. Abnormalities of the eye include short palpebral fissures, ptosis, strabismus and asymmetry. More serious abnormalities, including optic disk hypoplasia, absence of macula, cataract, persistent hyaloid artery and microcornea, have also been reported.

b. Hydantoin
Fetal hydantoin syndrome has been associated with microphthalmos, colobomas and persistent hyperplastic primary vitreous.

c. Lysergic acid diethylamide (LSD)
Severe eye abnormalities have been associated with maternal ingestion of LSD. These include microphthalmos, absence of an anterior chamber, cataracts and persistent hyperplastic primary vitreous.

d. Thalidomide
Maternal ingestion of thalidomide has been associated with anophthalmia, microphthalmos and colobomas.

8. Congenital cataracts. From 10 to 15% of blindness in children is caused by congenital cataracts. Cataracts may occur alone or in association with other eye abnormalities or other systemic diseases. They can be hereditary or sporadic. Cataracts are common in trisomies 13 and 21. Nonhereditary causes include rubella infection, birth trauma or hypoxia, and endocrine disturbances. Its associations with other ocular abnormalities include persistent hypoplastic primary vitreous, congenital retinal detachment and aniridia.

II. THE EAR

The ear develops from three different sources. 1) The inner (internal) ear develops from the epidermal otic placode that appears at the level of the future hindbrain. 2) The middle ear cavity forms from the first pharyngeal pouch, and the middle ear ossicles arise from the mesenchyme of the first and second pharyngeal arches. 3) The outer (external) ear develops from the first pharyngeal cleft and the six surrounding swellings, the auricular hillocks. The first pharyngeal membrane, which separates the first pharyngeal cleft from the first pharyngeal pouch, becomes the tympanic membrane. Its external surface is thereby ectodermal, the middle portion is mesodermal, and its internal surface is endodermal in origin.

A. The Inner (Internal) Ear

The inner ear is the first part to develop. At 3 weeks, the otic disks appear as a thickening of the surface ectoderm opposite the neural fold of the hindbrain. Studies of avians suggest that formation of the otic disk is influenced by the inductive action of the chondromesoderm (corresponding to the notochord in mammals). Migrating neural crest cells also influence inductive processes. Interactions between the otic disk and associated mesenchyme and between the otic disk and the neural fold of hindbrain are essential. The otic disk invaginates to form an otic pit, and its opening to the surface becomes progressively narrow. By this, the otic disk transforms into the otic vesicle, which shortly closes itself, migrates deep and loses its connection to the surface. The otic vesicle is a primordium for the membranous labyrinth of the inner ear and will give rise to semicircular ducts, cochlear ducts, endolymphatic ducts, the utricle and the saccule. The mesenchyme surrounding the otic vesicle condenses to form the otic capsule, which will give rise to the bony labyrinth, including the semicircular canals, cochlea, and vestibule (Fig. 17.3).

1. The Otic Vesicle (Otocyst)

Growth of the head causes the otic vesicle to shift to the level of the second pharyngeal arch. The otic vesicle elongates and becomes divided into the dorsal vestibular and ventral cochlear regions. A hollow diverticulum grows out from the dorsomedial aspect of the vestibular region. This diverticulum elongates and forms the endolymphatic duct. Later (seventh week) three other flattened hollow diverticula (resembling plates) also grow out of the vestibular region. The central portions of these plates collapse, and the epithelial layers from two sides fuse and degenerate, transforming plates into semicircular ducts. The part of the vestibular region where the semicircular ducts open is known as the utricle, and its remaining part is known as the saccule. One end of each semicircular duct dilates to form the ampulla (Fig. 17.3). At week 5, the cochlear part of the otic vesicle begins to elongate, curling into a spiral shape. It grows rapidly, spiraling to complete 2½ turns by the end of the embryonic period, thus forming the cochlear duct. During the early fetal period, the epithelium of the floor of the cochlear duct thickens to form the spiral organ (organ of corti) as the hair cells develop within it. These sensory (hair) cells undergo highly regulated differentiation, so that precise sound wave frequencies stimulate precise hair cells to transmit accurate signals to specific neurons in the brain. A tectorial membrane grows out, covering the hair cells. The neural crest cells migrate out of the otic vesicle to form, at least in part, the vestibular and cochlear (spiral) ganglia. Hair cells develop in the ampullae of the

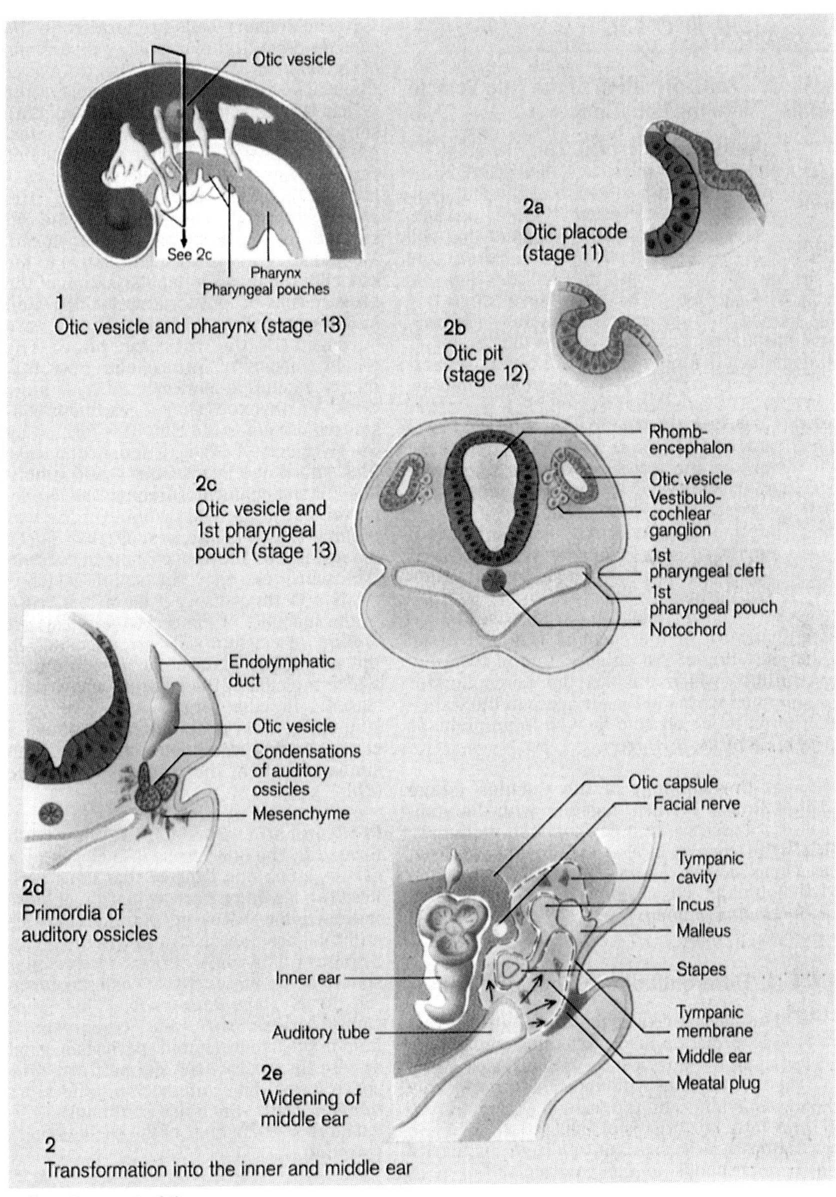

Figure 17.3
Derivation of the ear (see Color Plate 17.3). Reproduced with permission from Thieme Medical Publishers Inc., New York, 1995, *Color Atlas of Embryology*, Ulrich Drews, Chapter 5: Sense Organs.

semicircular ducts; the utricle and saccule and these hair cells are innervated by the vestibular ganglion. Fibers from the cochlear (spinal) and vestibular ganglia form the vestibulocochlear (VIII cranial) nerve. Contact of the nerve is not necessary for sensory cells differentiation; it is required for their maintenance.

2. The Otic Capsule

During the 9th week, the mesenchyme surrounding the membranous labyrinth, the otic capsule, undergoes chondrification. The membranous labyrinth induces chondrogenesis, and its presence is necessary to shape and control the cartilaginous model of the otic capsule. The otic capsule then joins the cartilaginous base of the skull. From weeks 12 to 16, the portion of the capsule adjacent to the membranous labyrinth undergoes vacuolization to form a cavity, the perilymphatic space, around the membranous labyrinth. The perilymphatic space becomes filled by perilymph. The precise shape of the perilymphatic space is created by removal of cartilage by chondroclasts (Fig. 17.4). From weeks 16 to 24, many centers of ossification appear in the remaining cartilaginous otic capsule to form the petrous portion of the temporal bone. Continued ossification later forms the mastoid process of the temporal bone.

During the third trimester, vibratory acoustic stimulation applied to the maternal abdominal wall results in a startle response, which can be seen in ultrasonic images of the fetal face. This test indicates normal development of the brain stem. It is also reported that the fetus can respond to sound by weeks 26–28, and it enables the fetus to influence and orientate its own position in the uterus.

B. The Middle Ear

The middle ear consists mainly of a cavity in the temporal bone that contains auditory ossicles. Its development is closely associated with development of the first pharyngeal pouch and first and second pharyngeal arches (Chapter 11).

1. The Middle Ear Cavity

The first pharyngeal pouch rapidly expands to form the tubotympanic recess. Its expanded part differentiates into the tympanic cavity, and its connection to the pharynx becomes the auditory or pharyngotympanic tube. The tympanic cavity also communicates with mastoid air cells through the mastoid antrum.

2. The Middle Ear Ossicles

The dorsal parts of the cartilage of the first and second pharyngeal arches are situated cephalic and caudal to the tubotympanic recess. It is likely that the head of the malleus and the body and short crus of the incus arise from pharyngeal arch 1 and the handle of the malleus. The long crus of the incus and the stapes arise from pharyngeal arch 2 (Chapter 11). It is suggested that ossicles may be derived from the neural crest cells. The cartilaginous models of ossicles are surrounded by loose mesenchyme that is removed by apoptosis (programmed cell death). The expanding tympanic cavity then surrounds and lines the ossicles. Thus, the endodermal epithelium lines the entire middle ear cavity, the ossicles and the internal surface of the tympanic membrane. Although during the ninth month the malleus becomes attached to the tympanic membrane, and the foot plate of the stapes is attached to the oval window, free movement of the ossicle is established 2 months postnatally. The ossicles begin to ossify during the second trimester.

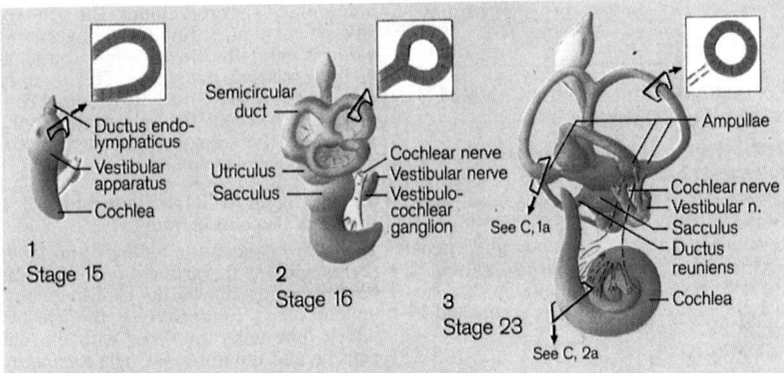

Transformation of the otic vesicle into the labyrinth

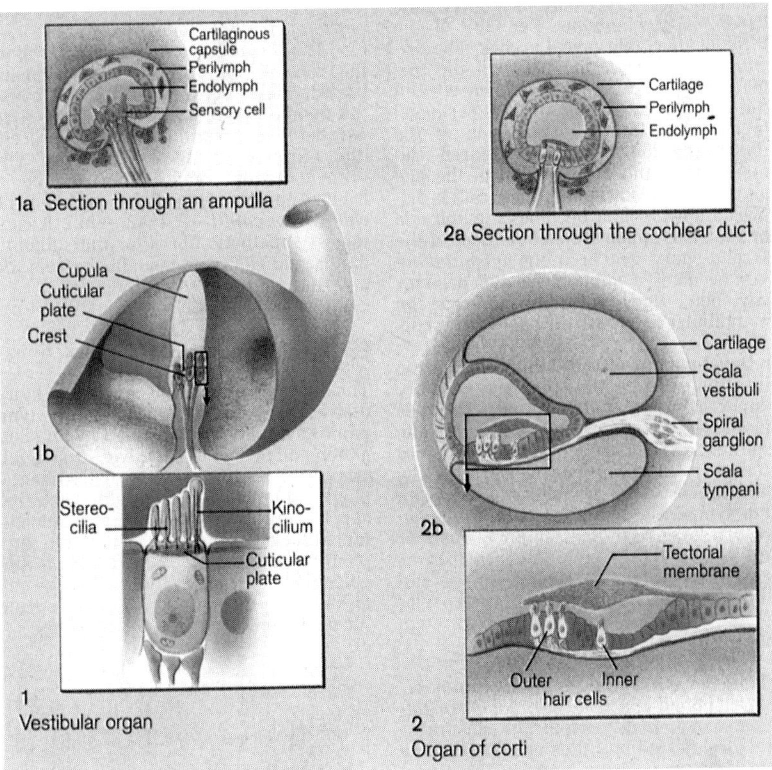

Differentiation of sensory cells

Figure 17.4
Inner ear (see Color Plate 17.4). Reproduced with permission from Thieme Medical Publishers Inc., New York, 1995, *Color Atlas of Embryology*, Ulrich Drews, Chapter 5: Sense Organs.

3. The Middle Ear Muscles

The tensor tympani, which arises from the first pharyngeal arch, and the stapedius, which arises from the second pharyngeal arch, become attached to the malleus and stapes, respectively.

4. The Tympanic Membrane

The expanding tympanic cavity is separated from the first pharyngeal cleft by a thin core of mesenchyme. This mesenchyme gives rise to the middle fibrous layer of the tympanic membrane. The endodermal epithelium of the tubotympanic recess lines the inner surface, and the ectodermal epithelium of the first pharyngeal cleft lines the outer surface of the tympanic membrane. Hence, the tympanic membrane is formed by all three germ layers.

C. The External (Outer) Ear

The external ear consists of the auricle and the external auditory meatus, and is derived from the first and second pharyngeal arches and the first pharyngeal cleft.

1. The Auricle

The mesenchyme from the first and second pharyngeal arches aggregates into a series of six elevations known as the auricular hillocks around the first pharyngeal cleft. These hillocks enlarge, coalesce and lose their identity. It is suggested that hillocks are only transitory structures and may not contribute to development of the auricle. Early, the auricles are situated ventrolaterally in the cranial region and later shift into their normal dorsolateral position. Because of their complex origin and close association with pharyngeal arches, the shape, size and location of the auricle are important indicators of abnormal development of the head and neck region. Abnormal location in relation to the external auditory meatus, so called "low-set" ears, are generally a subjective clinical impression. Abnormal auricles are factors of many abnormal syndromes.

2. The External Auditory Meatus

The external auditory meatus (the auditory canal) is formed by inward expansion of the first pharyngeal cleft. Its surface appears as a hole surrounded by auricular hillocks. The ectodermal epithelium proliferates, forming a meatal plug that completely fills the medial half of the auditory meatus. During the third trimester, recanalization in the plug occurs to form a new channel, the definitive auditory canal, which extends to the external surface of the tympanic membrane. Hence, the external surface of the tympanic membrane is lined by the ectoderm.

D. The Ear in Infants

In infants, the external auditory meatus is short and straight. The tympanic membrane faces caudally. The mastoid process is either not formed or is very short, leaving the facial nerve superficial and unprotected. The auditory tube is more horizontal, allowing pharyngeal infection to easily reach the tympanic cavity.

E. Congenital Anomalies

The auditory apparatus is composed of three separate units, derived from three separate origins. It is not surprising that malformation of one unit need not be accompanied by malformation in others. On the basis of studies in avian embryology, it has been suggested that many specialized mesenchymal structures of the pharyngeal arches, including the middle ear ossicles, are derived from neural crest cells. The neural crest elements are also important in aortopulmonary septation, and there is a close anatomical relation between these two neural crest elements. This may explain the association of some ear anomalies to cardiac lesions.

1. External (Outer) Ear Anomalies

a. Low-set ears

Low placement of the ear reflects a delay in development of the face and auricles. It may be present in infants without any other malformation. However, it is a common feature of many malformation syndromes. It generally is defined as attachment of a helix below the plane of the line joining the lateral angle of the eye and the external occipital protuberance or inion. In cases where the mandible is poorly formed or absent, upward migration of auricles may be retarded, and low-set ears often accompany lower jaw anomalies. When the mandible is absent, the two auricles may come together and fuse (synotia) below a rudimentary mouth.

b. Malformations of the auricle

Malformation of the auricles should draw attention to other possible problems. A detailed examination of all the auditory components should be performed.

For instance,

i. Thickening of the ear lobes may be associated with a curved, long process of the incus and absence of the head of the stapes.

ii. Cup ears, which are smaller than usual and deficient in cartilage, are associated with absence of incus and stapes in Mennonite kinders.

iii. Absence of superior crus in the auricle is associated with congenital fixation of the ossicles.

iv. Auricular malformations show great variability and have been reported in craniosynostosis, Klippel–Feil syndrome, cardiac and gonadal anomalies.

c. Anotia — complete absence of auricle

This may be associated with major malformation of the face and head. It occurs when the hillocks fail to form. The skin passes smoothly where the external auditory meatus should have formed. Middle ear ossicles are also abnormal, and there may be facial paralysis on the affected side.

d. Microtia — small auricle

This is usually associated with cup ear or overfolded auricle. In many cases, the external auditory meatus is absent. It has been reported in embryos affected by thalidomide and retinoic acid. Patency and adequacy of the external auditory meatus are important in hearing.

e. Auricular appendages

Auricular appendages and tags of skin and cartilage are most common. They are usually unilateral and vary in size. Unless they are associated with other anomalies, their importance is purely cosmetic.

f. Congenital auricular fistulas and sinuses

These are narrow tracts that end blindly, usually located anterior to the external auditory meatus. These lesions are harmless unless they become infected or form a cyst. Some fistulas may be associated with other auricular malformations.

2. Middle Ear Anomalies

Anomalies of the middle ear are generally associated with abnormalities of facial nerve and auditory ossicles.

a. Absence of incus and stapes have been described in diastrophic dwarfism. Absence of all ossicles was reported in cases of achondrogenesis, and fractured ossicles may be found in cases of osteogenic imperfecta.

b. Fixation of stapes, which is often accompanied by absence of the oval window, may be seen in congenital otosclerosis. This is primarily a disorder of the wall of the bony labyrinth (otic capsule). The foot of the stapes becomes fixed by fibrous tissue and results in interference with the conducting system.

c. Persistent stapedial artery
The stapedial artery passes through the foramen of the stapes in the fetus and disappears before birth. When persistent, it may interfere with conduction.

d. Delayed aeration of the middle ear cavity
In oligohydramnios, especially when caused by renal agenesis, mesenchymal resorption and aeration of the middle ear cavity are retarded. This anomaly may be variable.

3. Inner Ear Anomalies

Anomalies of the inner ear are generally combined with other malformations, are frequently hereditary, and involve either the sound conducting system or neurosensory system. These range from complete lack of development of the labyrinth to the presence of a reduced labyrinth.

a. Labyrinthine aplasia or hypoplasia

i. Michel's anomaly represents complete absence of the labyrinth.

ii. Mondini's anomaly shows presence of a single curved tube with reduced spirals and immaturity of the vestibule, affecting both membranous and bony components.

iii. The Bing–Seibenmann anomaly represents a poorly developed membranous labyrinth within an intact bony labyrinth.

iv. Scheibe's anomaly shows a normal vestibular apparatus, but the cochlear duct and saccule are underdeveloped.

b. Autosomal trisomies

i. In trisomy 13, underdevelopment of the cochlear duct and saccule have been described.

ii. In trisomy 18, an absence of spiral ganglia, anomalies of the tympanic cavity and malshaped ossicles have been described.

c. Prenatal infections

i. Rubella infection has been well documented as causing cochlear duct and saccular damage (Scheibe-type anomaly).

ii. Cytomegalovirus infection has been implicated in causing damage to the vasculature of the cochlear ducts. Inclusion-bearing cells have been demonstrated in the saccule and utricle.

iii. Toxoplasmosis has been known to cause otitis media and calcification of the spiral ligament.

iv. Hemorrhagic and necrotic lesions of the middle ear have been shown in herpes simplex 2 viremia.

TABLE 17.1. Summary of Development of the Eye and Ear

Age (days)	Embryonic structures
24	Optic and auditory placodes
28	Lens primordium; otic vesicle
35	Lens vesicle
42	Hyaloid artery, retinal pigment; fusion of auricular hillocks
49	Solid lens, eyelids; semicircular canals and cochlear duct
70	Fused eyelids
196	Open eyelids

REVIEW QUESTIONS — Chapter 17

Select the most appropriate answer.

1. The surface ectoderm during development of the eye gives rise to the:
   a. lens
   b. corneal epithelium
   c. conjunctival lining
   d. lacrimal gland
   e. all of the above

2. All of the following take part in the formation of the tympanic membrane except the:
   a. epithelial lining of the otic vesicle
   b. ectodermal lining of the first pharyngeal cleft
   c. the mesenchyme
   d. the endodermal lining of the first pharyngeal pouch

3. At 22 days, the thickened plates of surface ectoderm on each side of the developing hindbrain develop into the:
   a. external auditory meatus
   b. semicircular canals
   c. tympanic cavity
   d. membranous labyrinth

4. The pigmented epithelium of retina is derived from the:
   a. surface ectoderm
   b. neuroectoderm
   c. neural crest cells
   d. head and neck mesenchyme
   e. all of the above

5. The mesoderm associated with the optic vesicle contains neural crest cells that give rise to the:
   a. lens
   b. sclera
   c. retina
   d. optic stalk
   e. all of the above

6. Which of the following structures is not derived from the mesoderm?
   a. choroid
   b. muscles for eye movement
   c. dilators and constrictors of pupil
   d. the sclera
   e. the central artery of the retina

7. The origin of retina is from the:
   a. telencephalon
   b. diencephalon
   c. mesencephalon
   d. metencephalon
   e. myelencephalon

8. The muscles of the eye that develop from the neuroepithelium are:
   a. the extraocular muscles
   b. the ciliary muscles
   c. the dilator and sphincter papillae muscles
   d. all of the above
   e. none of the above

9. The cochlea of the inner ear is derived from the:
   a. ectoderm
   b. mesoderm
   c. endoderm
   d. head and neck mesenchyme
   e. all of the above

10. All of the following structures are derived from two pharyngeal arches except the:
    a. malleus
    b. incus
    c. stapes
    d. hyoid
    e. digastric

11. The part of the ear derived from the endoderm is the:
    a. external auditory meatus
    b. tympanic cavity
    c. tympanic membrane
    d. middle ear ossicles
    e. inner ear

12. The ciliary body of the eye develops from the:
    a. pigmented layer of the optic cup
    b. neural layer of the optic cup
    c. mesenchyme located at the edge of the optic cup
    d. mesenchyme located posterior to the lens placode
    e. none of the above

13. The iridopapillary membrane:
    a. is a continuation of the pigmented layer of the retina
    b. is a continuation of the optic cup
    c. is formed from the preotic myotomes
    d. should disappear completely before birth

14. The aqueous chamber of the eye:
    a. appears as a cavity in the mesenchyme associated with the developing iris
    b. is formed in the mesenchyme enclosed within the optic cup
    c. is formed by division of the mesenchyme by the lens placode
    d. appears in the pigmented layer of the ciliary body
    e. is a forward prolongation of the optic cup

15. The ganglionic cells of the retina:
    a. are derived from the diencephalon
    b. migrate to the optic cup from the neural crest
    c. originate from the lens placode
    d. are the microglia
    e. migrate through the blood vessels

16. Which of the following would cause congenital deafness?
    a. maternal rubella infection
    b. maternal goiter
    c. erythroblastosis fetalis
    d. maternal diabetes
    e. any of the above

17. Failure of complete closure of the retinal (choroid) fissure may result in
    a. ophthalmia (absence of eye)
    b. cyclopia
    c. coloboma
    d. congenital cystic eye
    e. any of the above

18. Malformation of auricles may be associated with:
    a. Down's syndrome
    b. craniosynostosis
    c. cardiac anomalies
    d. absence of one or more middle ear ossicles
    e. any of the above

19. The middle ear anomalies are generally associated with:
    a. facial nerve abnormalities
    b. first pharyngeal arch anomalies
    c. cretinism
    d. low-set ear
    e. any of the above

20. Rubella infection during pregnancy may cause:
    a. congenital cataracts
    b. congenital glaucoma
    c. cochlear duct damage
    d. corneal clouding
    e. any of the above

ANSWERS TO REVIEW QUESTIONS

1. e	2. a	3. d	4. b
5. b	6. c	7. b	8. c
9. b	10. c	11. b	12. c
13. d	14. a	15. a	16. e
17. c	18. e	19. a	20. e

CHAPTER

18

NERVOUS SYSTEM

During the third week of gestation, the notochordal process induces formation of the neural plate, which appears as a thickening in the epiblast on each side of the midline. This primary neural induction is believed to be caused by the action of two active diffusible signals (probably proteins). The initial signal induces formation of the rostral structures, and then induction proceeds caudally. However, these signals are not sufficient for all aspects of neural induction. Underlying tissue derived from the primitive streak is also required for neural tube formation. Nerve growth factor, epidermal growth factor, active inhibitors, noggins, follistatin and dorsalin have been implemented in the induction and maintenance of neural tissue. Many other proteins encoded by homologues of notch (xotch), Wnt and hedgehog genes have also been suggested to influence the early development of the nervous system.

Aside from induction, other major processes are needed, such as proliferation to build up the critical number of cells; pattern formation in response to genetic and environmental cues in forming subdivisions; cellular migration; cellular differentiation and signal transduction and adhesions; apoptosis and pruning of nonessential elements to finalize the integrated patterns of central nervous system function. Although the detail for each of these processes is beyond the scope of this book, some are explained here briefly.

I. NEURAL INDUCTION

In induction, one group of cells alters the behavior of another group of cells. As of 1995, it was poorly understood and its molecular mechanism not known. The majority of studies are done in lower animals, and then their findings are correlated to humans. During formation of the third layer (mesoderm), positioning of mesodermal cells between the ectoderm and endoderm initiates the inductive events that continue in later development. Two modes of signaling have been suggested. The notochord may vertically signal the overlying ectoderm, or mesodermal organizer signals may travel horizontally within the layer of the ectoderm. In amphibians, both types of signals have been suggested. A single inducer may act in both directions, but the underlying notochord is required for neural tube formation. Furthermore, the ectoderm (in that region) is also prepatterned for this induction. This prepattern can be demonstrated by the expression of an epidermal marker, epiI, in the nonneural ectoderm in the blastocyst. EpiI is not expressed in ectodermal cells of the future neural plate. This explains the differential response of the ectoderm to neural induction. This results in mediolateral patterning that is likely to rely on signals arising from the midline mesoderm (future notochord). This initial mediolateral pattern is later converted to a ventrodorsal pattern during neurulation by a secondary inductive interaction. Some neural-inducing factors are:

A. Follistatin
   This was originally isolated as an inhibitor of pituitary FSH. It is a natural antagonist of activin function. It is a specific inhibitor of activin function that is located in tissue that has neural-inducing activity during normal development. Follistatin is expressed in the notochord and later in the anterior nervous system.

B. Noggin
   This is implicated in neural induction. It may be the initial anterior inducer. It dorsalizes mesoderm and is expressed throughout the notochord in the later stages. Like follistatin, it is also expressed in anterior neural tissue, but noggin does not inhibit activin function. Furthermore, it neither stimulates nor is stimulated by follistatin. This suggests that these factors may represent redundant or independent inducers of anterior neural structures.

C. The notch (xotch) gene (encodes large transmembrane protein) is implemented in maintenance of an uncommitted state in precursors of neural and epidermal tissue. It plays a general role in cell fate decisions. Xotch acts specifically as a regulating factor in proportioning the uncommitted cells capable of responding to neural induction.

D. The wnt-1 gene-coded protein may function in regulating proliferation of specific regions of the developing neural tube.

E. Hedgehog (vhh-1), now referred to as sonic hedgehog (shh-1), is a segment-polarity gene. It is expressed in the notochord at the stage where it induces ventral cell types into the neural plate. The sonic hedgehog has been isolated also in the primitive node, neural tube floor plate, and posterior limb bud. Induction by it is a contact-dependent process.

F. Dorsalin-1
   A member of the TGF_{β} family, dorsalin-1 stimulates differentiation of dorsal tissue and inhibits differentiation of ventrolateral (motor neuron) tissue. In-

itially, the entire neural tube is competent to express dorsalin-1, but its expression is restricted to the dorsal third of the neural tube by diffusible notochord signals. Dorsalin-1 protein stimulates neural crest differentiation, and diffusion of this factor ventrally overrides ventral signals, thus setting the upper boundary for motor neuron differentiation. This demonstrates the involvement of opposing dorsal and ventral signals that specify cell fate along the dorsal–ventral axis.

Neural induction apparently involves both inducents such as follistatin and noggin, and modifiers of neural patterning like xotch, dorsalin-1 and sonic hedgehog. Nuclear factors are also implemented. It is shown that diffusible signals alone are not sufficient for all aspects of neural induction and that underlying notochord is required for neural tube formation. Further investigation indicates that floor plate induction is a contact-dependent process.

II. APOPTOSIS AND PROGRAMMED CELL DEATH

Homeostasis during embryogenesis, normal tissue turnover and metamorphosis are controlled by proliferation, differentiation and programmed cell death. In general, programmed cell death is preceded by apoptosis. Apoptosis is a process that includes nuclear condensation and segmentation, segmentation and fragmentation of the cytoplasm, and sometimes extensive segmentation of chromosomal DNA into nucleosomal units. In contrast, necrotic cell death is a pathologic form of cell death resulting from acute cellular injury that is characterized by cell swelling and lysis. During necrosis, cells lose their cell membrane integrity, resulting in leakage of cytoplasmic contents, leading to an inflammatory response. Apoptosis is characterized by controlled autodigestion of the cell. Cells appear to initiate their own suicide through activation of endogenous protease. The dying cell maintains its membrane integrity; however, alterations in the membrane signal the macrophage to engulf them to complete this degradation process. On the other hand, the cell that is not phagocytosed immediately breaks down into smaller membrane-bound apoptotic bodies. Apoptosis results in elimination of the dying cell without induction of an inflammatory response. The basic mechanism to carry out apoptosis is present in essentially all mammalian cells, but activation of cell death is regulated by many signals that originate intercellularly or extracellularly. Cell survival appears to depend on constant supply of survival signals provided by neighboring cells and the extracellular matrix. When cells are no longer needed or have become severely damaged, an intrinsic cell suicide program is activated to induce apoptosis. Diverse signals, like lineage information, extracellular survival factors and hormones, either suppress or promote activation of the cell death program, and the same signal may have an opposing effect on different cell types. During apoptosis an active gene-directed mechanism has been suggested.

Programmed cell death could be divided in four stages, 1) the decision of whether a particular cell will die or survive, 2) induction of apoptosis and the death of the cell, 3) elimination of the dead cell by phagocytosis or fragmentation of the dead cell, and 4) degradation of the phagocytosed cell. Studies in nematodes have shown that, once the decision has been made, the activity of ced_3 and ced_4 (cell death defective) genes is required for cell death to occur. However, the ced_9 gene protects the cell that should survive. Ced_9 encodes a protein that is homologous to the bcl-2 family of cell death regulators in mammalian cells. Human bcl-2 can inhibit cell death in nematodes. Furthermore, the expression of another antiapoptotic protein, P_{35}, from bacula virus can inhibit cell death in insects, worms and mammalian neurons. On the other hand, major aspects of apoptosis do not require transcription of new genes. They are always present in most mammalian cells, and their potentially lethal activities must be suppressed for the cells to survive.

The extracellular factors produced by neighboring cells are also important for the cell to survive. This social interaction appears to regulate the appropriate number of different cell types. Although the mechanism and cause of apoptosis are still unknown, the prominent changes seen in the nucleus at the early stages of apoptosis suggest endonucleolytic cleavage of DNA. Another suggested cause is accumulation of reactive oxygen species. Bcl-2 has been shown to protect against peroxidase-induced cell death. Some antioxidants also protect against death in response to cytokinase deprivation. Due to the overall similarities between apoptosis and the cell cycle, it has been suggested that apoptosis and mitosis may mechanically be related or even coupled. Some even suggest that apoptosis may be an aberrant mitosis, because genes like P_{53}, c-myc, Rb-1, E1A, cyclin DI, C-FOS and P_{34} (ced-2) kinase have been involved in both mitosis and apoptosis during regulation of cell proliferation. If any of the genes required for cell death (especially ced_3–ced_4) are inactivated, that cell that normally dies during development will survive and unwanted growth will occur. The family of cystine protease such as interleukin IB-converting enzyme (ICE), nedd-2/ich-1 and cpp32 are also implemented in apoptosis in mammalian cells. Multiple signaling pathways for activation of apoptosis converge on the "reaper" gene. Expression of reaper gene is sufficient to induce apoptosis, and it therefore is considered as a universal activator of apoptosis. It appears that the reaper is not part of the actual cell death program. It encodes a regulatory molecule (a 65-amino-acid polypeptide) rather than an apoptotic effector. It may inhibit negative regulators such as ced_9/bcl-2. It serves as an integrator that links different signaling pathways with the basic cell death program. There may be more discoveries in the near future toward further understanding of programmed cell death. Finally, it should be noted that inhibition of apoptosis by protective antiapoptotic proteins is of fundamental importance for cell survival, and it appears that most cells need a constant supply of these to survive.

III. SIGNAL TRANSDUCTION

Through signal transduction, cells are able to communicate with each other and respond to their external environment. As discussed above, this is critical for their survival. During transduction, an external signal must in some way pass through the lipid bilayer of the cell membrane. In most cases, the external signal does not directly enter the cell, but rather conveys its information to specific proteins located on the surface (external face) of the cell membrane. These proteins then transmit this information to other proteins located on the internal (intracellular) face of the cell membrane. The signal from these proteins is passed on to the additional pathways that involve post- or co-transitional modification (lipidation) by specific lipids. Thus, the lipids attached to the signaling molecule play crucial roles in their function. These lipid-modified proteins are classified on the basis of the attached lipids. Each type of lipid involved in protein modification adds a unique functional property to its protein host.

A. G Proteins

The signaling events initiated by extracellular ligands such as hormones, growth factors and neurotransmitters result in activation of specific G (trimeric GTP binding) proteins associated with the intracellular face of the cell membrane. This activation is through direct interaction with the G protein-coupled receptors (GPCRs) located on the surface of cell membranes. The G proteins are comprised of three subunit polypeptides (α, β and γ) and can have multiple lipids attached.

B. Protein Tyrosine Kinase

Protein tyrosine kinases are involved in many aspects of cell growth and development. These include receptors for growth factors as well as nonreceptor tyrosine kinases. Cytokines — including interleukins, interferons and colony-stimulating factor — are soluble mediators of cell-to-cell communication. Most of the cytokine receptors do not possess a protein tyrosine kinase domain, but receptor stimulation involves rapid tyrosine phosphorylation of intracellular proteins. It is now known that cytokine receptors are capable of recruiting and/or activating a variety of nonreceptor protein tyrosine kinases.

C. Small G Proteins

The members of the group of small G (monomeric) proteins are involved in membrane traffic in cells. This process includes secretion of neurotransmitters and other messengers in response to cell stimulation. Specific forms of these proteins are associated with specific membrane compartments and influence the membrane vesicles between distinct compartments. Some other small G proteins (A and f) function in vesicle budding from Golgi cisternae.

D. Glycosylphosphotidylinositol

Glycosylphosphotidylinositol (GPI) anchored proteins are involved in cell adhesion, nutrient uptake, catalysis and membrane signaling events. Notch proteins are implicated in a general mechanism for local cell signaling that includes interaction between similar or different cell types. It is suggested that notch signaling may involve direct signal transduction from the cell surface to the nucleus. Notch signaling plays an important role in differentiation of uncommitted cells. This signaling pathway modulates the ability of the cell to respond to other specific developmental signals. Notch proteins may also indirectly modulate a number of other signaling pathways.

E. Integrins

Integrins are the major cell surface receptors. They mediate attachment to the extracellular matrix and mediate cell-to-cell adhesive interaction. Adhesive interaction plays important roles in directing proliferation, differentiation and migration of cells. This interaction also regulates intracellular signal transduction pathways. These integrin-mediated adhesive interactions are involved in regulation of embryonic development, programmed cell death, hemostasis, tumor cell growth, metastasis, bone resorption, clot retraction and response of cells to mechanical stress. Integrins link extracellular matrix proteins on the surface of the cell membrane to cytoskeleton proteins and actin filaments on the intracellular face. Integrins also regulate changes in gene expression that are essential for developmental and proliferative response. Protein phosphorylation is one of the early responses to integrin stimulation. The interaction between growth factors and integrins signaling pathways can regulate cell proliferation, adhesion and migration. Several cytokine and immune response receptors stimulate integrin-dependent signaling pathways. It appears that numerous receptors systems can synergize with integrins to influence cell proliferation and migration.

F. Calcium

Neuronal activity causes increased concentration of calcium in the cytosol. Calcium binds to calmodulin and stimulates the activity of calcium–calmodulin kinases and other enzymes. Calcium thus functions as a second messenger to mediate a wide range of cellular responses. Depending on the route of entry into a neuron, calcium affects activity-dependent cell survival, modulation of synaptic strength and calcium-mediated cell death. Influx through the voltage-sensitive calcium channels leads to increased survival of the embryonic neurons. Calcium also appears to be a central mediator of plasticity in the nervous system.

IV. NORMAL DEVELOPMENT OF THE NERVOUS SYSTEM

A. Neurulation: Neural Plate and Neural Tube Formation

In humans, by about 16 days of gestation, a thickening appears in the epiblast along the midsagittal axis rostral to the primitive streak. This transformation of epiblast (ectoderm) into the nervous tissue is considered in five major events (Fig. 18.1).

1. Primary Neural Induction

The substances (listed under neural induction) secreted by the underlying prechordal plate diffuse to overlying prospective neural ectoderm. It has been known since the 1920s that contact between these two layers is essential. In response to primary neural induction, the cells in that region of the epiblast increase in height

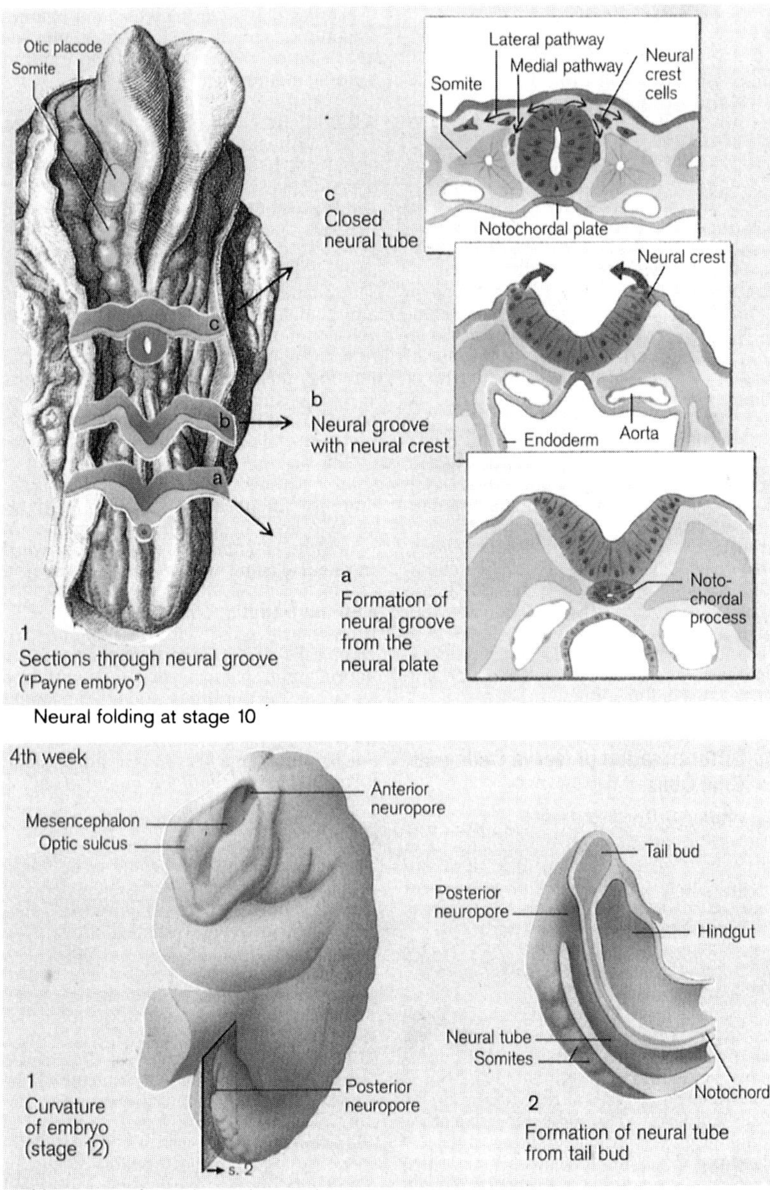

Figure 18.1
Neural tube (see Color Plate 18.1). Reproduced with permission from Thieme Medical Publishers Inc., New York, 1995, *Color Atlas of Embryology*, Ulrich Drews, Chapter 4: Nervous System.

and differentiate to become the neural plate. These transformed cells also are restricted in expression of N-cell adhesion molecules, whereas other ectodermal cells express both N-CAM and L-CAM. The neural plate appears first at the cranial end of the embryo and then differentiates in a rostrocaudal sequence.

2. Region-Specific Changes

Activation of specific genes causes cells to change shape. The taller cells are arranged on both sides of the middle smaller cells. The neuroepithelium appears as pseudostratified columnar. The neural plate now exhibits a median sulcus. The changes in the cells are believed to be produced by the cytoskeleton under further induction by the notochord.

3. Primary Neurulation

The median sulcus deepens, forming the neural groove. It results in elevation of the sides of the neural plate to form neural folds. These changes are the result of both intrinsic and extrinsic factors. The neuroepithelial cells react differentially to the inducing substances. The distance of cells from the notochord and the concentration of inducing substance reaching the neuroepithelial cells are determining factors. Expansion of the extracellular matrix of the mesenchyme may also influence the elevation of neural folds. Furthermore, the neural plate becomes narrower and longer. Thus, changes in cell shapes, differential growth and differential cell death are important factors in morphogenesis. A bend soon appears in the still unfused neural folds.

4. Closure of the Neural Groove

Closure of the neural groove begins by about 21 days. The initial mediolateral induction pattern is converted to a ventrodorsal pattern, which transforms the open neural plate into the closed neural tube. The underlying notochord also regulates this secondary inductive interaction. The notochord induces the floor plate and ventrolateral motor neuron formation in the adjacent neural tissue. The interaction between the notochord and floor plate is contact-dependent. On the other hand, dorsalin-1 stimulates differentiation of dorsal tissue. Closure of the neural tube begins almost midway, demarcating the junction between the future brain and spinal cord. The lateral lips of the neural plates come together, and the surface (somatic) ectodermal cells from two sides fuse and the neural tube becomes detached from the somatic ectoderm (Fig. 18.2).

5. Origin of Neural Crest Cells

Before fusion of the two sides of neuroectodermal cells, the cells located at the lateral lip (now located dorsally) detach themselves and migrate away from the neural tube. These special neural ectodermal cells are termed neural crest cells. They break through the basal lamina of the neuroepithelium, lose cell–cell adhesion, change their morphology and migrate widely along the cell-free pathways in the extracellular matrix to specific regions of the body. The neural crest develops in a rostrocaudal sequence, and its formation continues for at least four weeks in the cephalic region, and much longer in the caudal region. The derivatives of neural crest cells include the head and neck mesenchyme, melanoblasts, sympathetic and parasympathetic ganglia cells, sensory neurons, chromaffin cells, pial cells and arachnoid cells. Other migration of neural crest cells is discussed in other chapters.

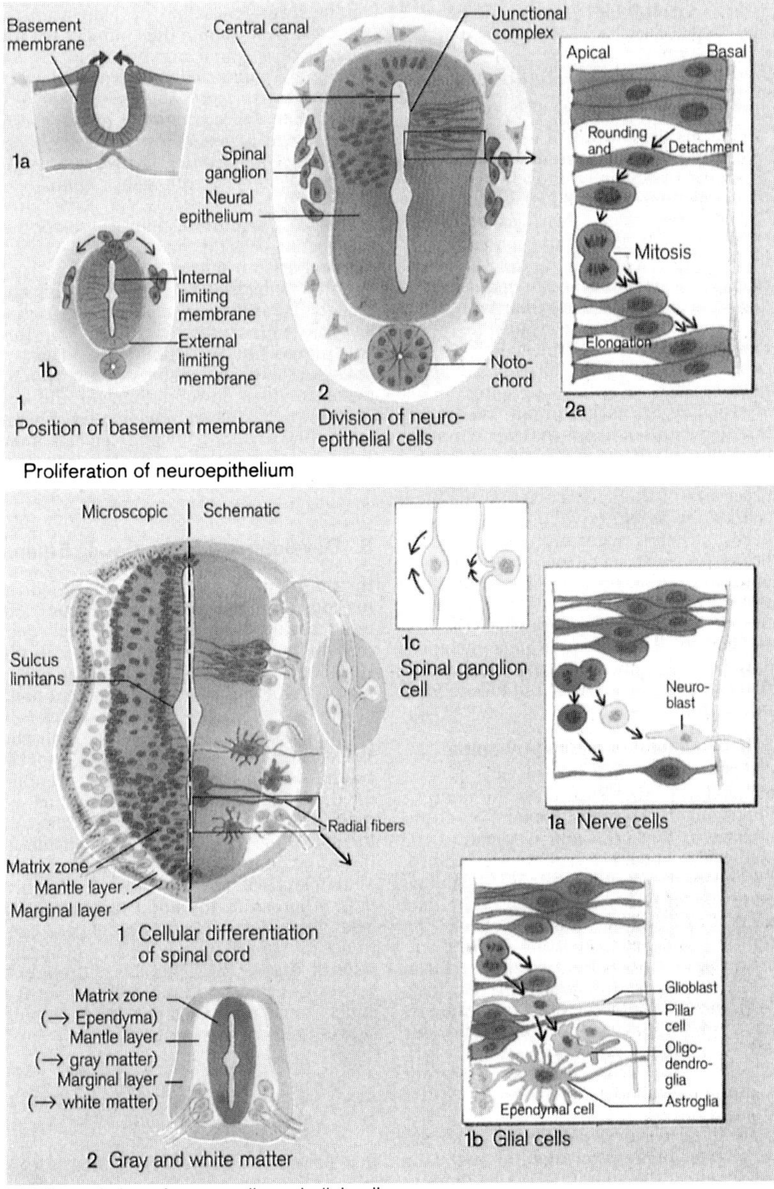

Figure 18.2
Spinal cord (see Color Plate 18.2). Reproduced with permission from Thieme Medical Publishers Inc., New York, 1995, *Color Atlas of Embryology*, Ulrich Drews, Chapter 4: Nervous System.

B. Secondary Neurulation

Closure of the neural tube proceeds both cephalically and caudally. The unclosed cephalic end is termed the anterior or rostral neuropore, and the unclosed caudal end is termed the posterior or caudal neuropore. The neural tube lumen communicates with the amniotic cavity through neuropores. The anterior neuropore closes at about 24 days and the posterior at about 26 days. After closure of the posterior neuropore, a condensation of the mesenchymal cells known as the neural cord forms under the surface ectoderm. A cavity soon appears in the neural cord that becomes continuous with the lumen of the caudal end of the neural tube. This extension of neural tissue from a nonneuroectodermal source is termed secondary neurulation, which is not very prominent in humans.

C. Primary Brain Vesicles

Even before neurulation, a bend, the mesencephalic flexure, appears in the neural plate. This and other indentations in the neural fold demarcate the three major divisions: 1) the prosencephalon (forebrain), 2) the mesencephalon (midbrain), and 3) the rhombencephalon (hindbrain). Rostrally, the neural tube shows two other bends: the cervical flexure at the junction of the brain and spinal cord, and the pontine flexure (in the opposite direction) in the hindbrain. The neural tube bends sharply at these flexures (Fig. 18.3).

D. Secondary Brain Vesicles

During the fifth week, the prosencephalon and rhombencephalon each subdivide into two distinct regions, converting three primary vesicles into five secondary vesicles. The prosencephalon divides into the rostral end portion, the telencephalon and the adjacent diencephalon, identified as having optic placodes. The rhombencephalon divides into the metencephalon and myelencephalon (Fig. 18.3).

E. Neuromeres: Further Subdivision of the Brain

Specific combinations of gene expression direct the unique development of each region. There are about 30 homobox genes that are implemented in directing the pattern of head development. The presence of constrictions in the wall of the neural tube causes the forebrain to subdivide into six segments, termed prosomeres. The prosencephalon could be grouped into large transverse subdivisions. Prosomeres P_1 to P_3 designate the diencephalon; prosomeres P_4 to P_6 designate the hypothalamus (secondary prosencephalon). The telencephalic region contributes to its dorsal aspect. Like the prosencephalon, the rhombencephalon is also subdivided into eight discrete rhombomeres (r_1 to r_8). Segmentation of the brain subdivides the central nervous system into functionally distinct domains. The constrictions in the wall correspond to the boundaries that partially restrict the intersegmental mixing of not only the cells but also the intercellular matrix. Furthermore, according to this prosomere model, the brain is also organized into longitudinal domains parallel to the longitudinal axis of the neural tube. Longitudinal domains are analogous to the roof, alar, basal and floor plates of the spinal cord. Segmentation within the neocortex may rely on signals from the boundary of each segment, which leads to region-specific gene expression, histogenesis and other early patterning. Once the new fibers arrive in a particular region, further refinement of the organization occurs (Fig. 18.4).

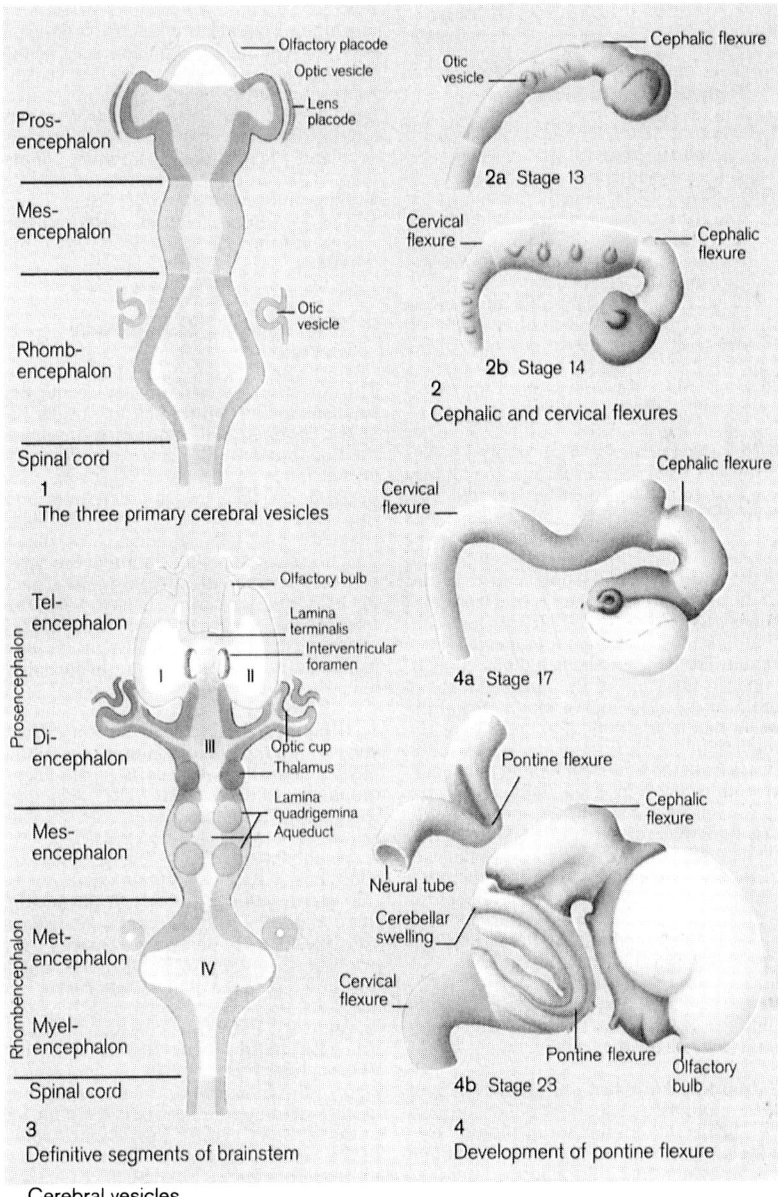

Figure 18.3
Brain vesicles (see Color Plate 18.3). Reproduced with permission from Thieme Medical Publishers Inc., New York, 1995, *Color Atlas of Embryology*, Ulrich Drews, Chapter 4: Nervous System.

F. Histogenesis — Cytodifferentiation of Neural Epithelium

Differentiation of the neuroepithelial cells forming the neural tube begins in the rhombencephalic region and proceeds rostrally and caudally as the neural tube

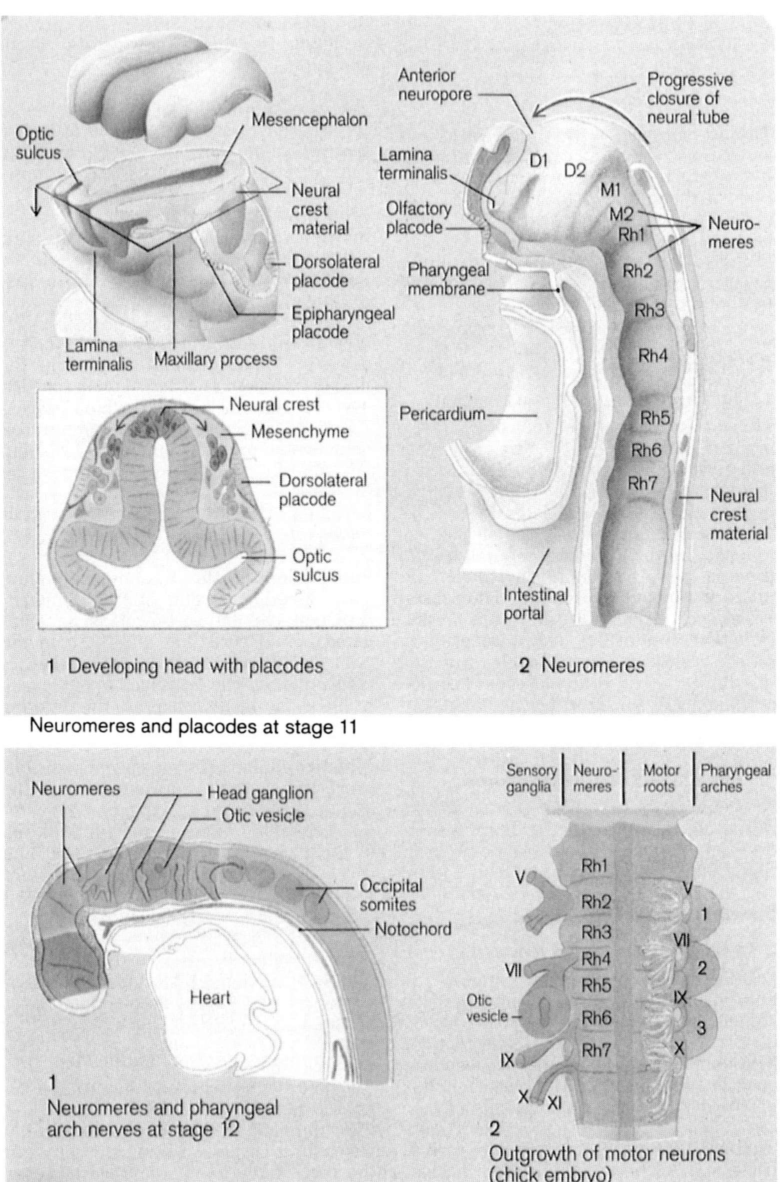

Figure 18.4

Neuromeres and placodes (see Color Plate 18.4). Reproduced with permission from Thieme Medical Publishers Inc., New York, 1995, *Color Atlas of Embryology*, Ulrich Drews, Chapter 4: Nervous System.

closes. Precursors of neurons and some types of glial cells are produced by proliferation of the neuroepithelial cells. The position of the nuclei in these cells is closely correlated with the stage of the cell cycle (Fig. 18.2):

1. In the G_1 period, the nuclei are toward the external limiting membrane (the basal lamina surrounding the neural tube), and the apical surface of the cells reaches to the lumen.
2. In the S period, the cells become wedge-shaped and the nuclei move further toward the outer zone. DNA replication occurs in these nuclei.
3. In the G_2 period, the nuclei begin to migrate toward the lumen of the neural tube.
4. During mitosis, the cells contract toward the lumen, where cell division is completed. The daughter cells thus formed either undergo cell division again or migrate out to form an intermediate (mantle) zone just beneath the external limited membrane. The cells in the mantle zone do not undergo cell division, and differentiate into neuroblasts, establishing the gray matter. Fibers from neuroblasts grow out, giving rise to the marginal zone (white matter). After production of neuroblasts ceases, the dividing cells become small and form the subependymal layer. The small cells divide and give rise to glial progenitor cells, which differentiate into two types of astrocytes: radial glial cells and some oligodendrocytes. Finally, the cells that remain lining the lumen differentiate into ependymal cells. Ependymal cells are responsible for secretion and absorption of cerebrospinal fluid. Microglial cells are mesodermal in origin and migrate to the central nervous system to act as macrophages.

G. Organization of the Central Nervous System

1. The Spinal Cord

The part of the neural tube caudal to occipital somites is termed the spinal cord. As mentioned above, it is formed by both primary and secondary neurulation. Because of active proliferation and differentiation, the neuroepithelium thickens and becomes organized in layers. The layer of cells lining the central canal (the lumen) forms the ventricular (ependymal) zone. Peripheral to the ventricular zone, a layer of cells forms the intermediate (mantle) zone. The cells of the intermediate zone differentiate into neuroblasts that give rise to processes (axons and dendrites). The processes (axons) extend outward to form the marginal zone. The marginal zone contains processes (fibers) and no cell bodies, and becomes the white matter. The intermediate zone becomes the gray matter of the spinal cord.

a. Division of the intermediate (mantle) zone into the alar and basal laminae

Although the median portion of the neural fold (closer to the notochord) is designated as the future basal plate and the lateral portion of the neural fold is designated as the alar plate, they are not distinct until after completion of neurulation. At about the end of the fourth week, the neuroblasts in the intermediate zone become organized into four distinct columns that run the length of the spinal cord. The dorsal pair is the alar laminae, and the ventral pair the basal laminae. These are separated on each side by the sulcus limitans, a groove in the central canal. The alar laminae are connected to each other by a thin roof plate, and the basal laminae by a thin floor plate.

b. Differentiation of motor neurons

The cells in the basal laminae differentiate into somatic motor neurons that will innervate motor structures such as skeletal muscles. The cells of the alar laminae differentiate into association neurons (interneurons) that interconnect the motor neurons to the processes of the sensory neurons. Both the contact of notochord and diffusible signals derived from the notochord are responsible for floor plate and motor neurons differentiation. The cells near the notochord differentiate into the floor plate, and the cells further away become motor neurons, indicating that diffusible signals may act at different concentration thresholds. It is now known that the notochord is responsible for motor neuron induction and that the floor plate may be involved in other aspects of ventral patterning. The floor plate helps to establish the left and right sides of the spinal cord and the brain. The midline is where decisions are made about differentiations of cell types. In addition, cell adhesive molecules and two new proteins — neutrin I and neutrin II — seem to influence the guidance of motor neuron axons. As mentioned above, dorsalin-1 transcripts are restricted to the dorsal region of the spinal cord.

c. The sensory neurons

After detaching from the neuroepithelium, neural crest cells migrate into the extracellular matrix. Further migration of these cells is influenced by intrinsic and extrinsic factors such as fibronectin and glycosaminoglycans. In general, the neural crest is divided into the cephalic neural crest and the trunk or spinal neural crest. The cephalic neural crest gives rise to diverse structures in the head and neck region. The cells from the spinal neural crest migrate in three directions:

i. Dorsolaterally between the ectoderm and somites, to become dermal melanocytes.

ii. Ventrally between the neural tube and somites, to contribute to the adrenal medulla and sympathetic ganglia.

iii. Ventrolaterally close to the anterior half of the sclerotomes and differentiate into segmental dorsal root ganglia. It has been shown that survival and differentiation of the dorsal root ganglion depend on brain-derived neural growth factor (BDNF), which is secreted by the adjacent neural tube.

The neural crest cells in dorsal root ganglia begin to form two processes. One grows medially toward the dorsal column of the spinal cord and synapses with the developing association neurons (interneurons). The other process of dorsal ganglion cells grows laterally, joins with the ventral root and continues within the spinal nerve to reach the end organ. Later the two processes of these neurons join together, transforming the bipolar shape to a pseudounipolar shape.

2. The Myelencephalon: The Medulla Oblongata

The caudal subdivision of the rhombencephalon differentiates into the medulla oblongata. It extends from the pontine flexure to the first spinal nerve. Expansion of the central canal causes the lateral walls to rotate outward. The roof plate

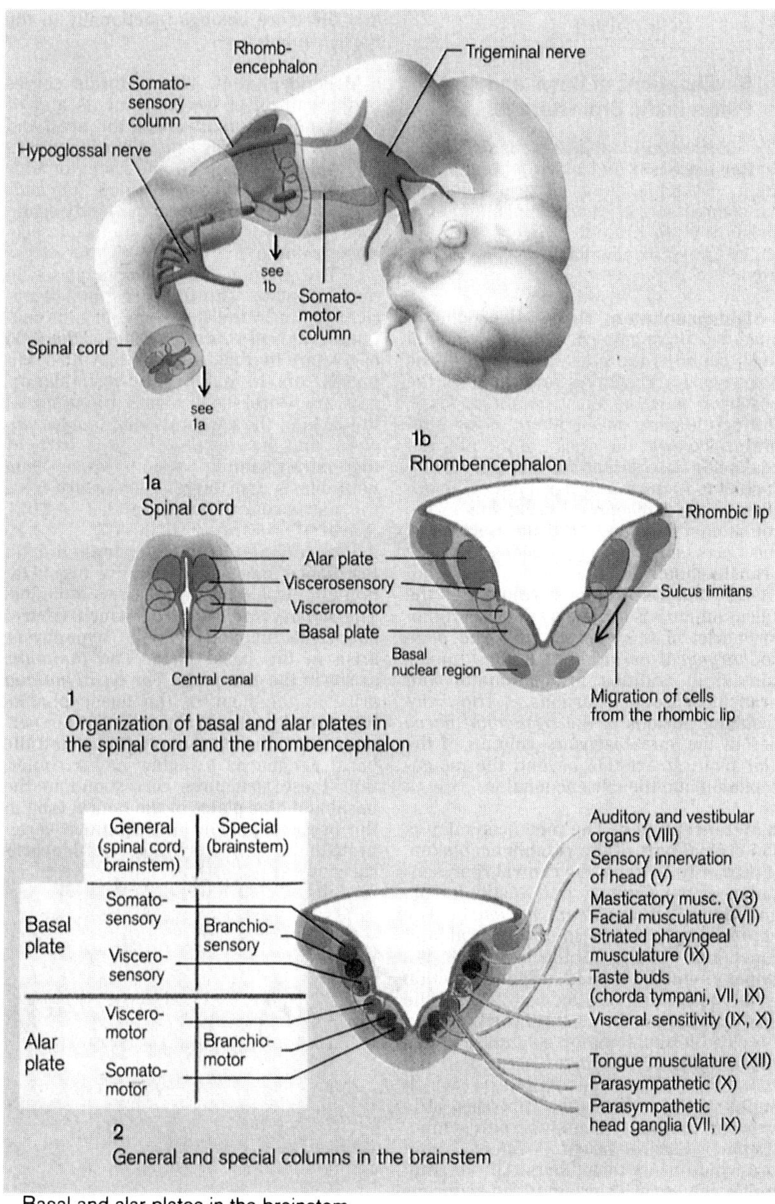

Figure 18.5
Basal and alar plates (see Color Plate 18.5). Reproduced with permission from Thieme Medical Publishers Inc., New York, 1995, *Color Atlas of Embryology*, Ulrich Drews, Chapter 4: Nervous System.

becomes stretched out. The alar laminae separate from each other. As a result, the alar laminae lie lateral to the basal laminae in the floor of the hindbrain. The thin roof plate becomes stretched and gives rise to the posterior choroid plexus (Fig. 18.5).

a. Alar laminae

The alar laminae nuclei are divided into three groups:

i. The most lateral group receives somatic afferent impulses from the ear and surface of the face by way of the vestibulocochlear (VIII) and bulbospinal (V) portion of the trigeminal nerve.

ii. The intermediate group forms the nucleus of the solitary tract and receives impulses from taste buds of the tongue, palate, oropharynx and epiglottis.

iii. The medial group receives interoceptor information from the lungs, heart, and aortic and carotid sinuses. Some cells from the alar laminae migrate ventromedially to form the olivary nuclei complex.

b. Basal laminae

The basal laminae nuclei can also be divided into three groups:

i. The medial group, which is a continuation of the anterior horn cell columns and innervates the skeletal muscles derived from the occipital somites, represented by the nucleus of the hypoglossal (XII) nerve.

ii. The intermediate group supplies the muscles derived from the third and fourth pharyngeal arches, represented by nuclei of the glossopharyngeal (IX), vagus (X) and accessory (XI) nerves. These nuclei are located in the floor of the fourth ventricle.

iii. The lateral group forms the dorsal nucleus of X and inferior salivatory nucleus of IX; it also supplies preganglionic parasympathetic fibers to the muscles of the respiratory tract, the heart, the GI tract and the parotid gland.

Because of descending axons from the motor cortex, the marginal layer on the ventral aspect becomes thick and forms the pyramids.

3. The Metencephalon

The metencephalon is the cephalic subdivision of the rhombencephalon. It consists of three parts: a) basal laminae of the primary portion and pons, b) alar laminae of the primary portion of the rhombic lips and c) the cerebellum (Fig. 18.6).

a. Basal laminae of the primary portion and pons

Its basic organization remains like that of the myelencephalon.

i. The medial group is a continuation of somatic efferent columns from the myelencephalon and gives rise to the abducent (VI) nucleus.

ii. The intermediate group contains trigeminal and facial nuclei that innervate the muscles derived from the first and second pharyngeal arches.

iii. The lateral group contains the superior salivatory nucleus. Fibers from this grow out to join the facial nerve to supply the salivary glands.

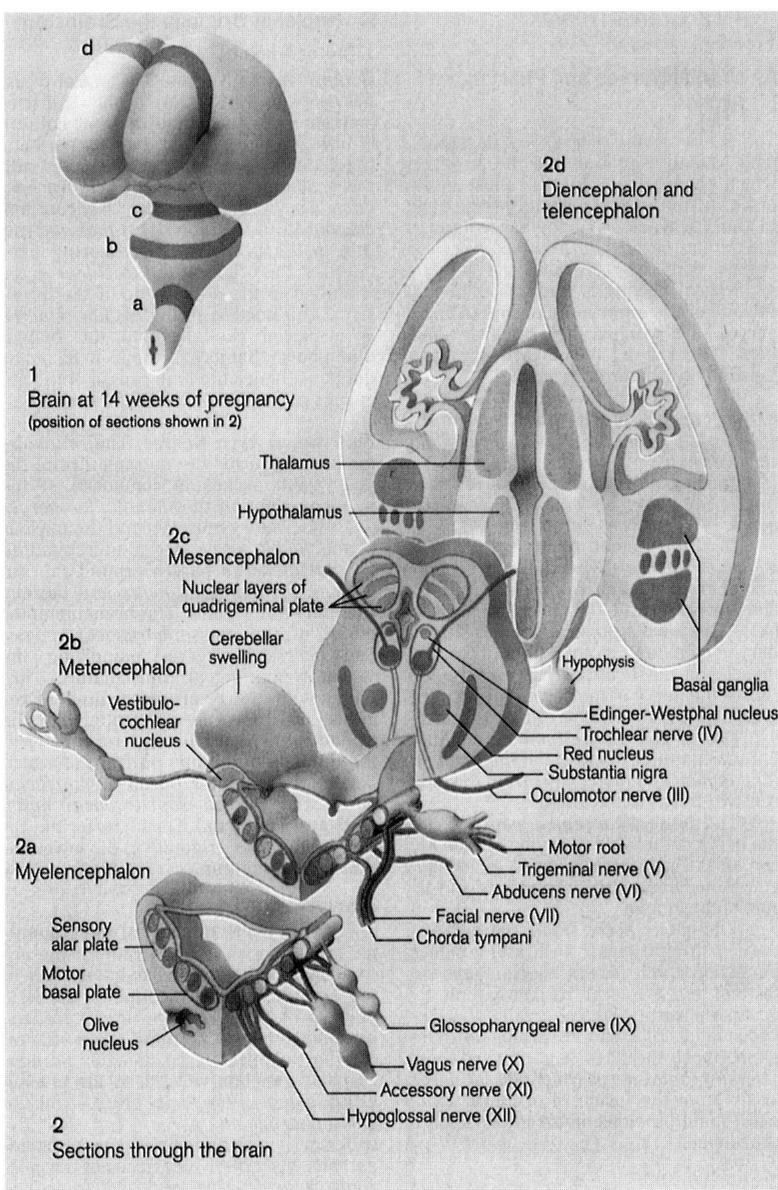

Figure 18.6
Brain stem (see Color Plate 18.6). Reproduced with permission from Thieme Medical Publishers Inc., New York, 1995, *Color Atlas of Embryology*, Ulrich Drews, Chapter 4: Nervous System.

b. Alar laminae of the primary portion

Its ventromedial portion contains the following afferents:

i. The lateral group contains the nucleus of the pontine portion of the vestibulocochlear complex (VIII).

ii. The intermediate group contains the cranial portion of the solitary tract nucleus.

iii. The medial group contains the most cranial portion of the dorsal sensory nucleus of the vagus (X).

Aside from the nuclei, the pons (bridge) contains fiber tracts that relay information between the cerebrum, cerebellum and spinal cord. These tracts arise from the marginal layer of the metencephalon.

c. The cerebellum

i. Development
The cerebellum arises bilaterally from the alar laminae of the metencephalon. The lateral walls of the rhombencephalon spread apart, stretching the roof plate. During week 4, proliferation in the most lateral aspect of this roof plate becomes intense, causing a marked thickness in this area. This dorsomedial area blends to form the rhombic lips. Continuous deepening of the pontine flexure, migration of neuroblasts to the rhombic lips and mitotic activity within the rhombic lips cause them to enlarge to form a pair of cerebellar primordia. By the second month, the two primordia unite dorsally across the midline, forming a single cerebellar plate that covers the fourth ventricle. By 12 weeks, the cerebellar plate shows a small midline portion known as the vermis, and two lateral portions known as the cerebellar hemispheres (Fig. 18.7). The lateral hemispheres expand rapidly and begin to acquire fissures. A transverse posterolateral fissure separates the developing cerebellum into cranial (main body) and caudal (flocculonodular) portions. The cranial portion grows much faster than the caudal portion and gives rise to the main bulk of the cerebellum. By the end of the third month, the primary fissure appears. As this fissure deepens, it divides the vermis and the hemispheres into a cranial anterior and a caudal posterior lobe. An additional transverse fissure appears that subdivides the lobes into lobules. The middle portion of the cerebellum grows more rapidly than the rostral and lateral portions, causing an intricate process of foliation, resulting in a mushroom shape. This process of fissuration and foliation continues throughout development, and it vastly increases the surface area of the cerebellar cortex. The cells from the alar plate of the metencephalon migrate ventrally to form the pontine nuclei. The fibers from the pontine nuclei grow transversely to enter the growing cerebellum, giving rise to transverse pontine fibers and the middle cerebellar peduncle. The cerebellum also becomes connected with the midbrain and medulla by the superior and inferior peduncles, respectively.

ii. Histogenesis of the cerebellar cortex
In the early stages, the cerebellar cortex develops from the germinal layer cells. As in the other regions of the neural tube, the neuroepithelium of the rhombic lips undergoes proliferation and gives rise to the ventricular, mantle and marginal layers. During the third month, cells from the rhombic lips migrate through the marginal region to the surface of the developing cerebellum, forming the external germinal layer. The ventricular zone (closer to the fourth ventricle) now is referred to as the internal germinal layer (Fig. 18.8). During the

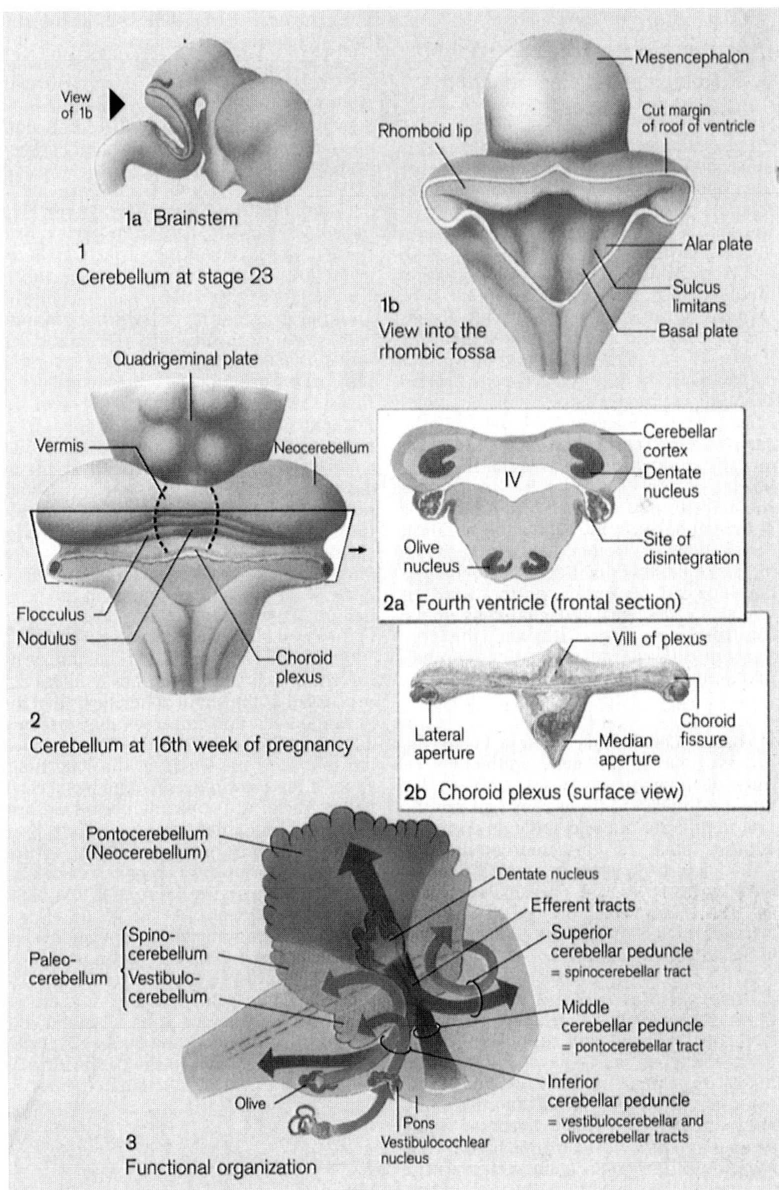

Figure 18.7
Cerebellum (see Color Plate 18.7). Reproduced with permission from Thieme Medical Publishers Inc., New York, 1995, *Color Atlas of Embryology*, Ulrich Drews, Chapter 4: Nervous System.

fetal period, the cells in the internal and external germinal layers undergo further proliferation and migration. The cells from the internal germinal layer migrate to form deep cerebellar nuclei (dentate, globose and emboliform nuclei), which are situated in the white matter. In addition, the internal germinal layer also gives rise to Purkinje and Golgi cells that migrate to the cortex. The

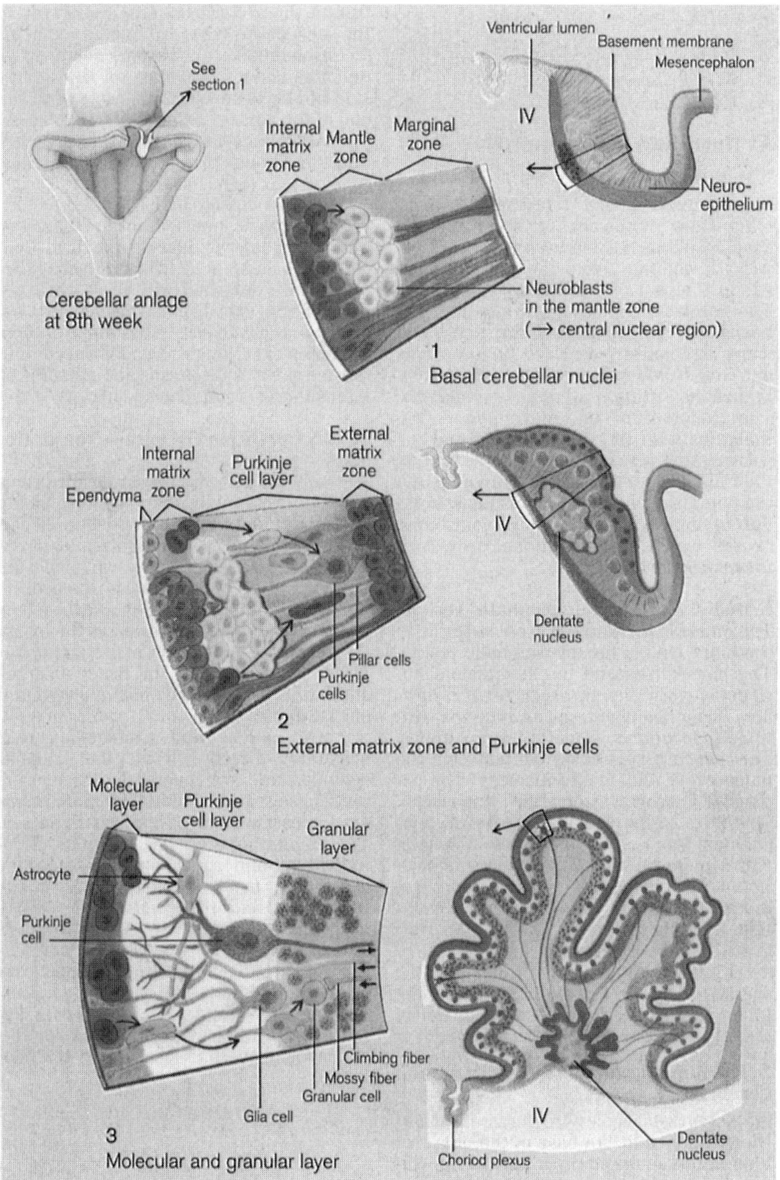

Figure 18.8

Cerebellar cortex (see Color Plate 18.8). Reproduced with permission from Thieme Medical Publishers Inc., New York, 1995, *Color Atlas of Embryology*, Ulrich Drews, Chapter 4: Nervous System.

axons from Purkinje cells synapse in the deep nuclei; from there the efferent fibers grow into the midbrain contributing to the superior cerebellar peduncle. The external germinal layer undergoes proliferation, and cells from this layer migrate in three waves. The first wave gives rise to basket cells, the second to granular cells, and the third to stellate cells. The granular cells and some basket cells migrate deeper to the Purkinje cells and form the granular layer of the cortex. The remaining basket cells and the stellate cells form the molecular layer of the cortex. After cessation of neuroblast formation, both internal and external germinal layers give rise to astrocytes and oligodendrocytes. In humans, the external germinal layer disappears at the end of the second postnatal year. The portions of the roof plate of the fourth ventricles situated in front of and behind the cerebellum give rise to the anterior and posterior medullary vela, respectively.

4. The Mesencephalon

The mesencephalon is the most primitive of all the brain vesicles. It consists of the alar laminae, the tectum, the basal laminae, the tegmentum and the peduncular region (white matter). The basal and alar laminae are separated by the sulcus limitans (Fig. 18.7).

a. Basal laminae, the tegmentum

Two groups of cells aggregate as the somatic motor oculomotor (III) nucleus and the general visceral efferent Edinger–Westphal nucleus. The trochlear and mesencephalic trigeminal nucleus originate in the metencephalon and become displaced into the tegmentum. The red nucleus and substantia nigra are also located here, but it is uncertain whether they derive from the basal laminae or migrate from the alar laminae.

b. Alar laminae, the tectum

Mesencephalic alar laminal cells proliferate and migrate medially into the roof plate. This migration of cells forms two thickenings that are separated by a midline groove. A transverse groove later appears and subdivides them into the superior and inferior colliculi. The superior colliculi receive fibers from the retina and form the ocular (visual) reflexes. The inferior colliculi serve as synaptic relay centers for auditory reflexes. The ventricle (neural canal) of the mesencephalon becomes narrow, giving rise to the cerebral aqueduct, which connects the third ventricle to the fourth ventricle.

5. The Diencephalon

The sulcus limitans does not extend into the forebrain; therefore, division into the alar and basal laminae is not clear. In fact, it is widely believed that the diencephalon lacks basal laminae and its walls are formed by the alar laminae (Fig. 18.9).

a. The thalamus

At the end of the 5th week, two pairs of swellings appear in the lateral wall of the third ventricle. These swellings are separated by the hypothalamic sulcus. The dorsal swellings grow disproportionately and become two thalami. The two thalami approach each other across the third ventricle and fuse at one or more places by interthalamic adhesions. The caudal part of the thalamus gives rise to medial and lateral geniculate bodies (metathalamus). The medial geniculate body is asso-

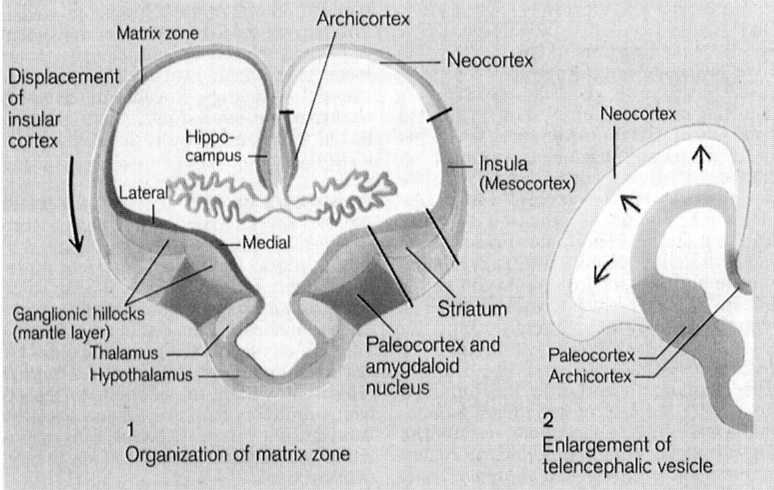

Figure 18.9
The paleo-, archi- and neocortex at the end of the embryonic period (see Color Plate 18.9). Reproduced with permission from Thieme Medical Publishers Inc., New York, 1995, *Color Atlas of Embryology*, Ulrich Drews, Chapter 4: Nervous System.

ciated with hearing, and the lateral geniculate body with sight. Many aggregations of cells in the intermediate (mantle) layer give rise to other thalamic nuclei. The thalamus receives all information en route to the cerebral cortex from the subcortical region and relays it to the appropriate cerebral areas.

b. The epithalamus

At about six weeks, a shallow groove, the sulcus dorsalis, appears in the dorsal (thalamic) swelling. It demarcates a small epithalamic area that differentiates into the epithalamus. Beside this dorsal rim, the epithalamus also develops in the adjacent roof plate. The median part of the roof plate evaginates and develops into the pineal gland. Anteriorly, the roof plate expands and becomes attenuated. The blood vessels from its outer surface push inward to give rise to the choroid plexus of the third ventricle. Other ependymal structures in the third ventricle form circumventricular organs that add neuropeptides to the cerebrospinal fluid. The epithalamus also gives rise to the habenular nuclei and posterior habenular commissures. Rapid growth of the thalamus displaces the epithalamus dorsally, obliterating the sulcus dorsalis.

c. The hypothalamus

The swelling in the lateral wall of the diencephalon ventral (lower) to the hypothalamic sulcus differentiates into the hypothalamus. The aggregation of cells in the mantle layer of the hypothalamus, especially in the floor plate, forms several nuclei. One of these nuclear groups enlarges rapidly into mammillary bodies that project into the ventral surface. Downward extension of the hypothalamic floor gives rise to the neurohypophysis. Development of the hypophysis (pituitary) is discussed in Chapter 16 (Endocrine System). The hypothalamus receives input from numerous structures of the central nervous system and acts as a regulatory center for emotions, hunger, thirst (electrolytes and water balance), sleep, body temperature and sex drive. Due to its neurosecretory nature, it serves as an interface between neural integration and the humoral status of the body.

d. The subthalamus

This is the other ventral structure. It is demarcated by the region between the thalamus and the cerebral peduncle. Growth of the thalamus shifts the subthalamus laterally.

6. The Telencephalon

Paired dorsolateral evaginations, the telencephalic vesicles (cerebral vesicles), appear on day 32, in the most rostral part of the neural tube. The rostral walls of these vesicles are connected in the median portion by the lamina terminalis. The telencephalic vesicles expand and extend rostrally, covering the median portion; vesicle caudal extensions cover the entire diencephalon and mesencephalon. The walls of the telencephalic vesicles begin to differentiate into two planes: a) ventrally the primordia of the corpus striatum, and b) dorsally the primordia of the pallium, the future cerebral cortex (Fig. 18.9).

a. The corpus striatum

The intermediate (mantle) layers of the ventrolateral walls of the telencephalic vesicles increase in size and bulge into the lateral ventricles (the cavities in the vesicles). Each has a striated appearance and is known as the corpus striatum. It expands posteriorly and appears as medial and lateral elevations. The elevations are separated by bundles of fibers to and from the cerebral cortex. These bundles of fibers are known as the internal capsule. The internal capsule partially separates the caudate nucleus from the putamen, and the thalamus from the globus pallidum.

i. The medial elevation is derived from the diencephalon and is involved in formation of the amygdaloid. Some of the telencephalic components later migrate into the amygdaloid.

ii. The lateral elevation is derived from the telencephalon and gives rise to the lentiform nucleus. Continuous growth of the cerebral hemispheres in the posterior region brings its medial walls closer to the lateral walls of the diencephalon, and they consequently fuse to each other. Some of the cells from the lateral elevations migrate into the thalamus. The corpus striatum appears first closer to the interventricular foramen, then elongates in a C-shaped manner.

b. The pallium: the cerebral cortex

Each telencephalic vesicle initially contains a large lateral ventricle (cavity). Due to thickening of the cortex, the ventricles progressively become smaller. However, the medial wall of the vesicles along the floor remains thin and at the junction with the diencephalic roof forms a longitudinal groove termed the choroid fissure. A choroid plexus forms along this fissure. The lateral ventricles extend the whole length of the hemispheres. Another prominent fissure, the rhinal (hippocampal) fissure, develops parallel to and above the choroid fissure, and divides the pallium into a) the hippocampal cortex (archipallium) and b) the neopallium (Fig. 18.9).

i. Hippocampal formation — the archipallium
Hippocampal formation includes the parahippocampal gyrus, the subiculum, and the dentate and hippocampus gyrus. An enlargement on the ventral surface of the cortex appears to form the olfactory stalk. The pallium in the region thickens and gives rise to the hippocampus. It follows the curve of the caudate nucleus into the temporal lobe. The efferent system of the hippocampus is represented in the fornix, which curves over the thalamus (Fig. 18.10).

ii. Neopallium
Most of the cerebral hemisphere visible from the surface is neopallium. Initially, the surface of the hemisphere is smooth. At about the 16th week, a small indentation, the lateral cerebral fossa, appears in the lateral wall of the cerebral hemisphere. The caudal end of the hemisphere lengthens and curves ventrally and rostrally to form the temporal lobe. The fossa deepens and becomes the lateral fissure (sulcus). The area of the cerebral cortex that becomes covered by the temporal lobe is known as the insula. The cerebral cortex then undergoes a complex pattern of folding, forming gyri. By 24 weeks, many other sulci appear to separate the gyri and demarcate lobes.

c. Commissures

During early development, two hemispheres develop separately, but at about week 12 bundles of fibers begin to cross from one cerebral hemisphere to the other through the lamina terminalis. The first set of these crossing over becomes the anterior commissure, which connects the two olfactory areas. The hippocampal commissure (the fornix), as mentioned above, connects both sides. The most important connection between the right and left cerebral hemispheres is the corpus callosum. Although it starts in the lamina terminalis, it expands into a broad band connecting a large part of the base of the cerebral hemisphere, thus forming an additional roof for the third ventricle (Fig. 18.11).

d. Histogenesis of the cerebral cortex

The neuroepithelium of the cerebral cortex is similar to that of the neural tube in the early stages. Later, the neuroblasts in the ventricular zone give rise to processes that form a primordial plexiform layer of the marginal zone. Many of the cells from the ventricular zone migrate superficially to form an intermediate zone between the ventricular and marginal zones. New neuroblasts produced in the

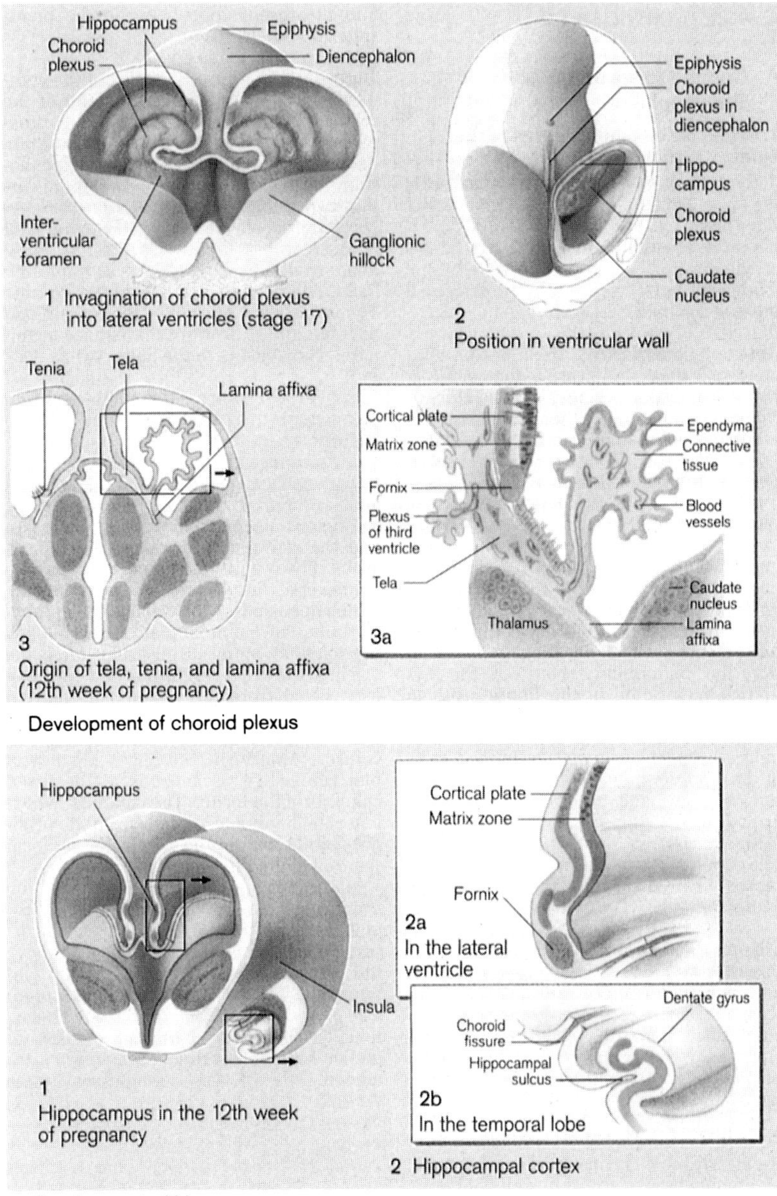

Figure 18.10

Choroid plexus and hippocampus (see Color Plate 18.10). Reproduced with permission from Thieme Medical Publishers Inc., New York, 1995, *Color Atlas of Embryology*, Ulrich Drews, Chapter 4: Nervous System.

ventricular zone and some cells from the intermediate zone migrate to establish a cortical plate between the marginal and intermediate zones. Some cells differentiate into pillar cells, which send out radial fibers stretching out between the fiber mass

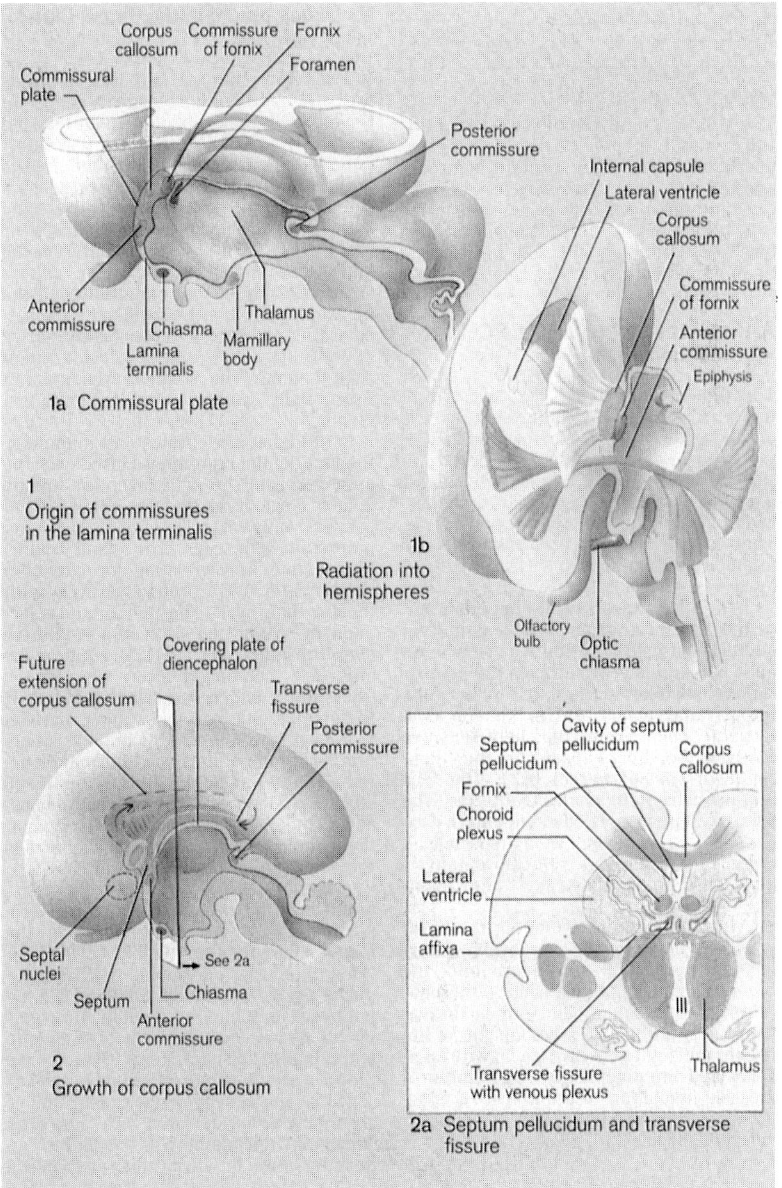

Figure 18.11

Commissures (see Color Plate 18.11). Reproduced with permission from Thieme Medical Publishers Inc., New York, 1995, *Color Atlas of Embryology*, Ulrich Drews, Chapter 4: Nervous System.

and the ventricular zone. Migrating neurons follow these radial fibers as guides. After establishment of the cortical plate, some cells from the ventricular zone migrate between the ventricular and intermediate zones to form the subventricular zone. Production of neuroblasts is taken over by the subventricular zone. The neuroblasts from the subventricular zone migrate outward and form a new subcortical plate, and some pass through the cortical plate and form its outer layer. In this manner, new neuroblasts come to rest external to their predecessors. Meanwhile, the intermediate zone becomes devoid of cell bodies of neurons and is converted into subcortical white matter. As production of neurons ceases, the ventricular zone produces some types of glioblasts and finally becomes the ependyma (Fig. 18.12).

Thus, the cerebral cortex is typically made up of six layers:

i. The external part of the original marginal zone is transformed into the molecular or plexiform layer.

ii, iii. The constituted neurocortex (cortical and subcortical plates).

iv. Subcortical white matter (intermediate zone).

v. The subventricular zone.

vi. The ependyma.

As mentioned above, pillar or radial glial cells guide the neuroblasts during their migration. It has been suggested that the human cerebral cortex is arranged into about 200 million discrete radial units. Each unit consists of radial glial cells and the neurons that are migrating on it. The number of units is determined by neural input. Although the sequence and timing of cellular migration differ in various regions of the cortex, at 8 months its cytoarchitecture is comparable to that of the adult brain.

e. Transcription factors involved in development of the cerebral cortex

A number of transcription factors have been identified that are expressed in the developing cerebral cortex. They likely control the differentiation and laminar arrangement of cortical neurons. These molecules (factors) regulate gene expression by binding to genomic DNA. The factors belong to the following classes:

i. Homeodomain, which includes

   a. The POU domain (Brn1, Brn2, Oct6, Brn5, mPOU). During development these genes are expressed in the subventricular zone, the intermediate zone and the cortical plate, but not in ventricular zone.

   b. Emx and Otx genes are expressed in the embryonic dorsal telencephalon. They control the identity of different regions of the brain.

   c. Pax-6 is the only member of this class expressed in the cortex. Its transient expression, especially in the ventricular zone, is essential for normal neuronal migration.

ii. Helix loop — helix factors

   These are important in neurogenesis, as they can either inhibit or activate transcription in the cortex. They include:

   a. MASH I, which is expressed in the ventricular zone of the dorsal telencephalon.

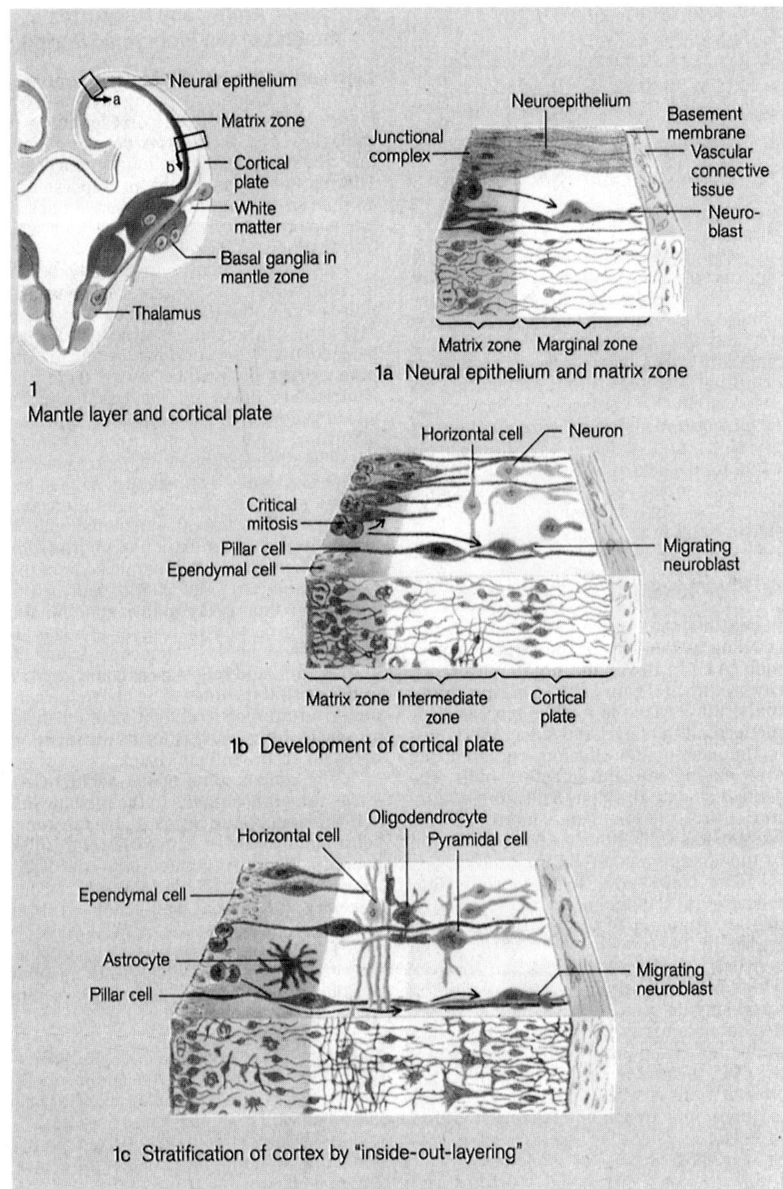

Figure 18.12

Cerebral cortex (see Color Plate 18.12). Reproduced with permission from Thieme Medical Publishers Inc., New York, 1995, *Color Atlas of Embryology,* Ulrich Drews, Chapter 4: Nervous System.

b. N-myc7, which is expressed in the ventricular zone of the dorsal and ventral telencephalon.

c. Id-2 codes for inhibitory proteins. These lack a DNA-binding domain and inactivate other HLH proteins by forming a heterodimer with them. Id-2 is expressed at high levels in laminated regions of the cortex, such as the cerebral cortex, the olfactory bulb and the cerebellum. In early embryos, it is expressed in the ventricular zone, later in the intermediate zone and the cortical plate and not in the ventricular zone. These results suggest that HLH proteins are needed during cortical lamination.

iii. MEF-2 genes
MEF-2c is expressed in postmitotic neurons located in the cortical plate of the developing cortex. These genes appear to have overlapping patterns of expression. MyoD, MEF-2 and HLH factors may be part of the network that control neuroblast differentiation.

iv. Zinc-finger genes
These genes share a cysteine–histidine domain that binds with zinc. These are early genes that are expressed in alternating rhombomeres, and include Krox-20 (EGR-2), EGR-1 and NGF1-A.

v. Leucine zipper genes
These factors dimerize through a leucine zipper domain. They include the fos and jun families. A transient fos-c expression has been reported in the subplate (subcortical plate). These factors are best known for their role in normal function of the mature brain. The transcription factors known to be expressed during development of the cerebral cortex are also expressed in other tissues. No factor has been identified as specific for a specific cell type or specific zone. The combination of expression of these factors may be essential during development. The interaction between transcription factors has been shown to be necessary in cortical development.

f. Other factors that regulate histogenesis

Normal histogenesis is also regulated by:

i. Interaction between neuroblasts and glial cells.

ii. Neurotrophic factors, such as nerve growth factor, brain-derived growth factor, apolipoprotein E and purpurin.

iii. Extracellular matrix molecules, such as laminin, fibronectin, N-cell adhesion molecules, heparin sulfate proteoglycans and collagen.

iv. Sex hormones.

v. Other growth factors such as epidermal growth factor, fibroblast growth factor, insulin and insulin-like growth factors.

g. Axonal guidance

Santiago Ramon y Cajal described terminal swellings in embryonic fibers that show progressive growth and active neoformation. He called them the "cone of growth" (growth cone). He also pointed out that one neuron contacts another by sending out a single fiber, an axon. The growth cone leads the way for an axon to reach its target. Once at the target, the growth cone transforms into a presynaptic terminal. Both local factors and long-range diffusible factors might steer growth cones in the direction of the appropriate target. Growth cones have been shown to follow the nerve growth factor gradient. Molecules such as N-cadherin, N-CAM and L1 help cells and outgrowing fibers (neurites) attach to one another. Furthermore, a target might release a diffusible factor that would promote its own innervation. In addition to CAMs and diffusible chemoattractants, the growth of axons may be influenced by other contact-mediated repulsive or inhibitory cues present on the surface of the cell or in the extracellular matrix. In this case, when an axon comes into contact with a nontarget cell, the cone collapses. After a short while, the cone reforms and moves in another direction. Defining the midline helps to establish the right and left sides of the developing spinal cord and brain. Decisions on axon guidance are made in the midline. For instance, while some axons are guided across the midline from one side to the other, other axons are directed to avoid the midline. The chemoattractants, chemorepulsants and CAMs may provide the mechanism for axonal guidance. Some of these factors are as follows:

i. Netrins
   Netrins are bifunctional proteins. Growth cones of floor plate axons bind to the netrin molecule and signal axons to grow toward the concentration of these proteins. Netrins are found to guide the commissural axons circumferentially around the spinal cord to the floor plate. Depending on the kind of receptors on the growth cone, one type of receptor interprets the protein as an attractant and another may see it as repulsant.

ii. Collapsin (also known as semaphorin-sema 1)
   Collapsin causes the growth cone to collapse. It may be a diffusible repulsive cue and may provide short-range inhibition to prevent axons from splitting into multiple branches before reaching the target.

iii. Connectin
   Connectin is a CAM and provides attractive cues to motor neurons. Connectin is also bifunctional. It may attract some motor neurons toward the ventral region, where it is most concentrated, and rebuff other axons, causing them to move to the dorsal side.

V. CONGENITAL ANOMALIES OF THE NERVOUS SYSTEM

The incidence of malformation of the nervous system ranges from 8 in 1000 to 10 in 1000 births. It is believed that more than 90% of defective embryos are lost early in pregnancy. No two anomalies are alike, as they vary in type, degree and other associations. Neural tube defects are more frequent in embryonic period than later. Teratogens are defined in Chapter 7 (Teratogenesis). The most common teratogens during development of the human brain are believed to be isotretinoin (vitamin A analogue used for treating acne), radiation, methyl mercury, maternal alcoholism, viral infection, parasitic infection, hypoxia, and hereditary and chromosomal disorders. It should be noted that there is no consistency in findings in fetal alcohol syndrome. The mechanism for this is not fully understood, and not all fetuses of alcoholic mothers are born with fetal alcohol syndrome.

A. Neural Tube Defects (NTDs)

These malformations involve a disturbance in neurulation (neural tube closure), or postneurulation. For the sake of simplicity, these are divided into defects that are related to the spinal region and defects related to the cranial region.

B. Neural Tube Defects Related to the Spinal Cord

1. Spina Bifida

This implies a skeletal defect in the spinal region. Failure of closure of part of the neural tube causes a disruption in differentiation of the nervous tissue and induction of vertebral arches. Vertebral arches remain open along the dorsal midline. This can result in various anomalies, which, depending on skeletal, meningeal or neural involvement, can be divided into four main types (Chapter 8, Fig. 8.7):

a. Myeloschisis
The neural groove remains open at one or several levels, especially in the thoracolumbar region. It is a rare but serious condition, and results in paralysis.

b. Myelomeningocele (Spina Bifida Cystica)
This appears as a dome-shaped cyst covered by an inadequate membrane that leaves a central red, wet, oozing area. Within the cyst neural tissue, leptomeninges and disposed glial neuronal tissue are present. The most frequent site is the lumbar region. Covered meninges and neural tissue may ulcerate after birth, leading to meningitis. Various forms of spina bifida cystica can be associated with hydrocephaly, urinary incontinence, clubfoot and Arnold–Chiari malformations.

c. Meningocele
Only the meninges are present in the cyst. Although part of the neural tube is intact, it may be abnormal. Aberrant nerve roots may be present in the sac of the cyst. In both meningocele and myelomeningocele, the neural arches of the vertebrae in the affected region are defective. Meningocele may be attributed to faulty migration of mesenchyme during neurulation.

d. Spina Bifida Occulta
The vertebral arches in the affected area are underdeveloped and fail to fuse in the dorsal midline. Although the vertebral canal is open, the meninges and the spinal cord remain entirely within the vertebral canal. The defect is covered by the skin. One-fifth to one-fourth of spina bifida occurrences are symptomless and may never be detected. But others can be associated with fibrous bands, aberrant nerve roots, and adhesions. In many cases, abnormalities of the lower limbs are found.

Neural tube defects, especially open defects, are associated with high levels of a-fetoproteins in amniotic fluid. It results from direct or indirect communication between cerebrospinal fluid and amniotic fluid. Although it is used for screening, it is not specific for this condition and must be confirmed by ultrasound or other techniques. A false positive result could be produced by a hydatidiform mole, omphalocele and multiple pregnancy.

2. Neural Tube Defects in the Cranial Region

a. Anencephaly

Anencephaly has a striking appearance. Major parts of the brain and overlying cranial vault are absent. Only the caudal part of the rhombencephalon may be present. Results obtained from animals indicate three stages of development:

i. Failure of closure of the anterior pore of the neural tube.

ii. Protrusion of the developed brain.

iii. Degeneration of the exposed part of the brain.

The primary cause is attributed to disturbance in the interaction of neural ectoderm and associated mesoderm. Prominent eyes are present at the base of the skull, but the optic nerves may or may not be present. A small adenohypophysis is present but separated from the brain. Varying degrees of an open spinal canal may be present. Depending on the severity, varying amounts of identifiable brain stem are present. Although this condition is regarded as incompatible with postnatal life, two anencephalic children recently lived for about 3 years.

b. Encephalocele

This is a protrusion of the brain and meninges outside the cranial cavity. These are for the most part mesodermal defects that occur after neurulation. The vast majority occur in the occipital region, but 10 to 20% are found in the frontal region. Appearance depends on the size of the encephalocele. Small defects may result in an almost normal brain. Large defects are associated with a small skull or hydrocephaly. Lesions are usually covered by skin, and meninges are present beneath the skin. In large defects, herniation of the cerebellar hemisphere and brain stem are found. The portion of neural tissue located in the herniated sac is occasionally abnormal. Various abnormalities of the cerebellum, brain stem and associated blood vessels have been described. If merely the meninges protrude through the defect, it is called a cranial meningocele.

C. Anomalies of the Brain

1. Holoprosencephaly

The spectrum of holoprosencephaly covers a series of facial and cerebral malformations. Depending on the severity of the disturbance, it ranges from cyclopia to mild facial anomalies. It should be remembered that there are important interactions between the prechordal mesenchyme, neuroectoderm and mesencephalic neural crest. Disturbance in early reciprocal induction can result in these abnormalities. A correlation between facial and cerebral abnormalities led to the saying that "the face predicts the brain." With some exceptions, the more severe the facial defect, the more severe the brain defect.

In typical holoprosencephaly, the prosencephalon remains more or less undifferentiated and incompletely cleft. The basal part of the telencephalon and diencephalon is missing. Other brain anomalies are directly proportional to the severity of brain malformation. The olfactory bulbs and tracts are absent. In some cases, the optic chiasma, neurohypophysis and eyes may be missing. Depending on the stage of development at the time of disturbance, the spectrum of cyclopia or synophthalmia may occur. In typical cases, due to involvement of the frontonasal prominence, agenesis of the nasal placode and intermaxillary processes are found. A short upturned nose, a long upper lip without philtrum and a short retracted lower jaw are seen. It has been suggested that, if the disturbance occurs at about 20 days, it results in the spectrum of cyclopia; at about 22 days, it may result in synophthalmia. In general, all malformations of this type occur at or before the 4th week of gestation. The most common cause of these anomalies can be associated with consumption of alcohol during the 3rd and 4th weeks of pregnancy. Holoprosencephaly is the most severe manifestation of fetal alcohol syndrome. Chronic small amounts or a single large amount of alcohol consumption could result in this condition.

2. Arnold–Chiari Malformation

This is characterized by a displacement of the cerebellum and medulla through the foramen magnum into the upper cervical canal. These herniated structures become constricted and show sclerosis when examined under a microscope. Almost all cases are associated with myelomeningocele and/or hydrocephaly.

Although Arnold–Chiari malformation has been well described, there is no agreement on its pathogenesis. Two hypotheses are:

a. An abnormal elongation of the hindbrain. Failed formation of the pontine flexure is an important factor.

b. An open neural tube defect that prevents normal growth of the 4th ventricle, which in turn leads to underdevelopment of the mesenchyme of the posterior cranial fossa. Many other anomalies, especially skeletal abnormalities, i.e., shortness of the posterior fossa and occipital and atlantoaxial abnormalities could be explained by this theory.

3. Dandy Walker Syndrome

This is characterized by a cystic dilation of the fourth ventricle, underdevelopment of the vermis of the cerebellum and hydrocephaly. The primary cause is disturbance of growth of the roof of the fourth ventricle. The posterior cranial fossa is enlarged. These cases are usually compatible with life.

4. Hydrocephaly

Hydrocephaly is characterized by a greatly enlarged head with widely separated sutures and very large fontanelles. The volume of cerebrospinal fluid in the brain is increased because of obstruction of CSF flow. The ventricles are increased in size. Hydrocephaly can occur prenatally or postnatally. Depending on the time of its onset, the cerebral mantle is affected. If the obstruction occurs during the early fetal or embryonic period, the cerebral hemispheres are present as a thin plate. If the obstruction occurs late, the condition can be alleviated surgically by shunting the CSF into the right atrium or peritoneal cavity. The obstruction usually occurs in the ventricular system, most frequently in the cerebral aqueduct. The obstruction can be caused by viral infection or toxoplasmosis, and could also occur outside the ventricular system.

D. Chromosomal Abnormalities

1. Trisomy 21 (Down's Syndrome)

The brain of these children is about three-quarters the normal weight and size. It is short in anteroposterior length. Most lobes are comparatively small. At later age, a reduction in dendritic spines is found. These changes are not specific for trisomy 21 and are found in other syndromes. As the gene location for β-protein of amyloid is located on chromosome 21, neurofibrillar tangles and plaque could be found in older Down's syndrome individuals.

2. Trisomy 18

Holoprosencephaly, various neural tube defects, absence of corpus callosum, and cerebellar malformation have been found in trisomy 18.

3. Trisomy 13

The brain may be slightly smaller or of normal weight. Only one-third show various degrees of anomalies associated with the forebrain.

E. Infections

1. Purulent Meningitis

Most commonly it is caused by *E. coli* or streptococci B. The appearance of the brain is normal. There is a subarachnoid purulent exudate with congested meningeal blood vessels. Microscopic examination of the exudate shows neutrophils and other inflammatory cells. The major complication of this is vascular thrombosis. The ependyma of the ventricles may be involved. In this case, the exudate may be seen covering the choroid plexus, and may cause obstruction of CSF flow, resulting in hydrocephalus.

2. Cytomegalovirus Infection

Cytomegalovirus generally causes a systemic infection and can cause widespread damage to the brain. Infection usually occurs late in pregnancy, initially causing necrosis. The inclusions mostly are present in the nuclei of glial and endothelial cells, resulting in increased cytoplasm and nuclear size.

3. Rubella Virus

Intrauterine rubella infection can result in severe microcephaly. Infection during the early fetal period may result in meningocephalitis with focal necrosis.

4. Herpes Simplex Virus

This is also a systemic infection. If the brain is involved during early development, it becomes a bag of necrotic liquid material. Histological examination shows necrotic tissue and inflammatory cells.

5. Toxoplasmosis

Infection by toxoplasma causes extensive areas of necrosis randomly distributed throughout the brain. The infection also involves the ventricles and frequently leads to hydrocephalus. Marked inflammatory filtrate is present in the brain and meninges; as the lesions increase in size, the inflammatory reaction becomes less intense, and necrotic areas become more evident. Chorioretinitis in the eyes is a feature of toxoplasmosis.

REVIEW QUESTIONS — Chapter 18

Select the most appropriate answer.

1. Formation of neural tube is induced by two signals that are diffused from:
   a. notochordal processes
   b. the primitive streak mesenchyme
   c. the overlying ectoderm
   d. the underlying endoderm
   e. all of the above

2. All of the following neural tube inducing factors are expressed in the notochord except the:
   a. follistatin
   b. noggin
   c. notch (xotch)
   d. Hedgehog (shh-1)

3. Apoptosis:
   a. is a pathological form of cell death
   b. results from acute cellular injury
   c. is characterized by cell swelling and lysis
   d. is involved in programmed cell death for normal tissue turnover
   e. all of the above are correct
4. The neurulation, neural plate and neural tube formation is caused by:
   a. changes in the shape and size of the neuroepithelial cells
   b. the secretion of N-CAM molecules by neuroepithelial cells
   c. the reaction of neuroepithelial cells similar and equal to the inducing substances
   d. inhibition of programmed cell death
   e. all of the above are needed
5. During early development, the neural tube is bent at three locations, and the pontine flexure is located in the:
   a. telencephalon
   b. diencephalon
   c. mesencephalon
   d. metencephalon
   e. myelencephalon
6. Segmentation of the brain into functionally distinct domains is known as:
   a. primary neurulation
   b. secondary neurulation
   c. primary brain vesicles
   d. secondary brain vesicles
   e. neuromeres
7. During early development, the position of the nuclei of the neuroepithelial cells changes. The nuclei are located near the lumen in:
   a. mitosis
   b. phase G1
   c. S phase
   d. phase G2
8. The CNS cells that share lineage with monocytes are the:
   a. cerebral granular cells
   b. microglia cells
   c. oligodendroglia cells
   d. astrocytes
   e. Schwann cells
9. During cytodifferentiation of the neural epithelium, glial progenitor cells arise from the:
   a. mantle zone
   b. marginal zone
   c. subependymal layer
   d. mesoderm
   e. all of the above

10. Microglial cells originate from:
    a. ependymal layer neuroblasts
    b. mantle layer neuroblasts
    c. subependymal layer neuroblasts
    d. the mesoderm
    e. the ectoderm
11. In the developing spinal cord, association neurons (interneurons) differentiate in the:
    a. basal laminae
    b. alar laminae
    c. floor plate
    d. roof plate
    e. none of the above
12. Large thickenings in the lateral walls of the rhombencephalon form the rhombic lips, which in turn give rise to the:
    a. basal laminae of the metencephalon
    b. pons
    c. alar laminae of the metencephalon
    d. cerebellum
    e. cerebellar peduncles
13. Derivatives of the neurocrest include all of the following except the:
    a. sensory ganglia
    b. oligodendroglia
    c. Schwann cells
    d. head and neck mesenchyme
    e. melanoblasts
14. The cerebellar cortex of the early stages contains an internal germinal and an external germinal layer. The internal germinal layer cells give rise to:
    a. deep cerebellar nuclei
    b. basket cells
    c. granular cells
    d. stellate cells
15. The Purkinje cells of the cerebellum arise from its:
    a. external germinal layer
    b. internal germinal layer
    c. both
    d. none
16. The sulcus limitans, which demarcates the alar plate and the basal plate in the neural tube, extends in all of the following regions except the:
    a. myelencephalon
    b. metencephalon
    c. mesencephalon
    d. diencephalon

17. The human cerebral cortex is arranged into discrete radial units. Each unit consists of the:
    a. pillar cell and the neurons that are migrating on it
    b. neuroblasts related to the primordial plexiform layer
    c. neuroblasts produced in the specific region of the ventricular zone
    d. neuroblasts produced in the specific region of the marginal zone
    e. there is no such unit

18. Closure of the neural pores:
    a. begins at the 4th somite and then progresses rostrally and caudally
    b. is essential for further differentiation of the neural tube
    c. causes formation of brain vesicles
    d. is influenced by the notochord
    e. all of the above

19. The myelin sheath of a peripheral nerve fiber is formed by the cells that originate from:
    a. mesenchymal cells
    b. neuroblasts
    c. the neural crest
    d. glioblasts
    e. primitive blood cells

20. Hydrocephaly results from:
    a. aqueduct stenosis
    b. choroidal plexus papilloma
    c. chronic meningitis
    d. hydromyelin
    e. subarachnoid hemorrhage

21. In spina bifida occulta:
    a. the vertebral arches in the affected area are underdeveloped but the defect is covered by the skin
    b. the neural groove fails to close
    c. a cyst, containing neural tissue, leptomeninges and glial tissue, appears
    d. only the meninges are present in the cyst
    e. all of the above are present

22. Anencephaly:
    a. in most part is a mesodermal defect
    b. primarily is attributed to disturbance in the interaction of neural ectoderm and associated mesoderm
    c. is characterized by displacement of the cerebellum and medulla through the foramen magnum

d. is a type of holoprosencephaly
e. contains a cystic dilated fourth ventricle

23. The a-fetoprotein levels in amniotic fluid are increased in:
    a. neural tube defects
    b. a hydatidiform mole
    c. an omphalocele
    d. multiple pregnancies
    e. all of the above

24. The chromosomal abnormality associated with anomalies of the brain most likely is:
    a. monosomy (45,XO)
    b. Klinefelter's syndrome (44,XXY)
    c. trisomy 21
    d. cat-cry syndrome
    e. supermale (44,XYY)

25. Intrauterine rubella infection can result in:
    a. vascular thrombosis
    b. hydrocephaly
    c. purulent meningitis
    d. microcephaly
    e. chorioretinitis

ANSWERS TO REVIEW QUESTIONS

1. a	2. c	3. d	4. a
5. d	6. e	7. a	8. b
9. c	10. d	11. b	12. d
13. b	14. a	15. b	16. d
17. a	18. e	19. c	20. a
21. a	22. b	23. e	24. c
25. d			

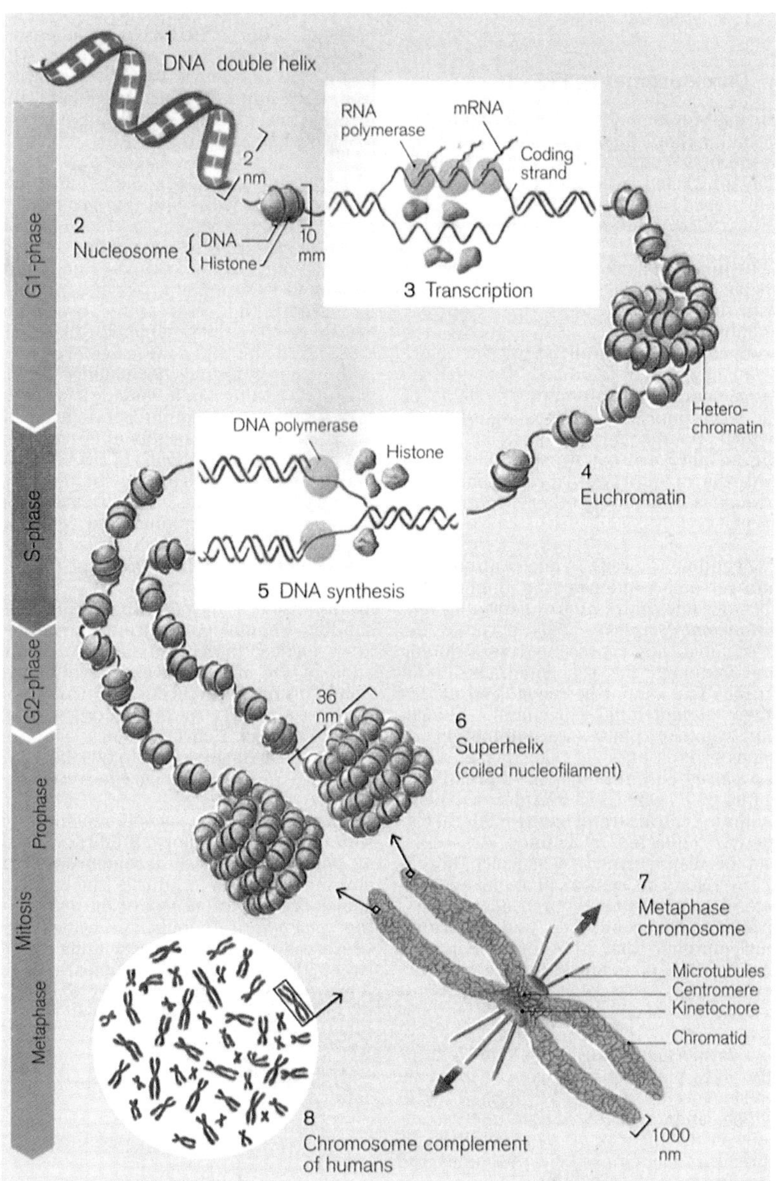

Structure of chromosomes

Plate 1.1

Chromosomes. Reproduced with permission from Thieme Medical Publishers Inc., New York, 1995, *Color Atlas of Embryology,* Ulrich Drews, Chapter 1: Reproduction.

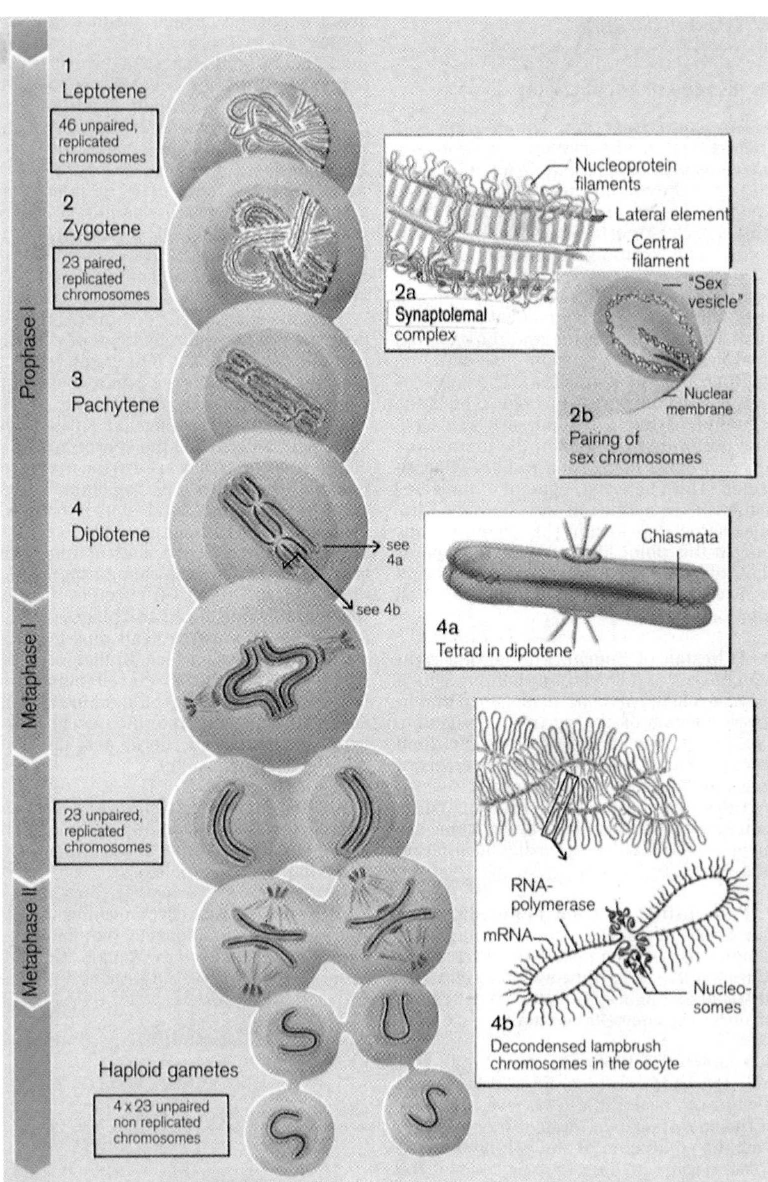

Chromosomes in meiosis

Plate 1.2

Meiosis. Reproduced with permission from Thieme Medical Publishers Inc., New York, 1995, *Color Atlas of Embryology,* Ulrich Drews, Chapter 1: Reproduction.

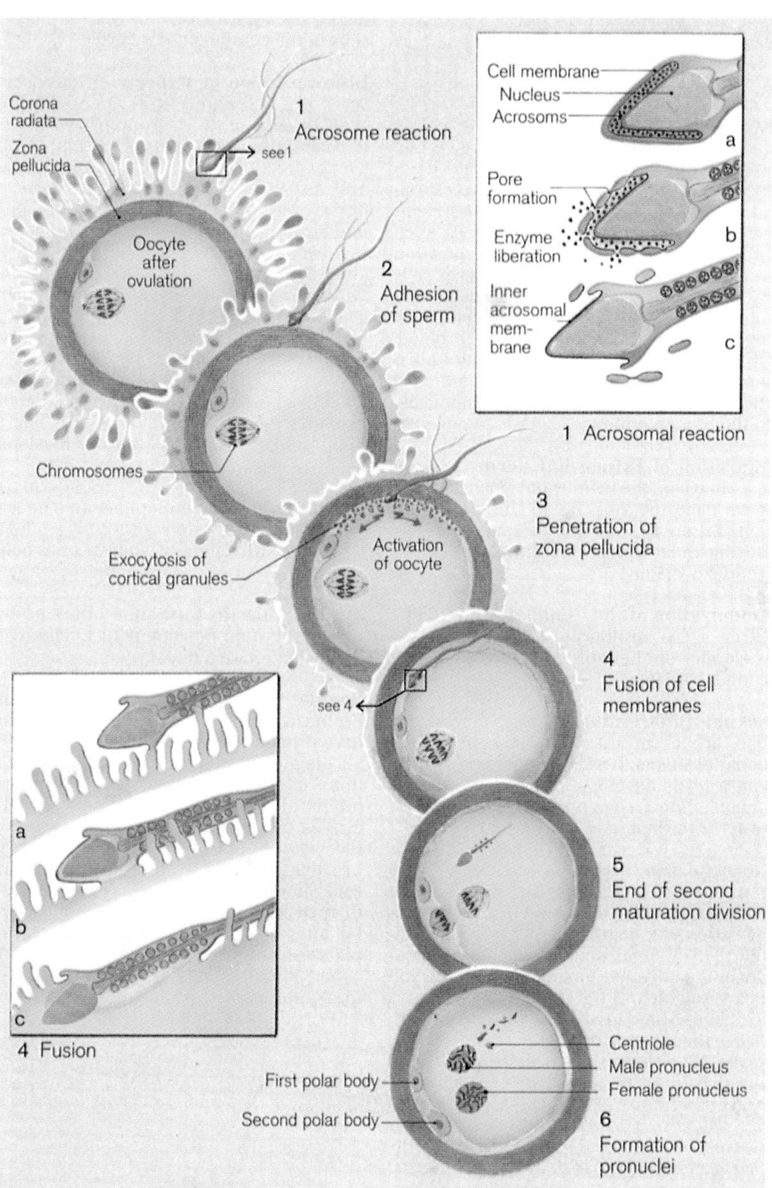

Plate 3.1

Fertilization. Reproduced with permission from Thieme Medical Publishers Inc., New York, 1995, *Color Atlas of Embryology,* Ulrich Drews, Chapter 1: Reproduction.

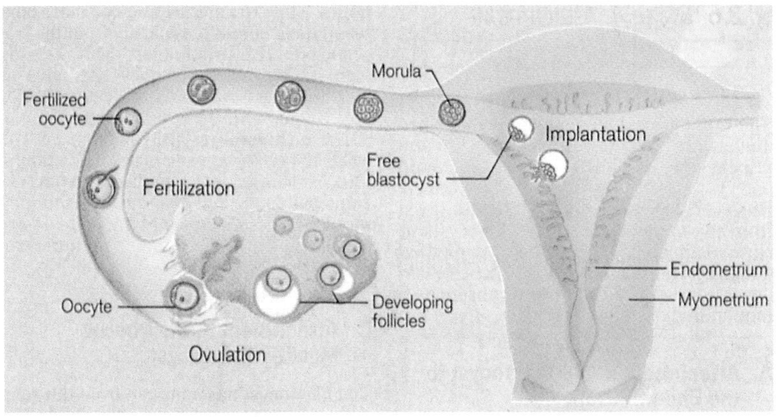

From ovulation to implantation

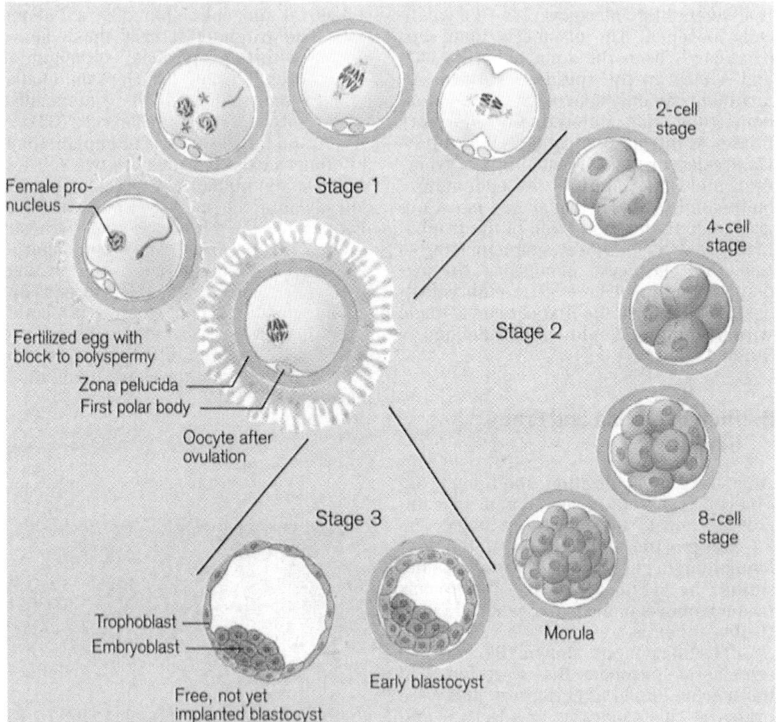

Development of oocyte to blastocyst

Plate 3.2

Overview: "first week." Reproduced with permission from Thieme Medical Publishers Inc., New York, 1995, *Color Atlas of Embryology*, Ulrich Drews, Chapter 2: Human Development.

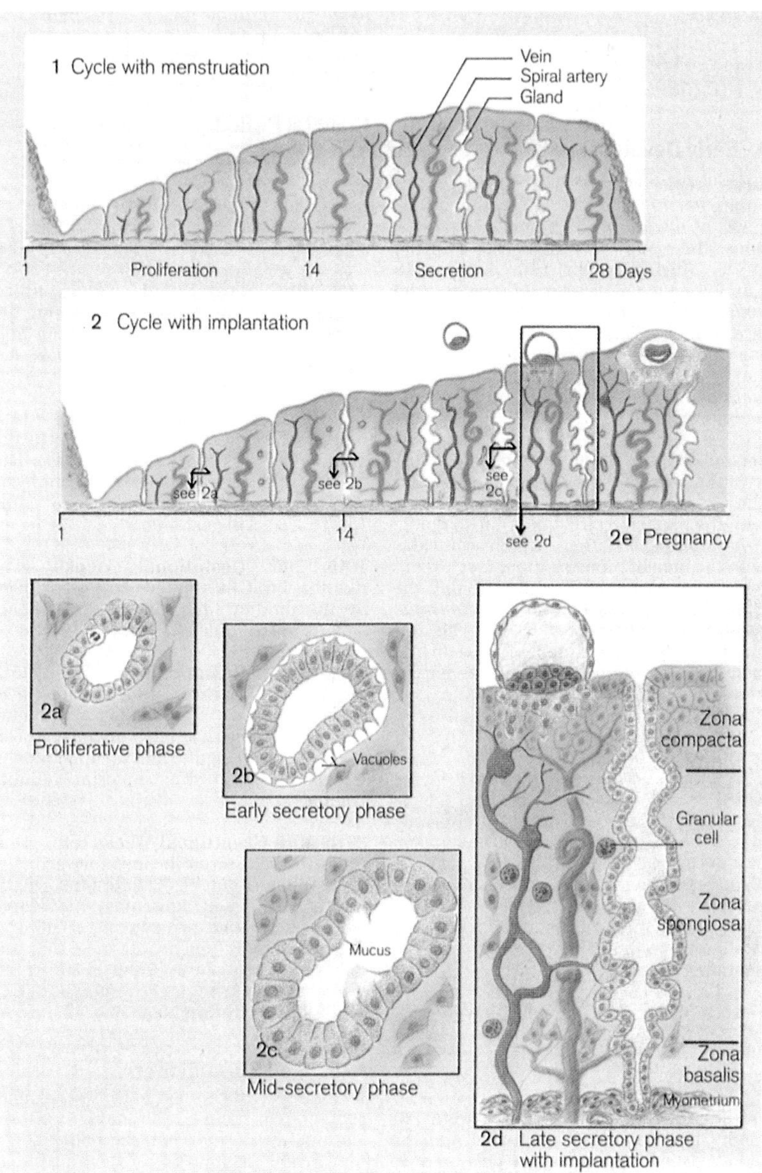

Preparation of endometrium for implantation

Plate 4.2

Endometrium. Reproduced with permission from Thieme Medical Publishers Inc., New York, 1995, *Color Atlas of Embryology,* Ulrich Drews, Chapter 1: Reproduction.

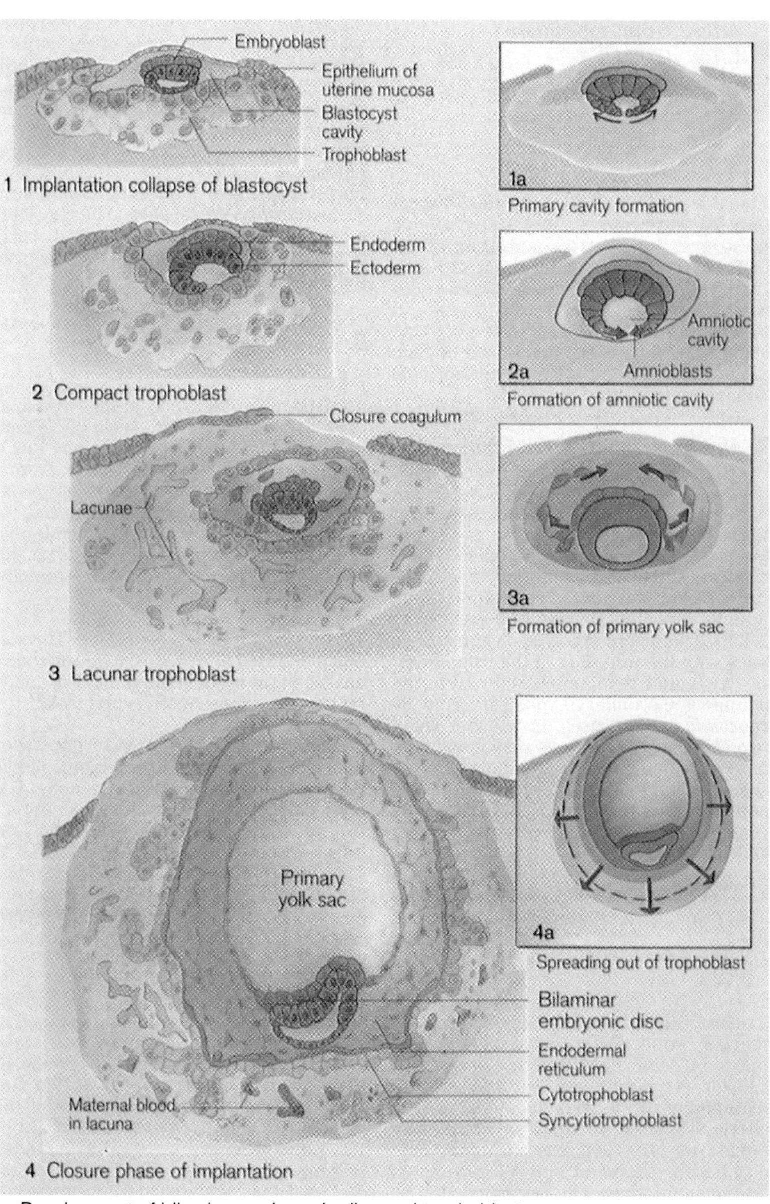

Development of bilaminar embryonic disc and trophoblast

Plate 5.1

Stage 5: bilaminar embryonic disk. Reproduced with permission from Thieme Medical Publishers Inc., New York, 1995, *Color Atlas of Embryology,* Ulrich Drews, Chapter 2: Human Development.

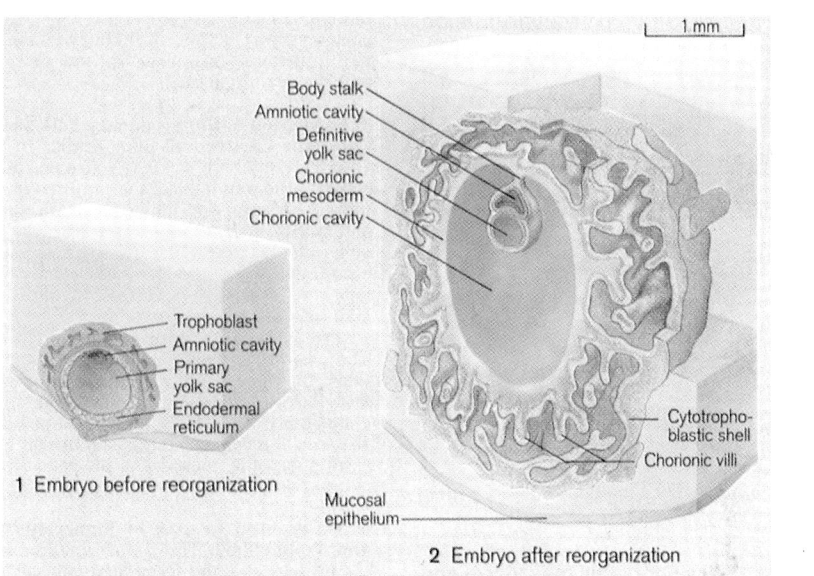

Reorganization of the embryonic cavities

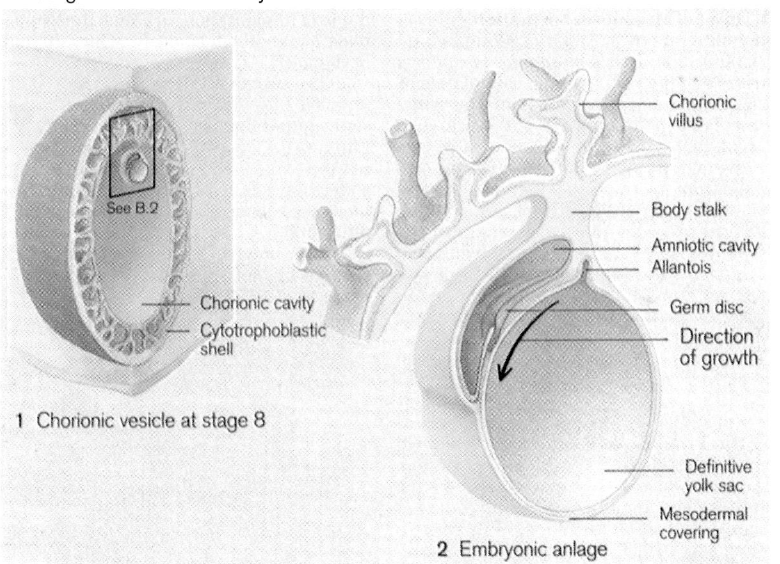

Direction of growth of germ disc

Plate 5.2

Overview: third week. Reproduced with permission from Thieme Medical Publishers Inc., New York, 1995, *Color Atlas of Embryology*, Ulrich Drews, Chapter 2: Human Development.

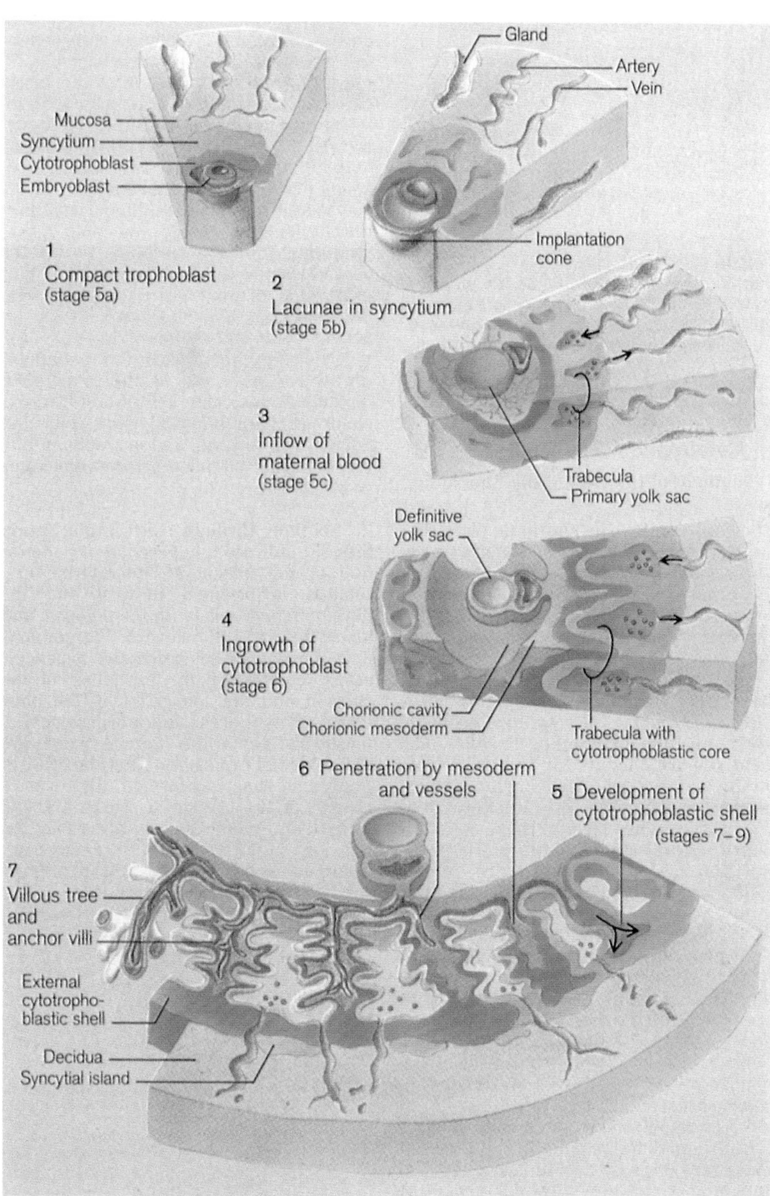

Development of chorionic villi

Plate 5.3

Overview: chorion and chorionic villi. Reproduced with permission from Thieme Medical Publishers Inc., New York, 1995, *Color Atlas of Embryology,* Ulrich Drews, Chapter 2: Human Development.

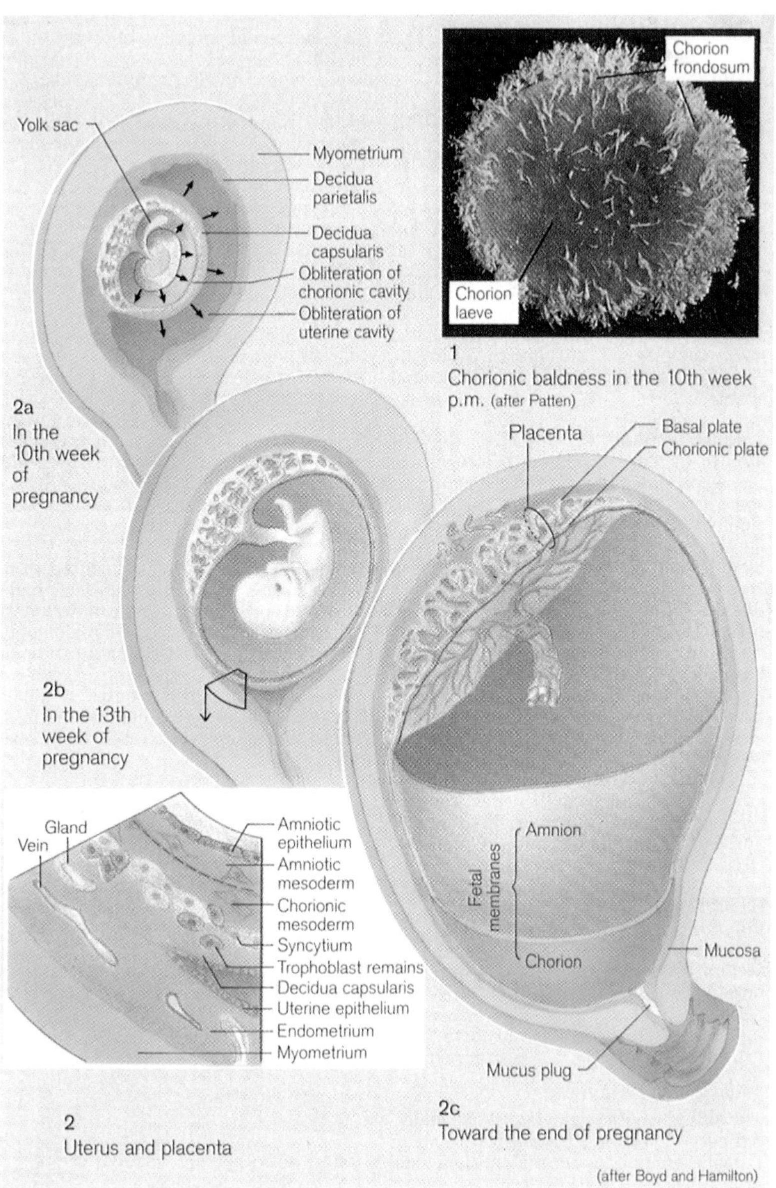

Development of placenta and fetal membranes

Plate 5.4

Overview: placenta and fetal membranes. Reproduced with permission from Thieme Medical Publishers Inc., New York, 1995, *Color Atlas of Embryology,* Ulrich Drews, Chapter 2: Human Development.

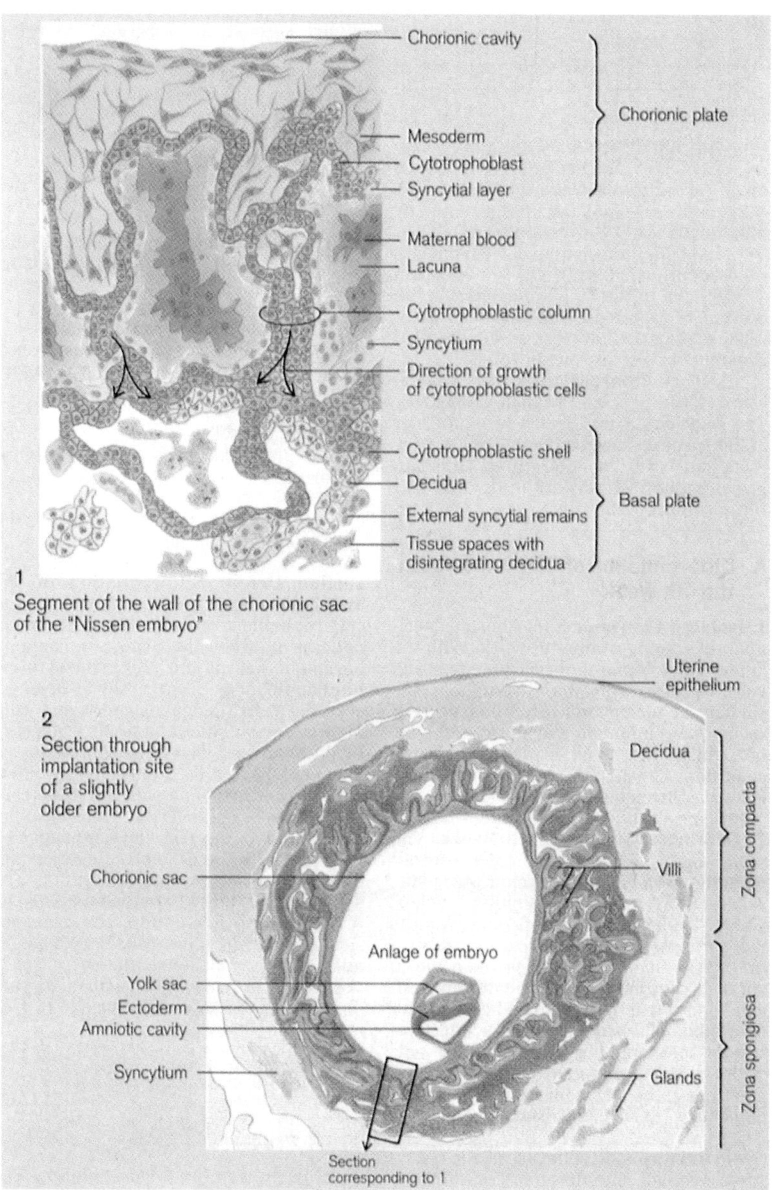

Development of the cytotrophoblastic shell at stage 7

Plate 5.5

Cytotrophoblastic shell. Reproduced with permission from Thieme Medical Publishers Inc., New York, 1995, *Color Atlas of Embryology*, Ulrich Drews, Chapter 2: Human Development.

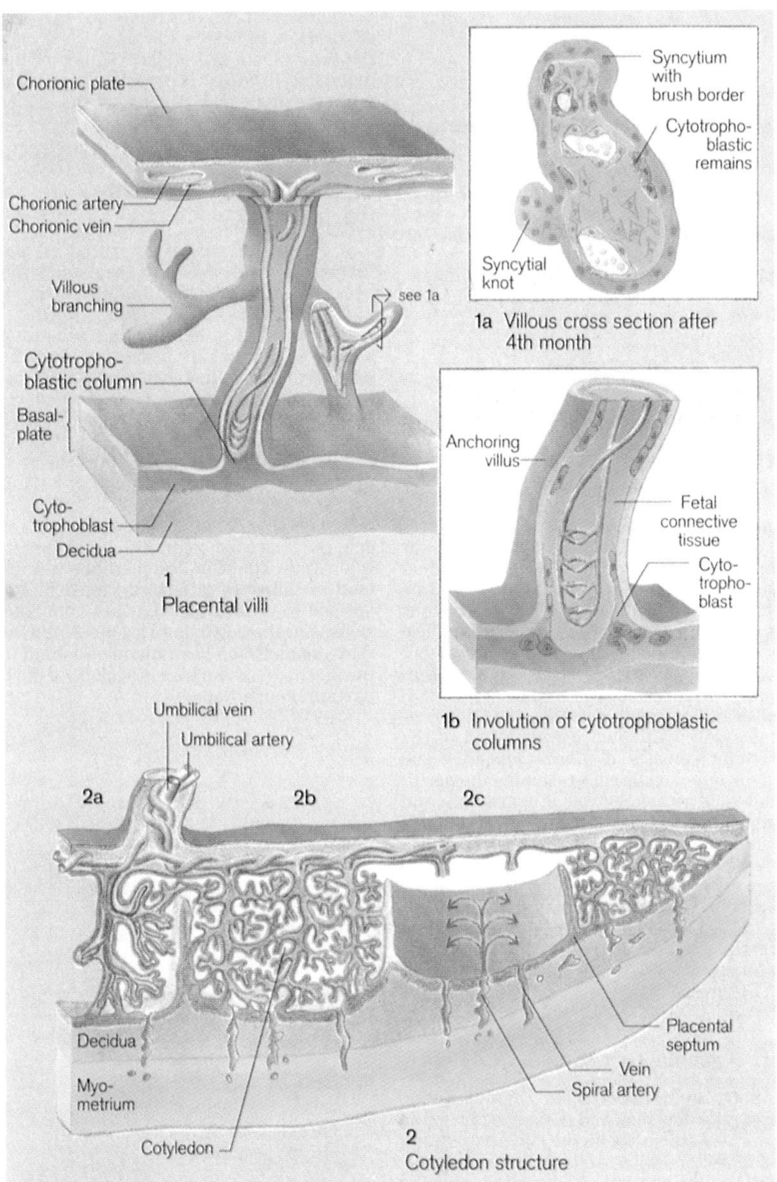

Functional structure of the placenta

Plate 5.6

Structure of the placenta. Reproduced with permission from Thieme Medical Publishers Inc., New York, 1995, *Color Atlas of Embryology*, Ulrich Drews, Chapter 2: Human Development.

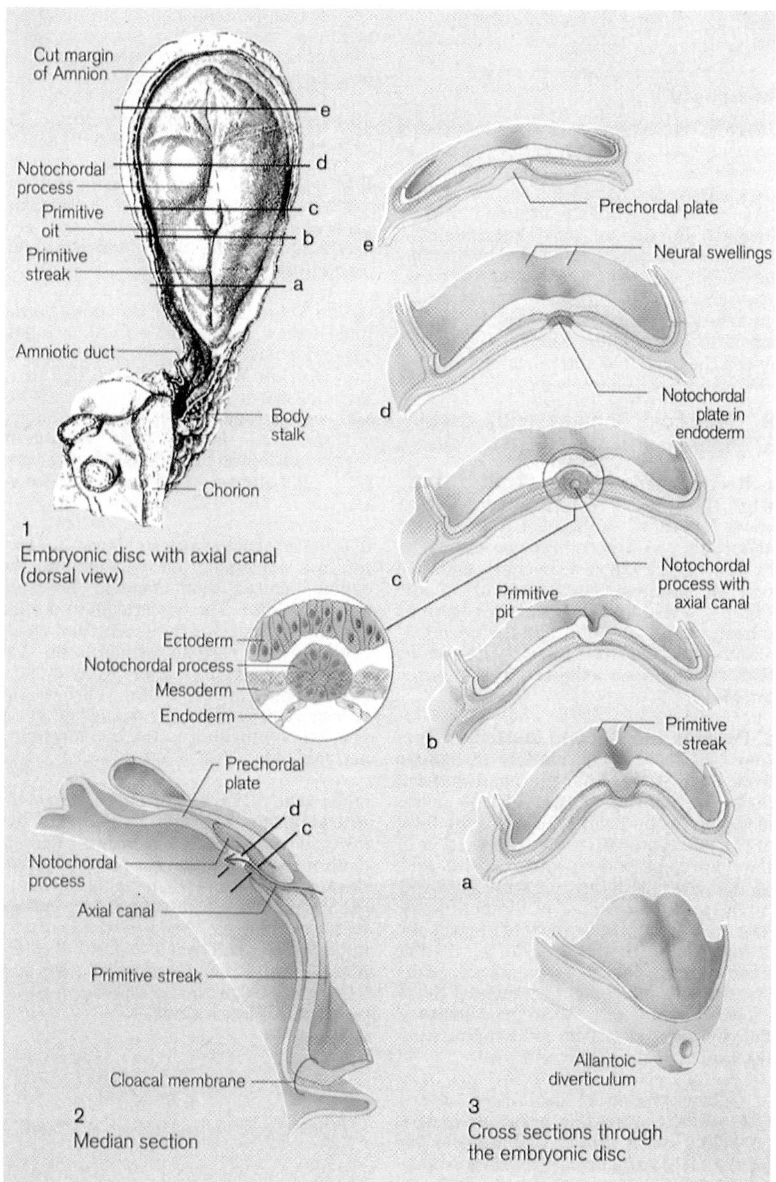

Axial canal and intercalation of notochord into roof of gut

Plate 6.1

Stage 8: axial canal. Reproduced with permission from Thieme Medical Publishers Inc., New York, 1995, *Color Atlas of Embryology,* Ulrich Drews, Chapter 2: Human Development.

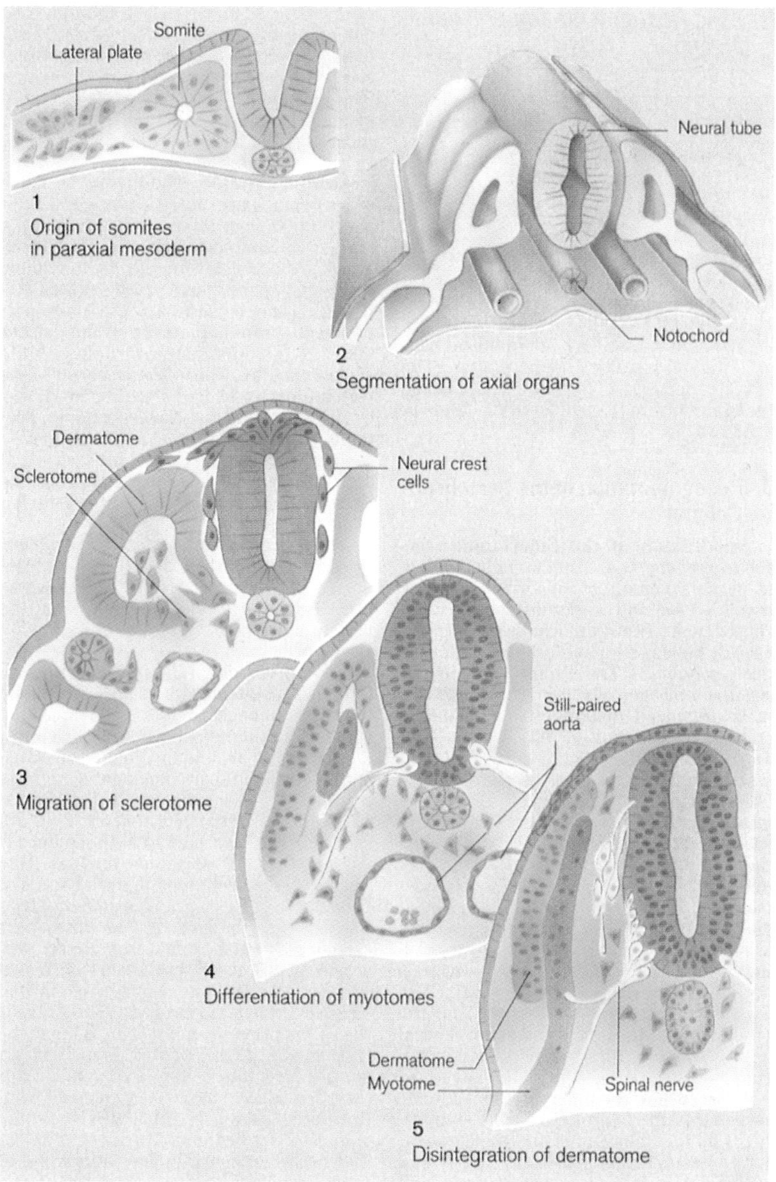

Differentiation of the somites

Plate 6.2

Differentiation of somites. Reproduced with permission from Thieme Medical Publishers Inc., New York, 1995, *Color Atlas of Embryology*, Ulrich Drews, Chapter 3: General Embryology.

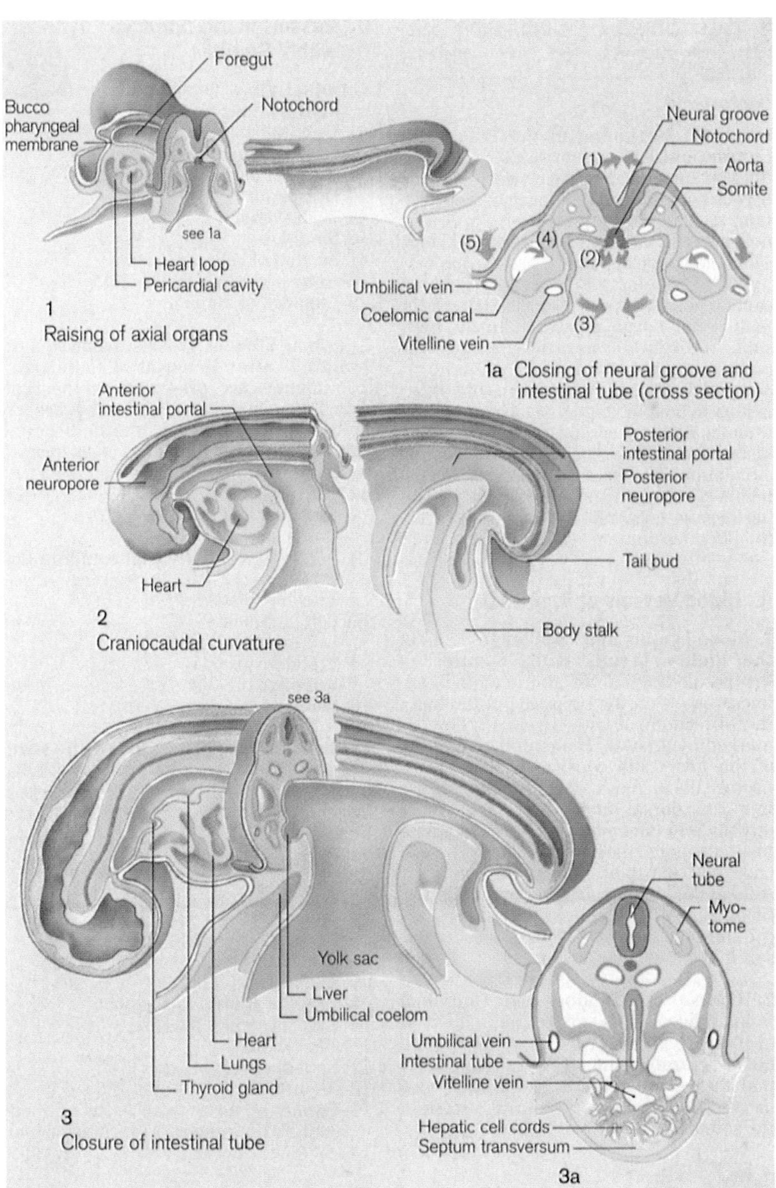

Folding from yolk sac and development of body form

Plate 6.3

Overview: folding. Reproduced with permission from Thieme Medical Publishers Inc., New York, 1995, *Color Atlas of Embryology*, Ulrich Drews, Chapter 2: Human Development.

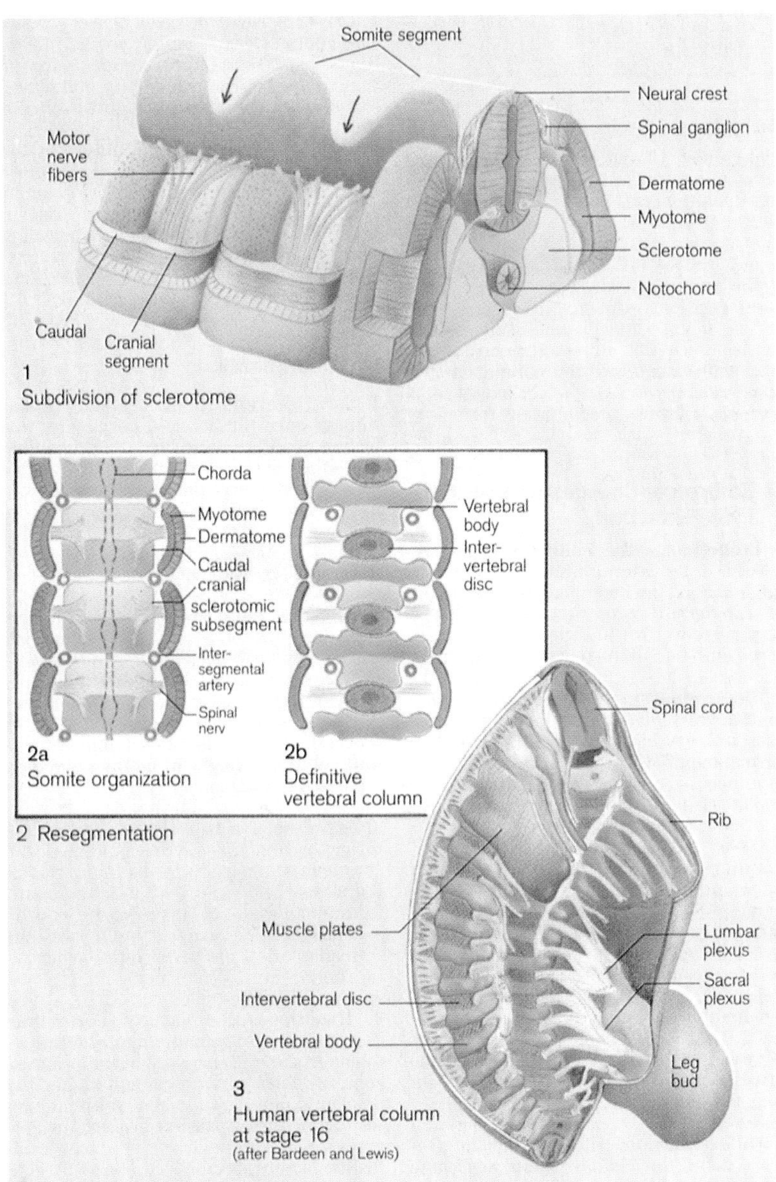

Resegmentation of axial organs

Plate 8.1

Resegmentation of axial organs. Reproduced with permission from Thieme Medical Publishers Inc., New York, 1995, *Color Atlas of Embryology,* Ulrich Drews, Chapter 3: General Embryology.

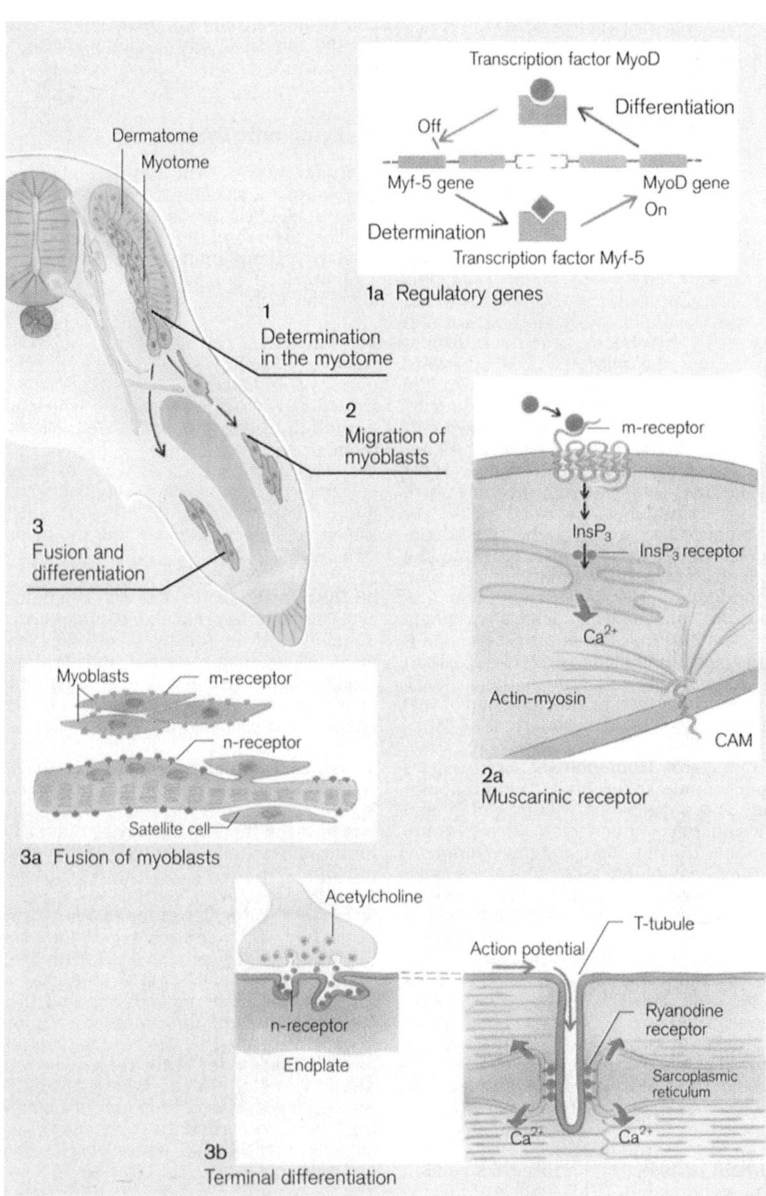

Development of skeletal musculature

Plate 8.2

Skeletal musculature. Reproduced with permission from Thieme Medical Publishers Inc., New York, 1995, *Color Atlas of Embryology*, Ulrich Drews, Chapter 3: General Embryology.

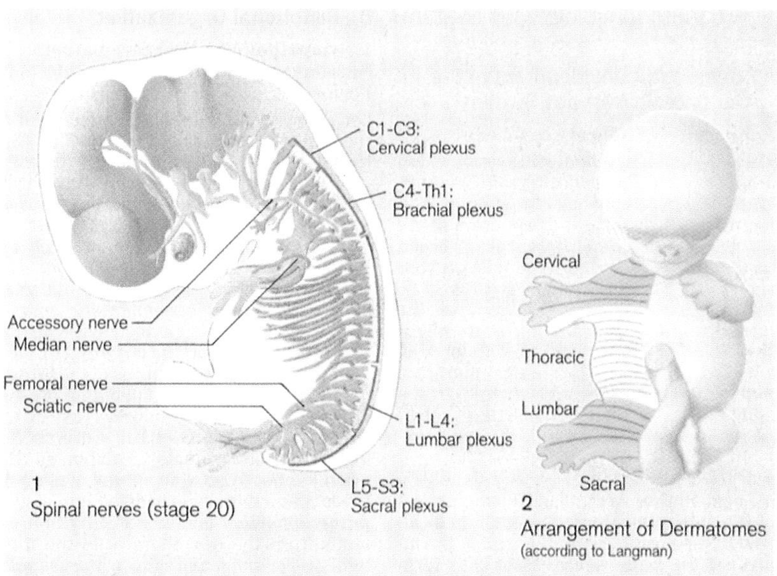

Spinal nerves and dermatomes

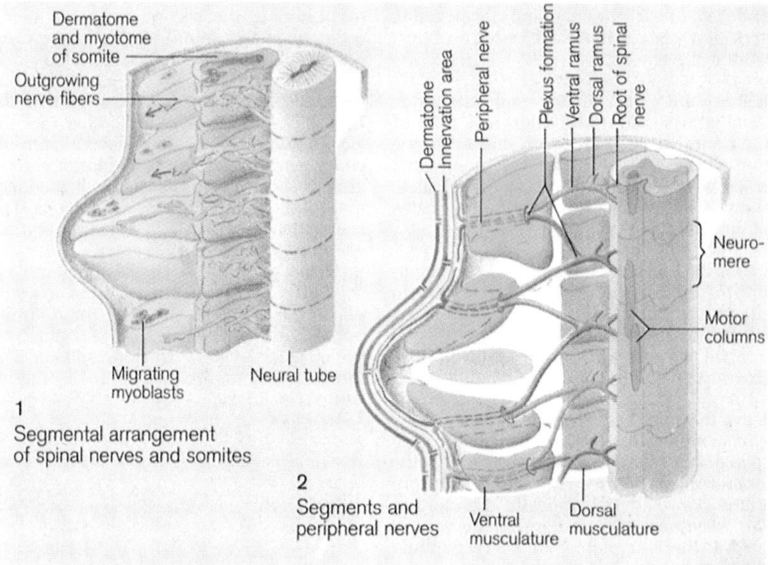

Peripheral nerves and dermatomes

Plate 8.6

Dermatomes. Reproduced with permission from Thieme Medical Publishers Inc., New York, 1995, *Color Atlas of Embryology,* Ulrich Drews, Chapter 4: Nervous System.

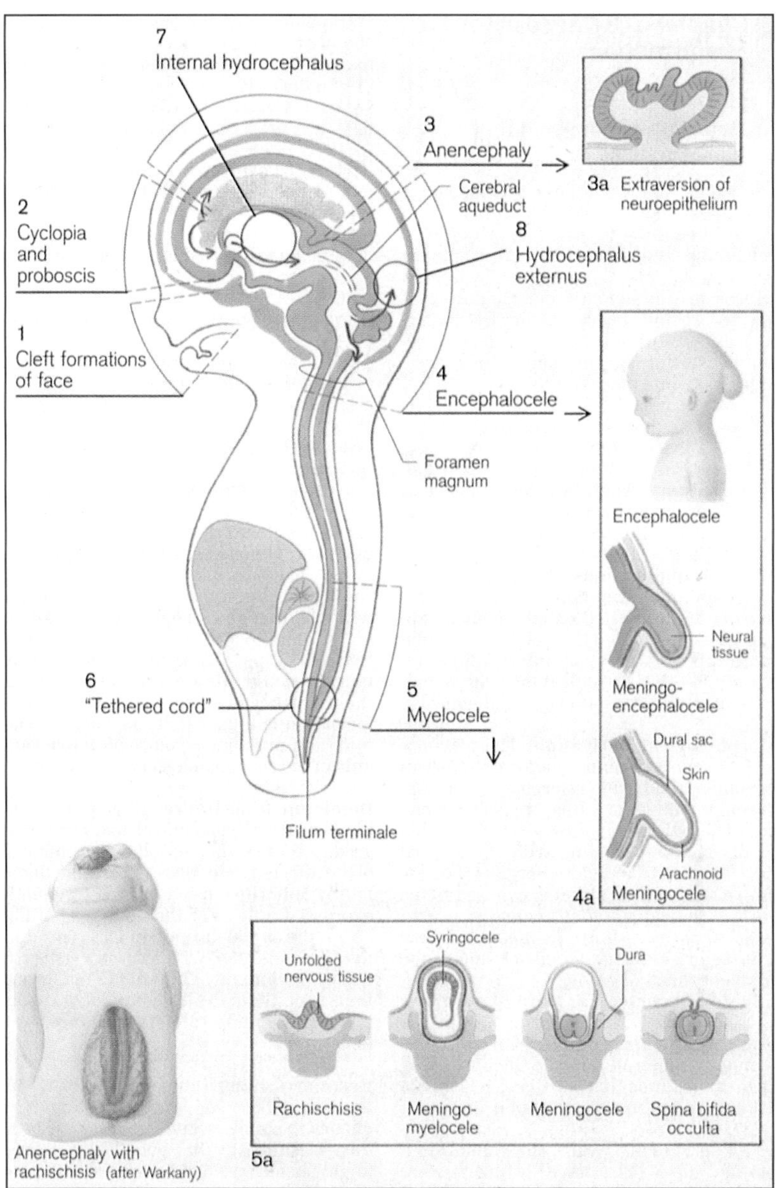

Malformations of head and CNS

Plate 8.7

Malformations. Reproduced with permission from Thieme Medical Publishers Inc., New York, 1995, *Color Atlas of Embryology,* Ulrich Drews, Chapter 9: Head.

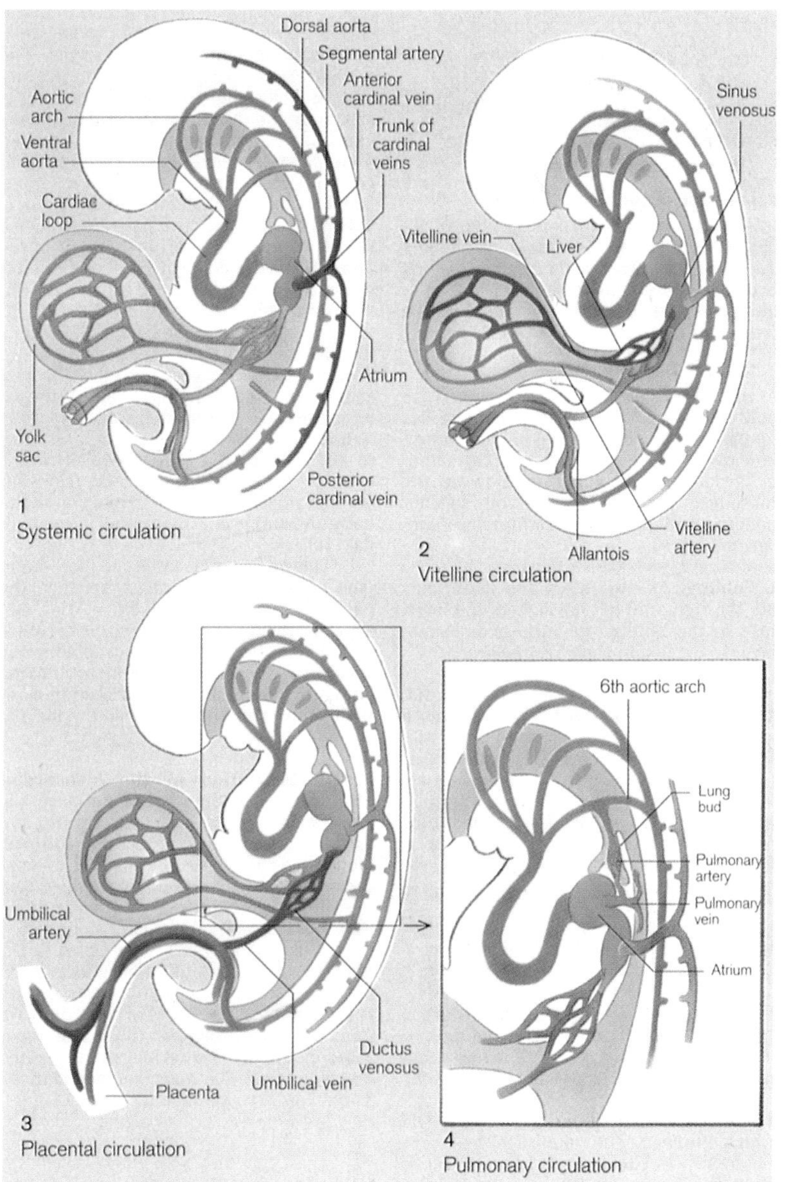

Circulatory systems in the embryo

Plate 9.1

Circulatory systems in the embryo. Reproduced with permission from Thieme Medical Publishers Inc., New York, 1995, *Color Atlas of Embryology,* Ulrich Drews, Chapter 6: Heart and Vessels.

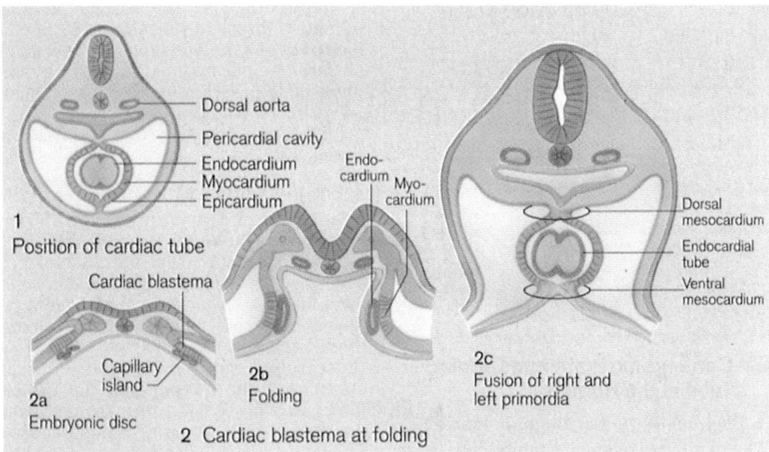

Origin of cardiac tube in the visceral mesoderm

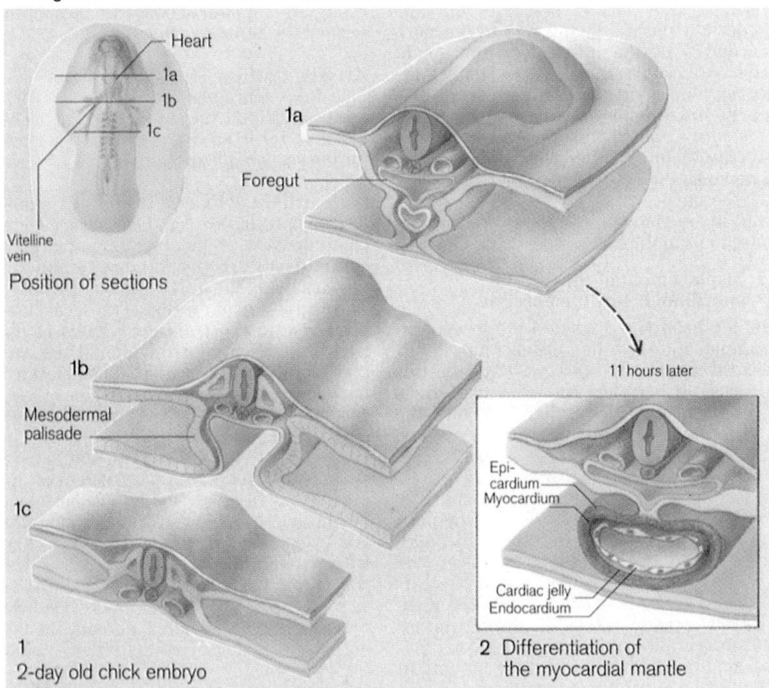

Development of the chick heart during embryonic folding

Plate 9.2

Formation of the heart tube. Reproduced with permission from Thieme Medical Publishers Inc., New York, 1995, *Color Atlas of Embryology,* Ulrich Drews, Chapter 6: Heart and Vessels.

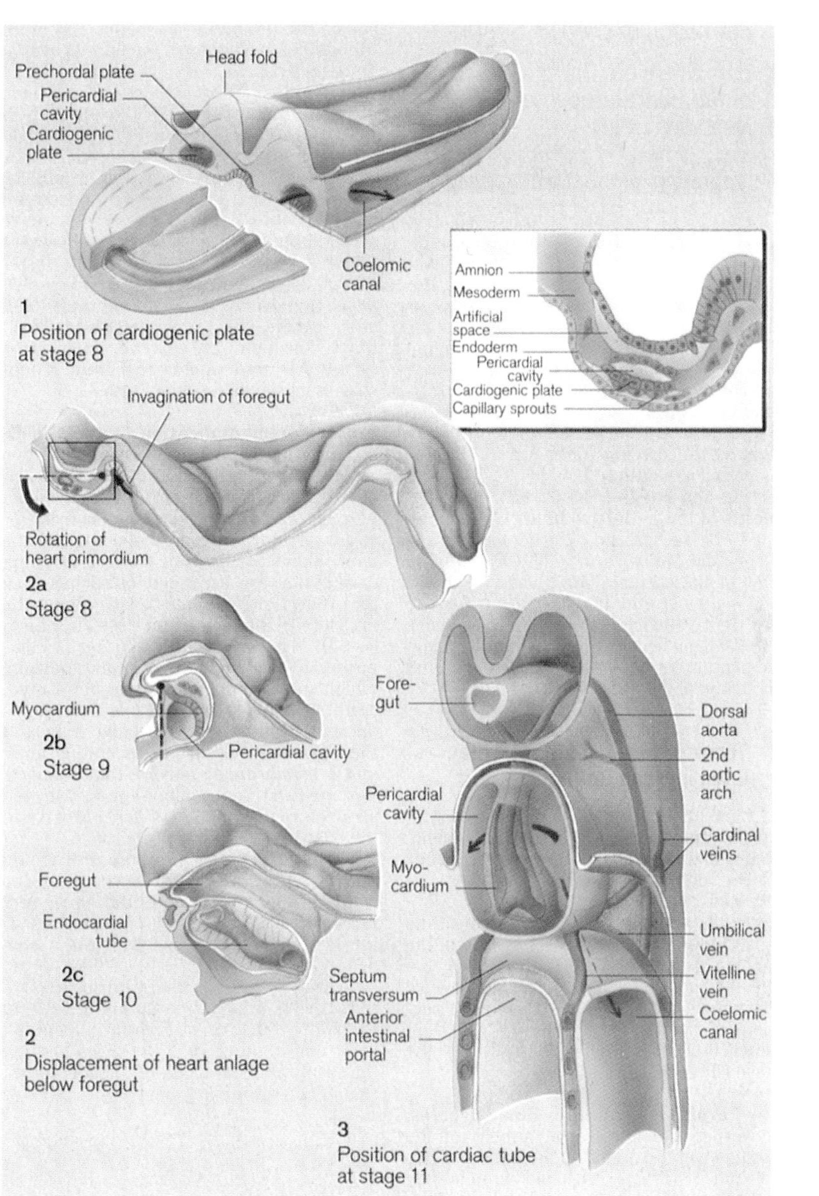

Cardiogenic plate and cardiac tube in the human

Plate 9.3

Cardiac tube in the human. Reproduced with permission from Thieme Medical Publishers Inc., New York, 1995, *Color Atlas of Embryology,* Ulrich Drews, Chapter 6: Heart and Vessels.

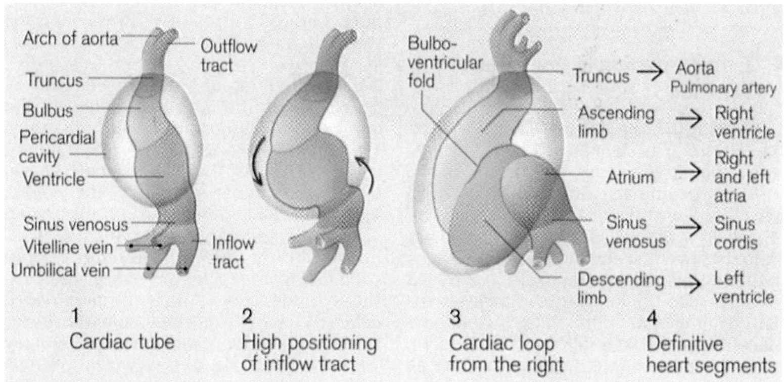

Segments of the cardiac loop

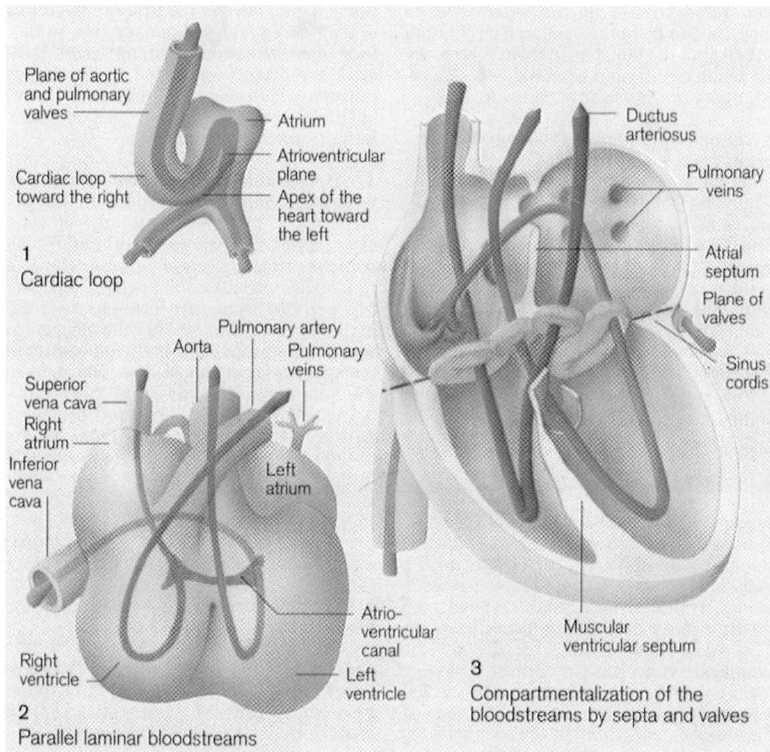

Transformation of the cardiac loop into the definitive heart

Plate 9.4

Overview: transformation of the heart loop. Reproduced with permission from Thieme Medical Publishers Inc., New York, 1995, *Color Atlas of Embryology*, Ulrich Drews, Chapter 9: Head.

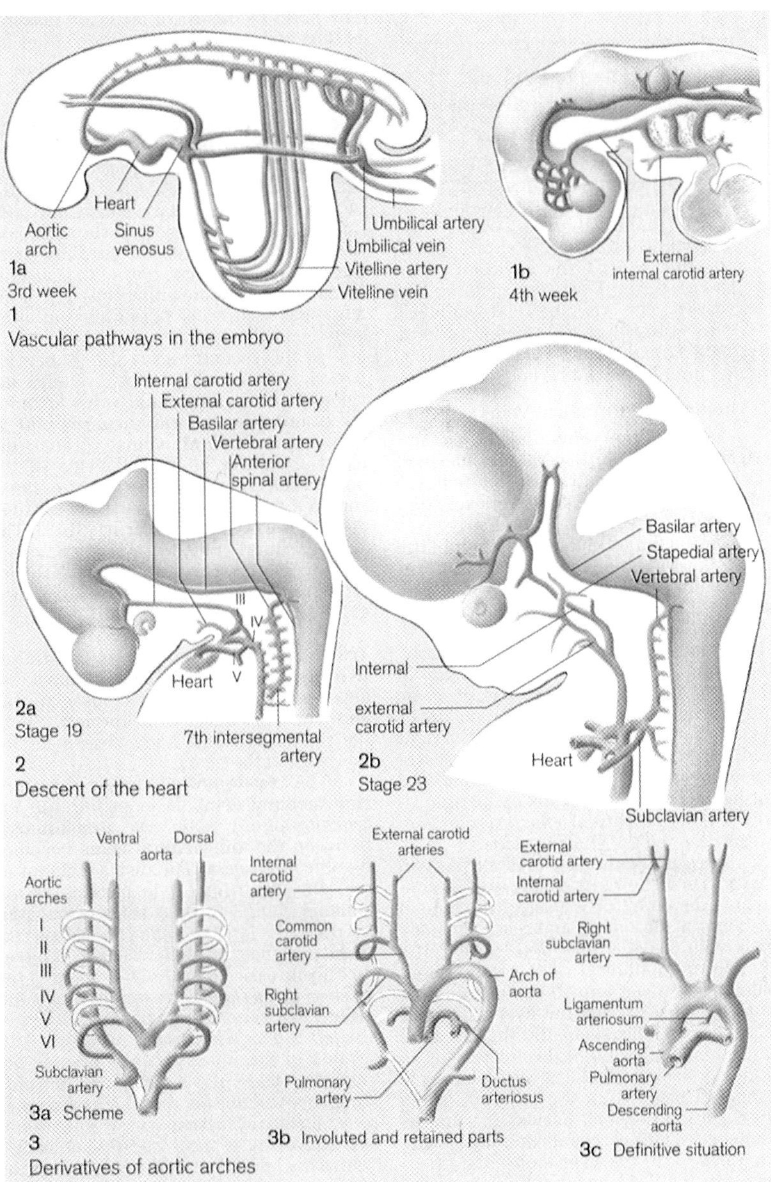

Development of arteries

Plate 9.7

Arteries. Reproduced with permission from Thieme Medical Publishers Inc., New York, 1995, *Color Atlas of Embryology*, Ulrich Drews, Chapter 6: Heart and Vessels.

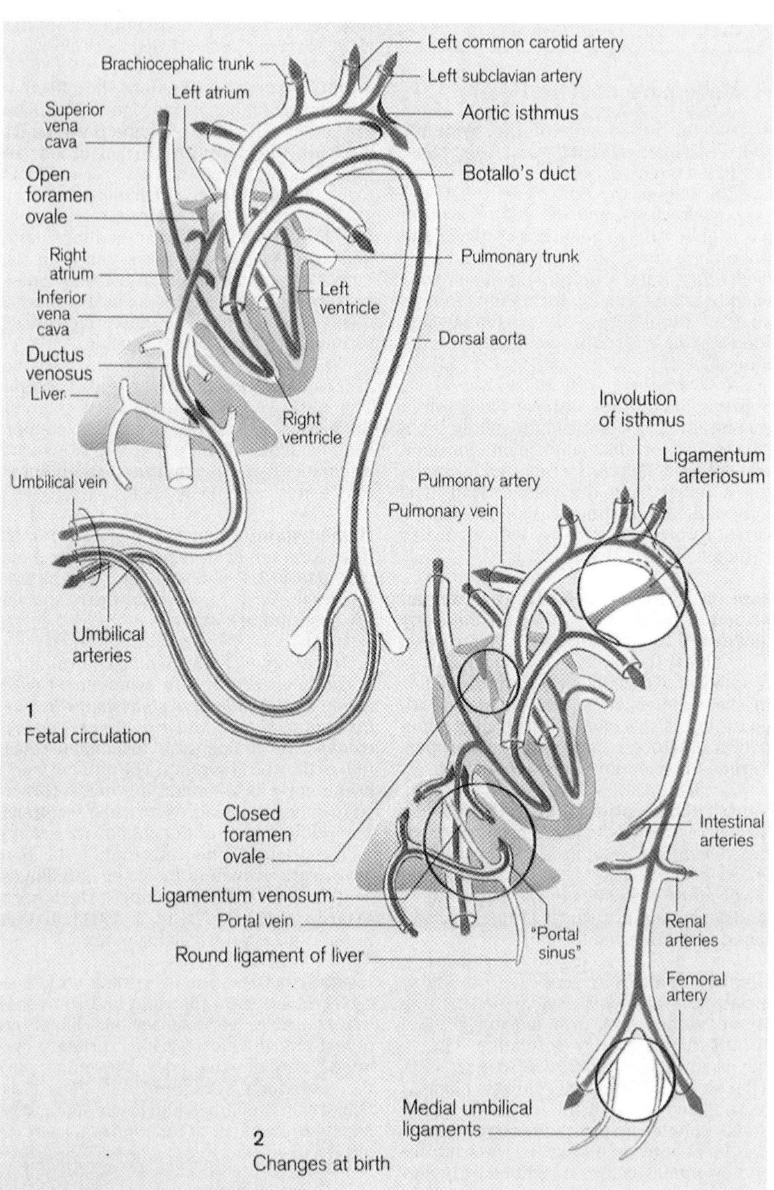

Changes in the circulation at birth

Plate 9.10

Changes in the circulation at birth. Reproduced with permission from Thieme Medical Publishers Inc., New York, 1995, *Color Atlas of Embryology*, Ulrich Drews, Chapter 6: Heart and Vessels.

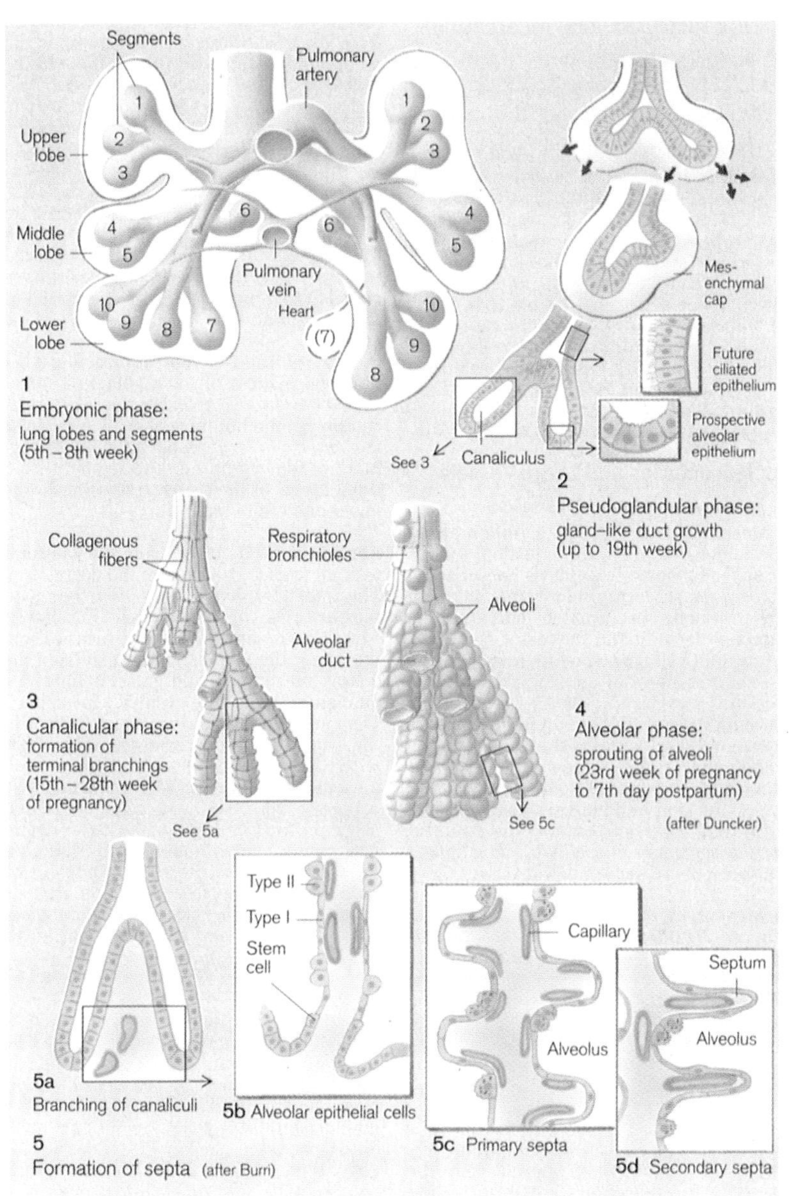

Differentiation of lungs

Plate 10.2

Differentiation of the lungs. Reproduced with permission from Thieme Medical Publishers Inc., New York, 1995, *Color Atlas of Embryology*, Ulrich Drews, Chapter 7: Gastrointestinal Tract.

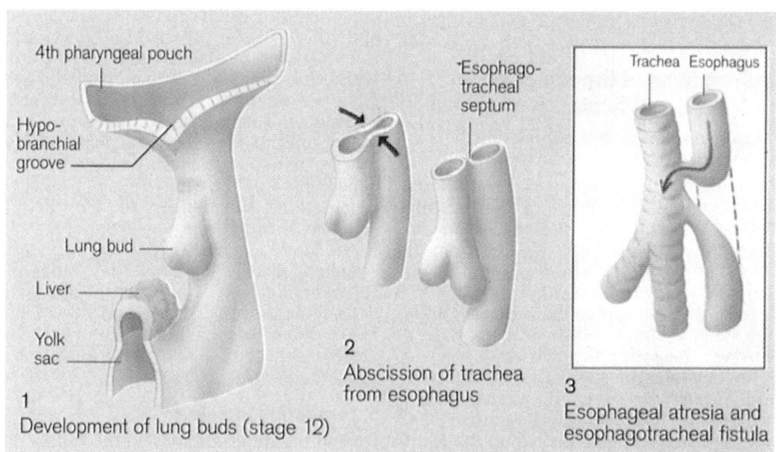

Development of lung buds

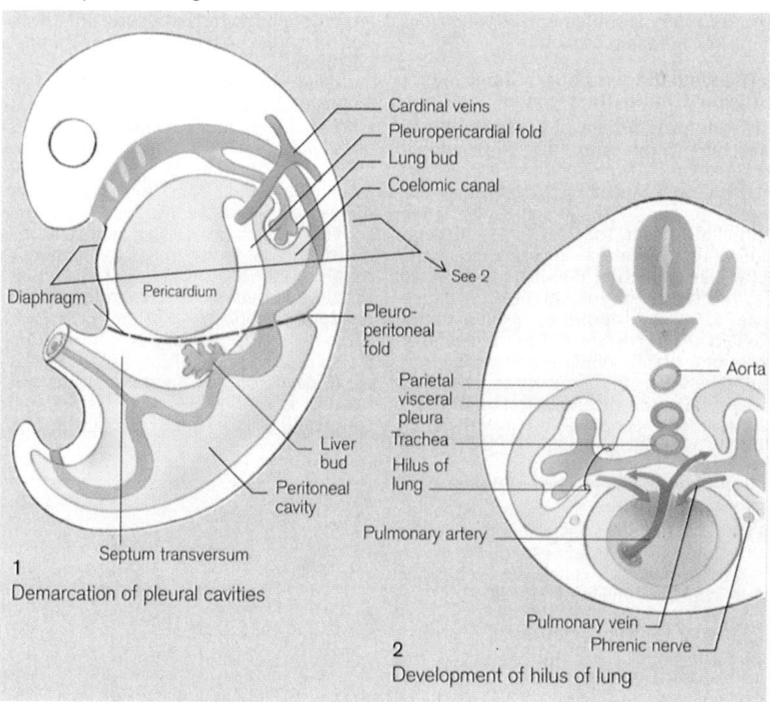

Growth of lungs into the coelomic canals

Plate 10.3

Lung buds. Reproduced with permission from Thieme Medical Publishers Inc., New York, 1995, *Color Atlas of Embryology*, Ulrich Drews, Chapter 7: Gastrointestinal Tract.

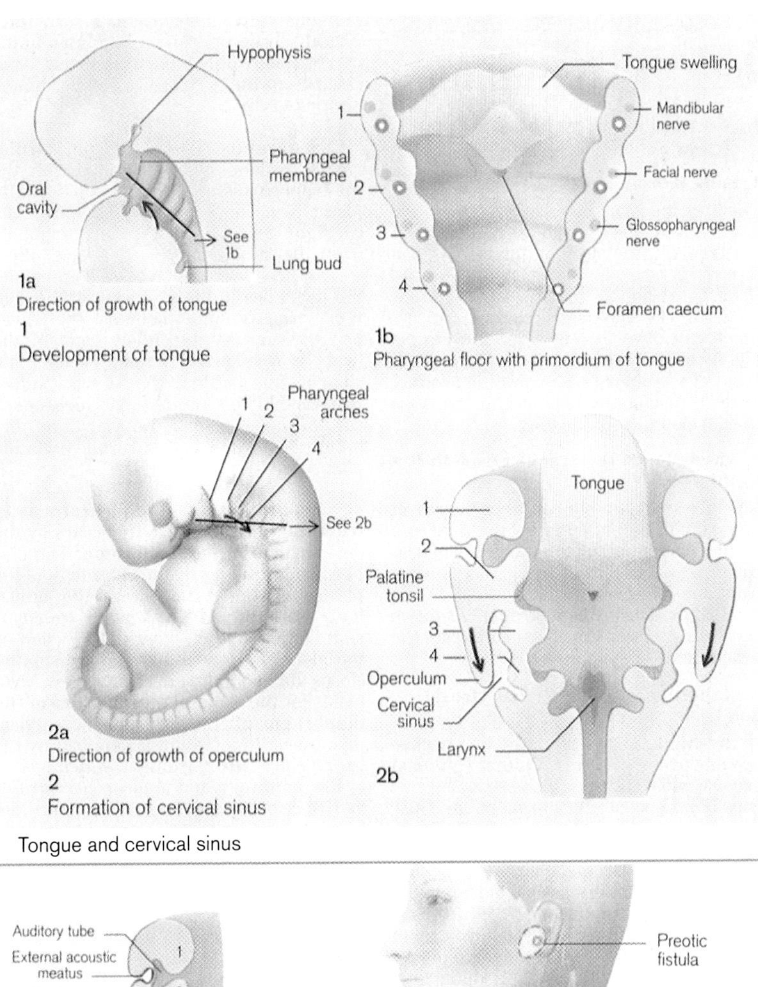

Tongue and cervical sinus

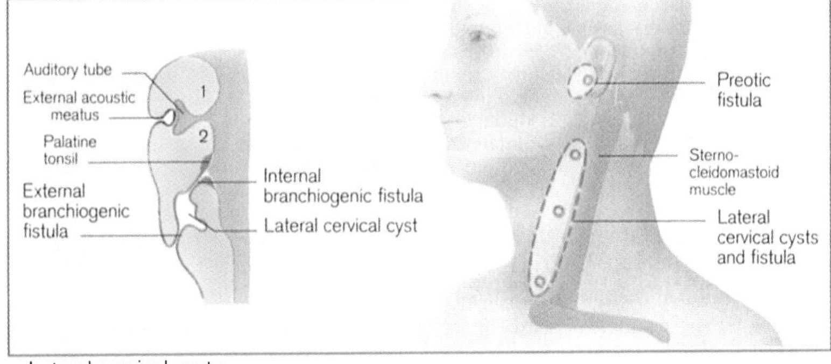

Lateral cervical cysts

Plate 11.1

Cervical sinus and cervical cysts. Reproduced with permission from Thieme Medical Publishers Inc., New York, 1995, *Color Atlas of Embryology,* Ulrich Drews, Chapter 7: Gastrointestinal Tract.

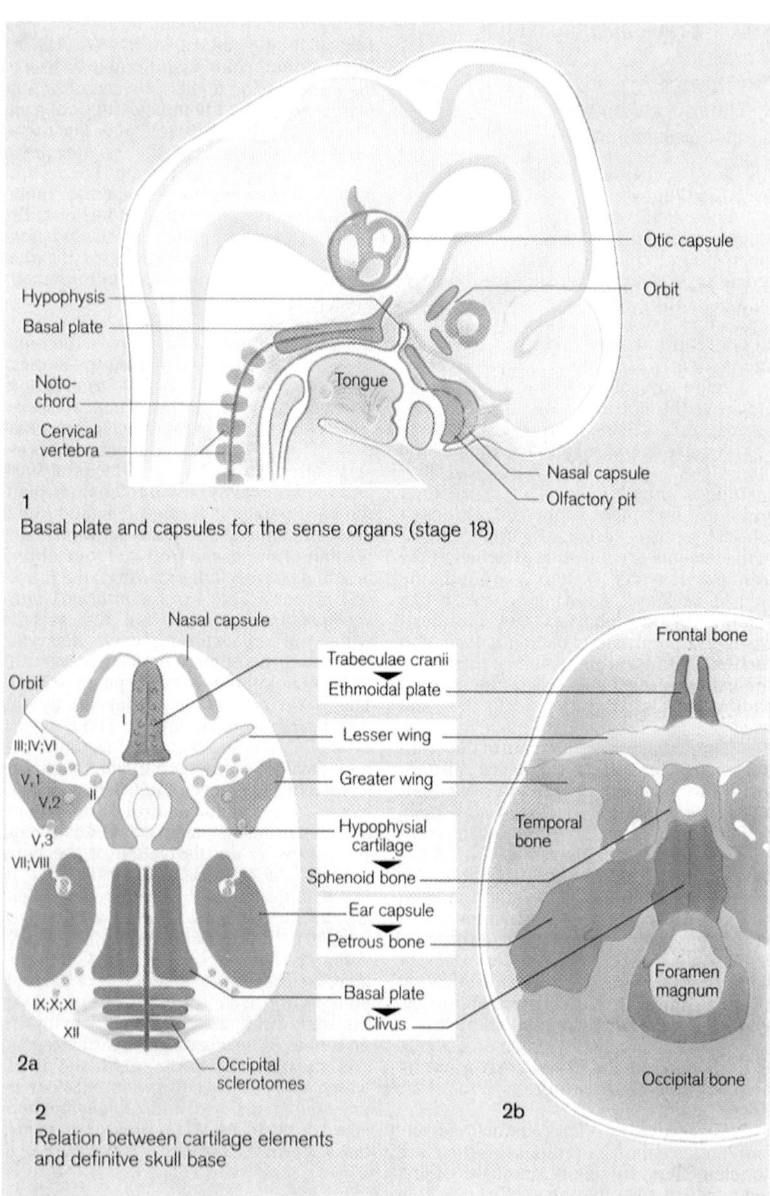

Development of the skull: chondrocranium

Plate 12.1

Skull base. Reproduced with permission from Thieme Medical Publishers Inc., New York, 1995, *Color Atlas of Embryology*, Ulrich Drews, Chapter 9: Head.

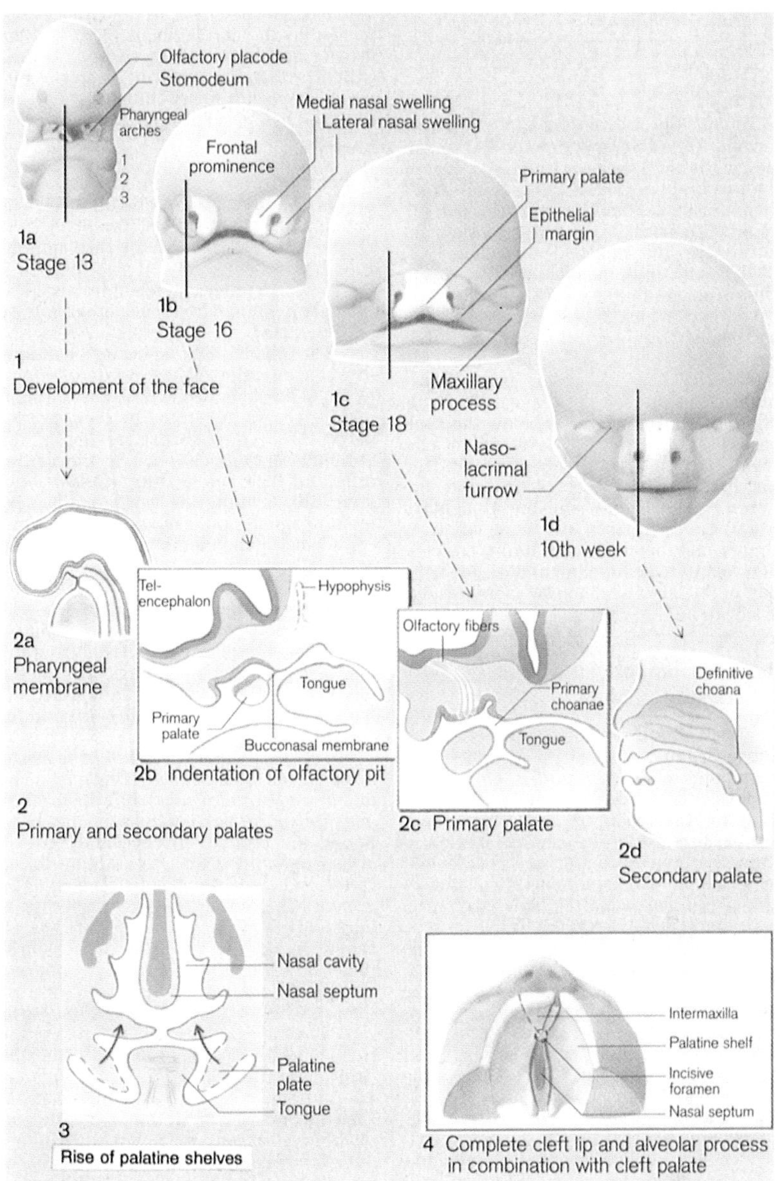

Facial swellings and palates

Plate 12.2

Facial swellings. Reproduced with permission from Thieme Medical Publishers Inc., New York, 1995, *Color Atlas of Embryology,* Ulrich Drews, Chapter 9: Head.

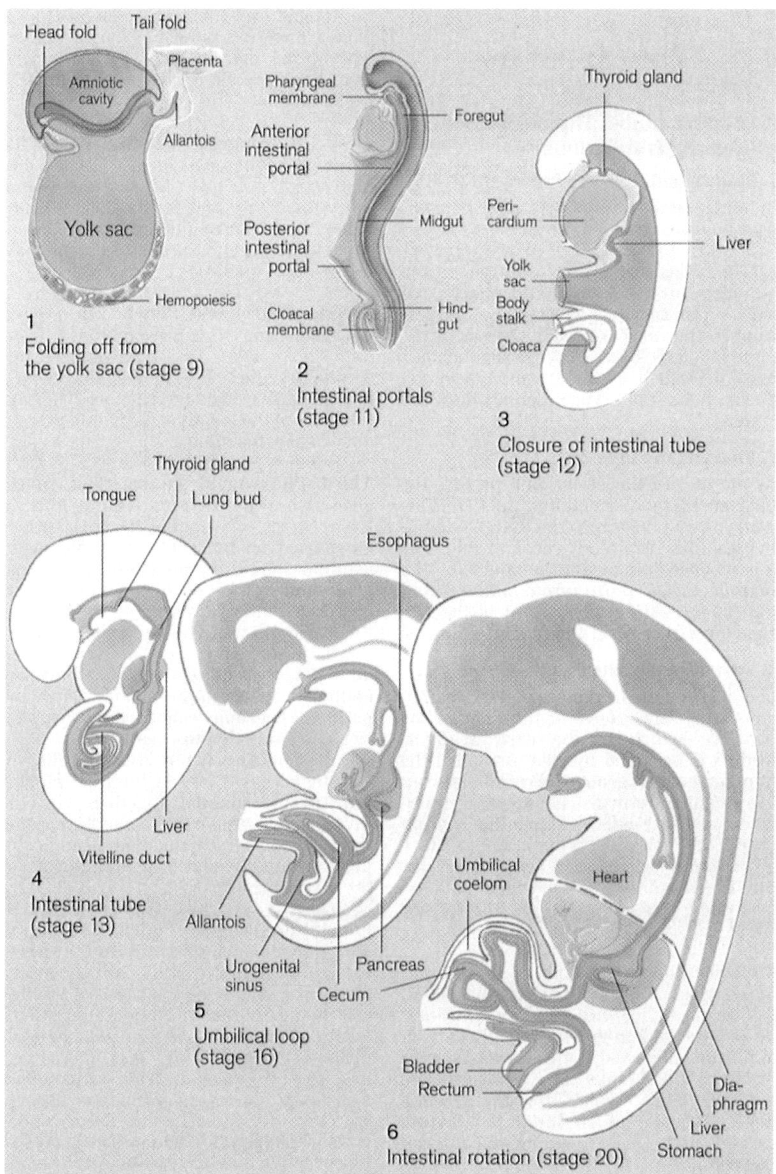

Development and organization of the digestive tube

Plate 13.1

Overview: digestive tube. Reproduced with permission from Thieme Medical Publishers Inc., New York, 1995, *Color Atlas of Embryology,* Ulrich Drews, Chapter 7: Gastrointestinal Tract.

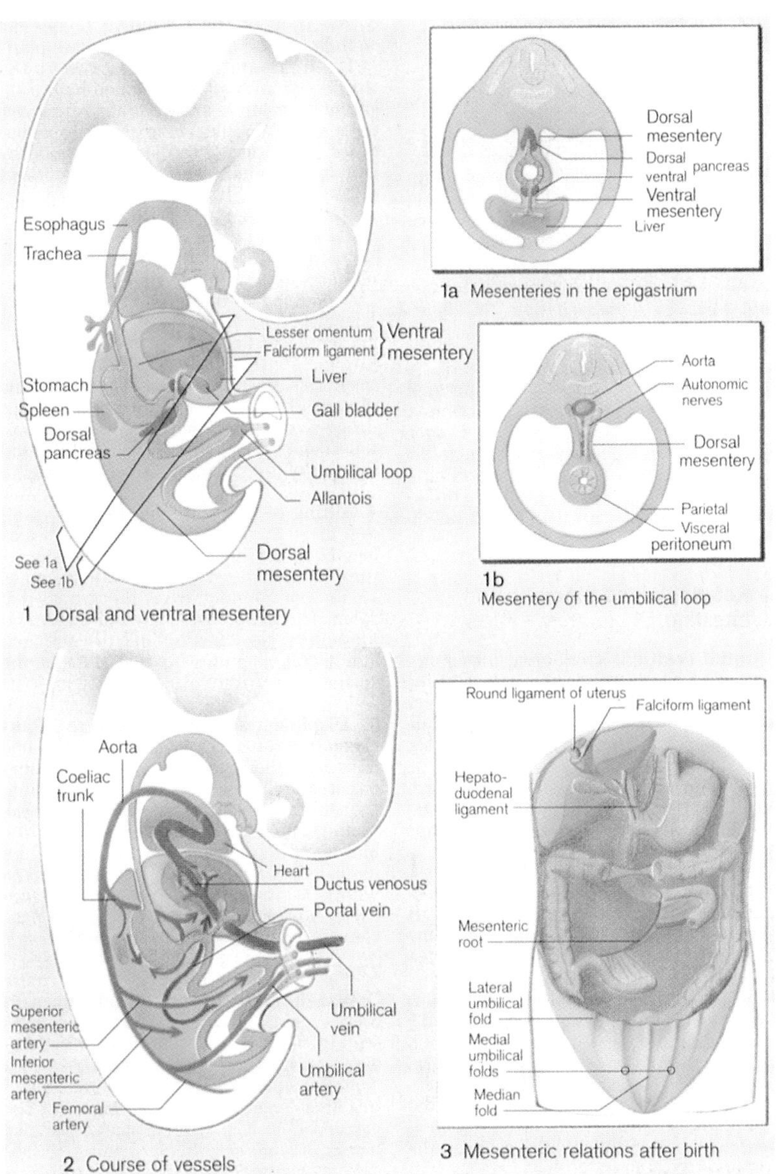

Mesenteries and large vessels

Plate 13.2

Overview: mesenteries. Reproduced with permission from Thieme Medical Publishers Inc., New York, 1995, *Color Atlas of Embryology,* Ulrich Drews, Chapter 7: Gastrointestinal Tract.

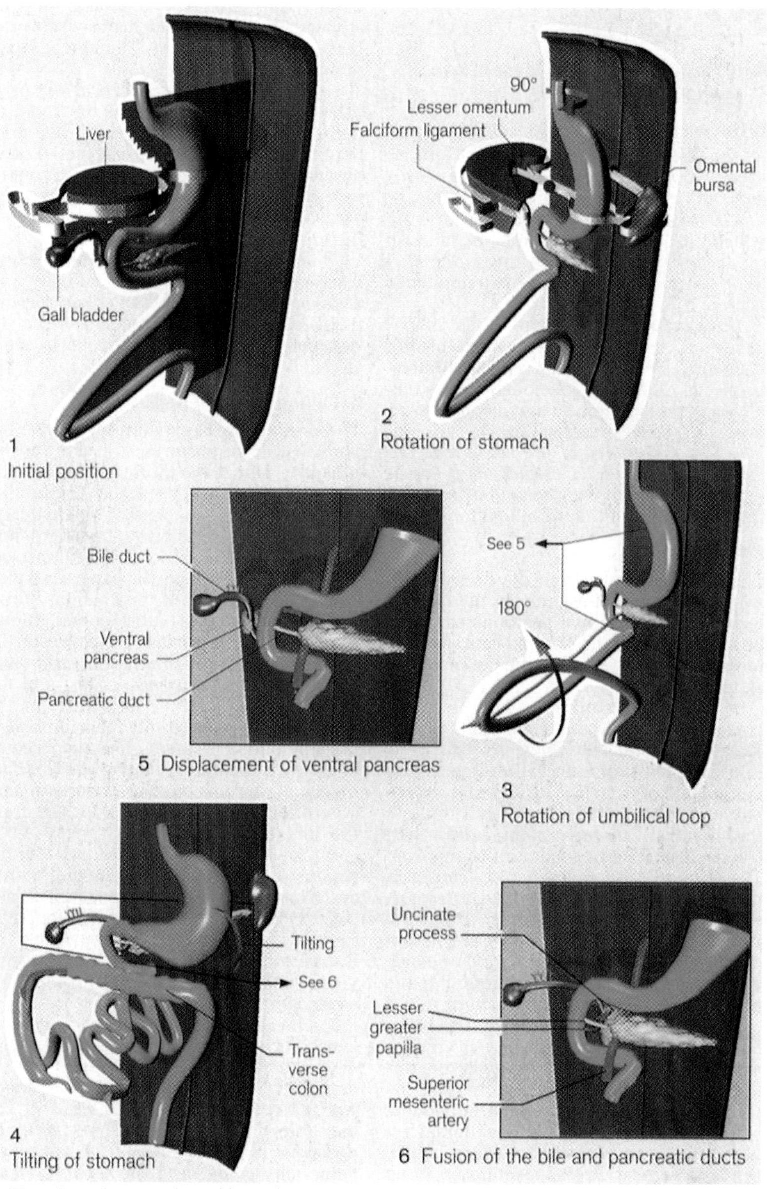

Rotation of stomach and intestine

Plate 13.3

Gastrointestinal rotation. Reproduced with permission from Thieme Medical Publishers Inc., New York, 1995, *Color Atlas of Embryology*, Ulrich Drews, Chapter 7: Gastrointestinal Tract.

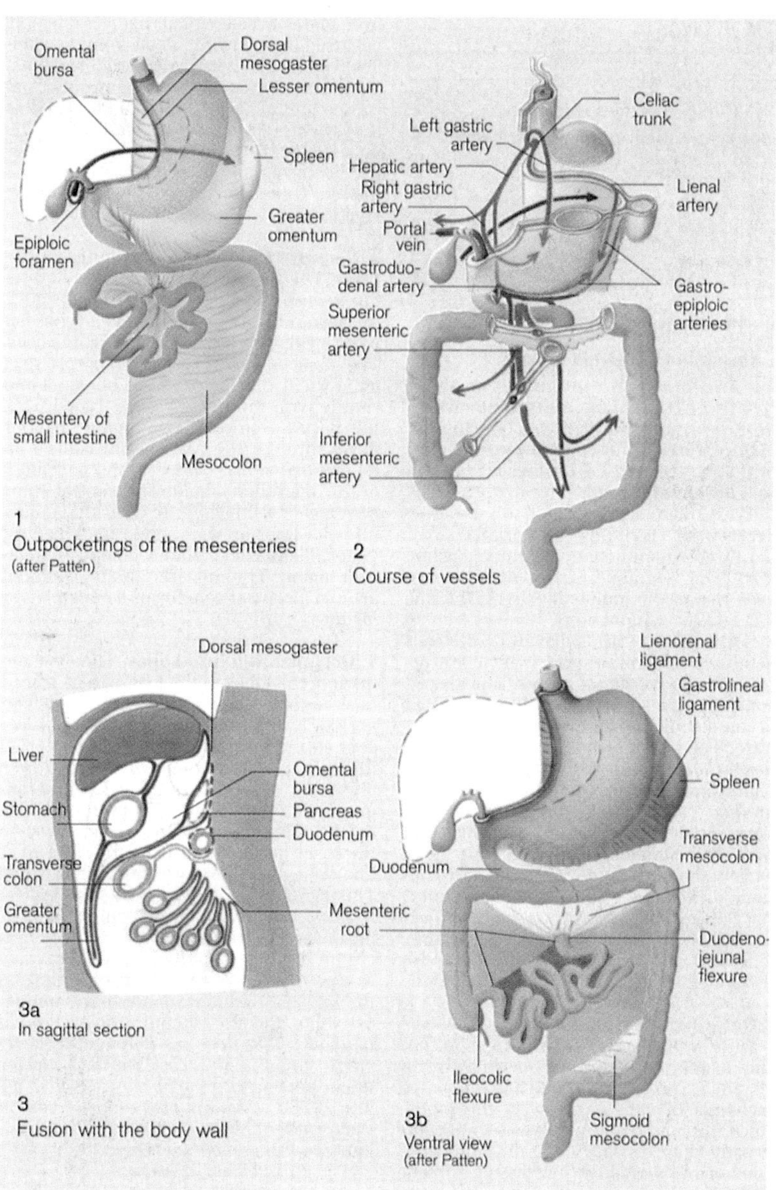

Derivation of definitive situs

Plate 13.5

Derivation of situs. Reproduced with permission from Thieme Medical Publishers Inc., New York, 1995, *Color Atlas of Embryology*, Ulrich Drews, Chapter 7: Gastrointestinal Tract.

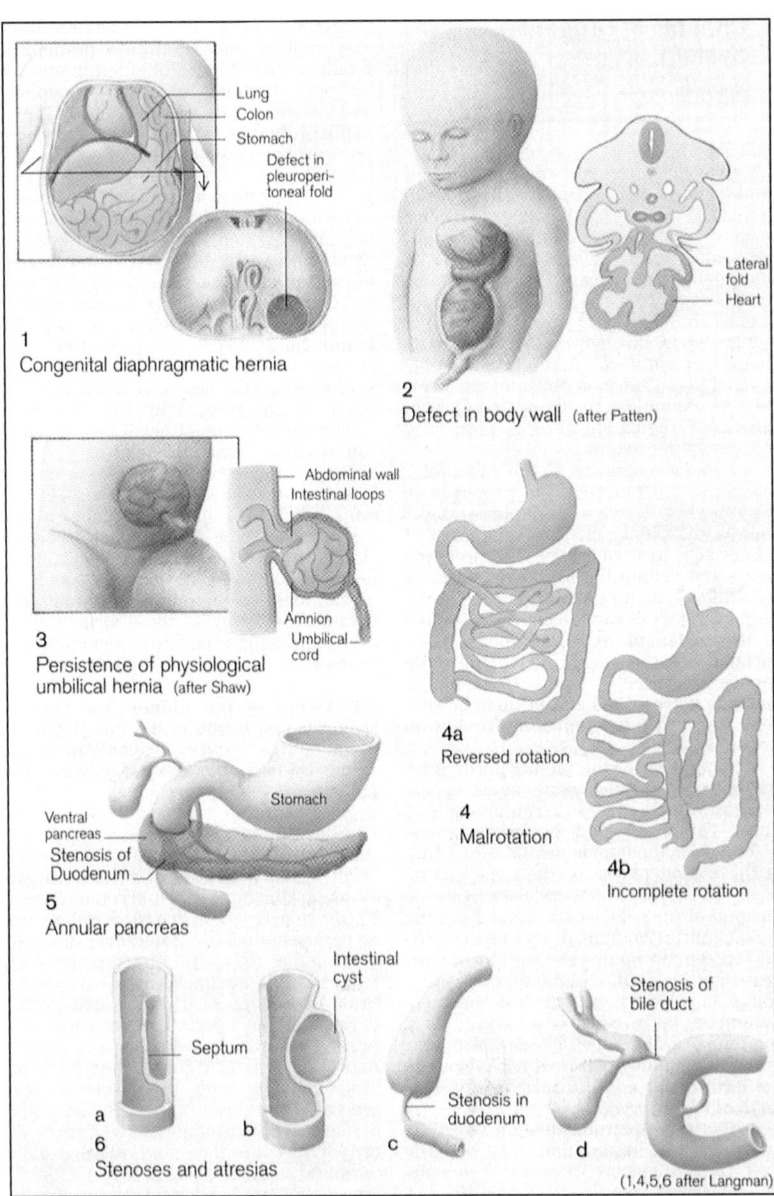

Malformations of the gastrointestinal tract

Plate 13.7

Malformations. Reproduced with permission from Thieme Medical Publishers Inc., New York, 1995, *Color Atlas of Embryology,* Ulrich Drews, Chapter 7: Gastrointestinal Tract.

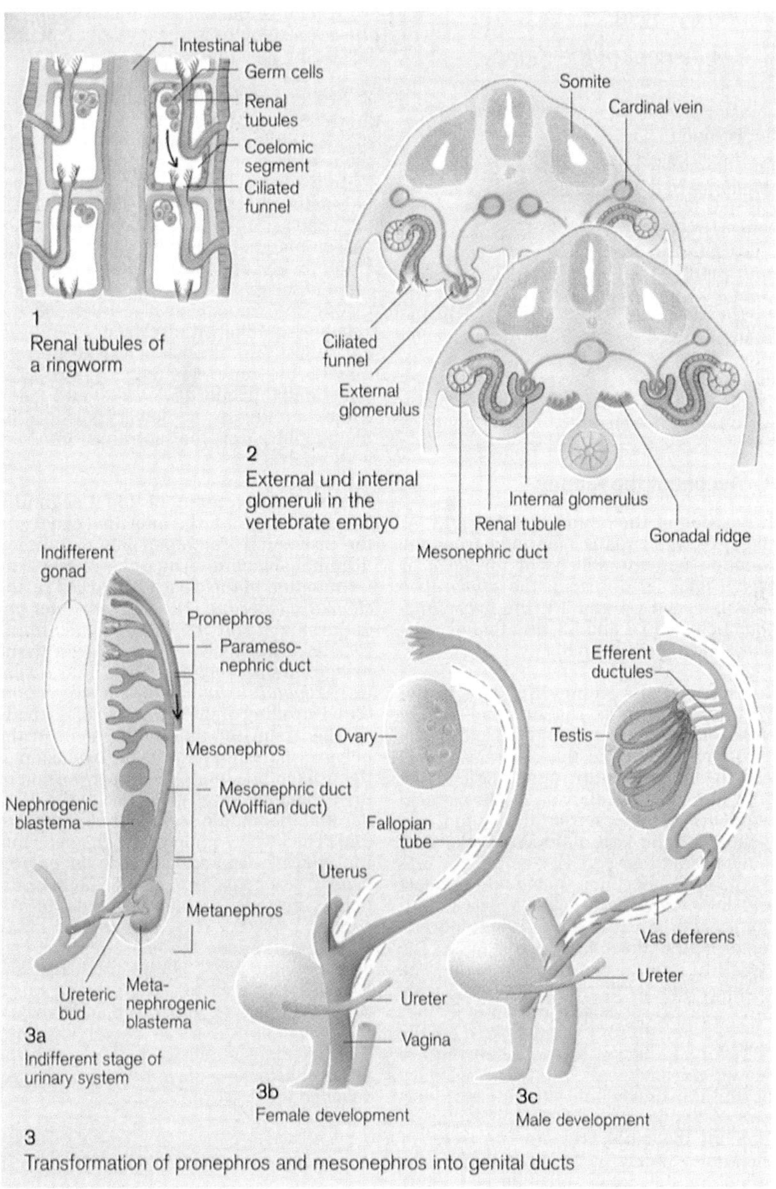

Relation between renal tubules and genital ducts

Plate 14.1

Overview: urogenital system. Reproduced with permission from Thieme Medical Publishers Inc., New York, 1995, *Color Atlas of Embryology,* Ulrich Drews, Chapter 8: Urogenital System.

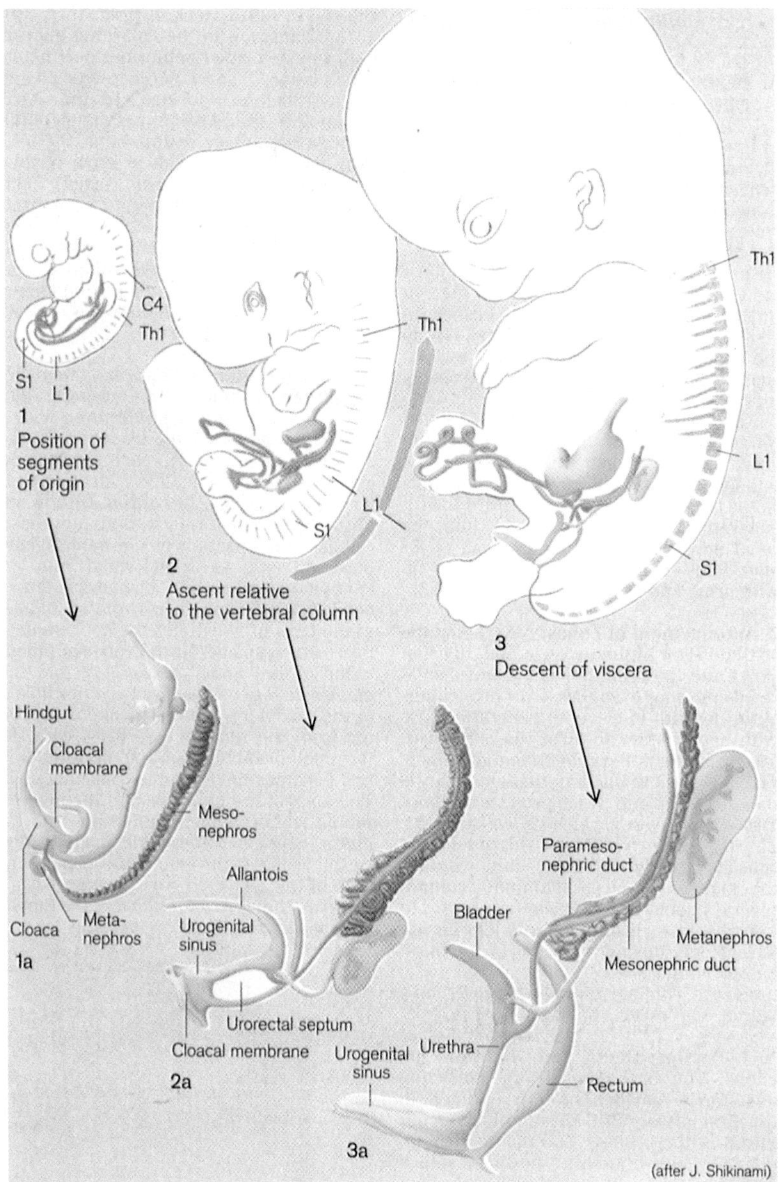

Ascent of kidneys

Plate 14.2

Ascent of kidneys. Reproduced with permission from Thieme Medical Publishers Inc., New York, 1995, *Color Atlas of Embryology,* Ulrich Drews, Chapter 8: Urogenital System.

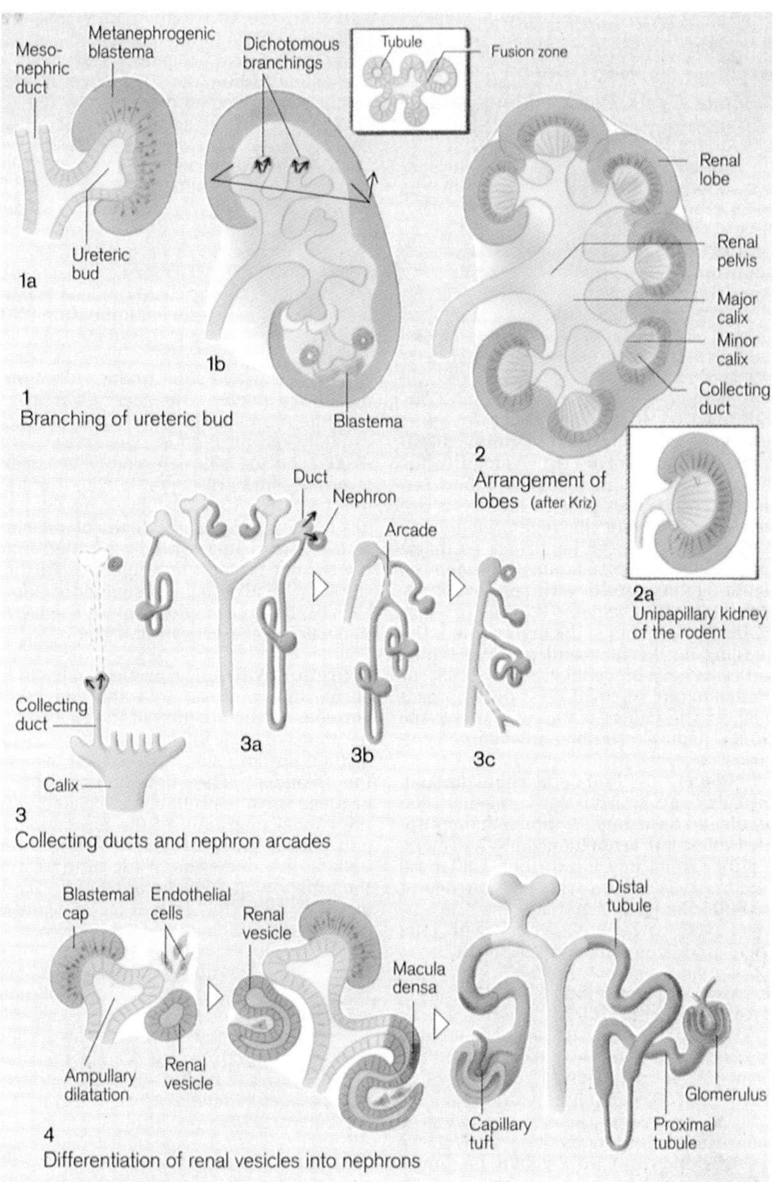

Differentiation of the metanephros

Plate 14.3

Differentiation of the kidney. Reproduced with permission from Thieme Medical Publishers Inc., New York, 1995, *Color Atlas of Embryology,* Ulrich Drews, Chapter 8: Urogenital System.

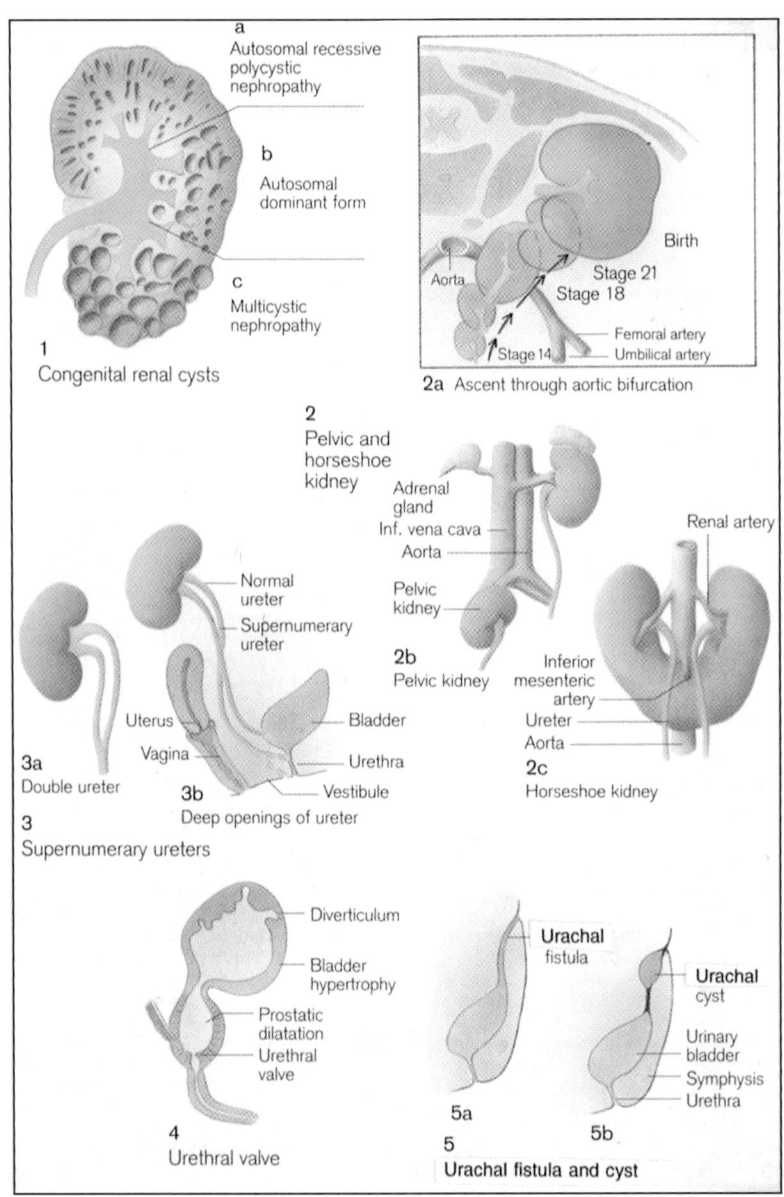

Malformations of kidney and urinary passages

Plate 14.4

Malformations of the urinary system. Reproduced with permission from Thieme Medical Publishers Inc., New York, 1995, *Color Atlas of Embryology*, Ulrich Drews, Chapter 8: Urogenital System.

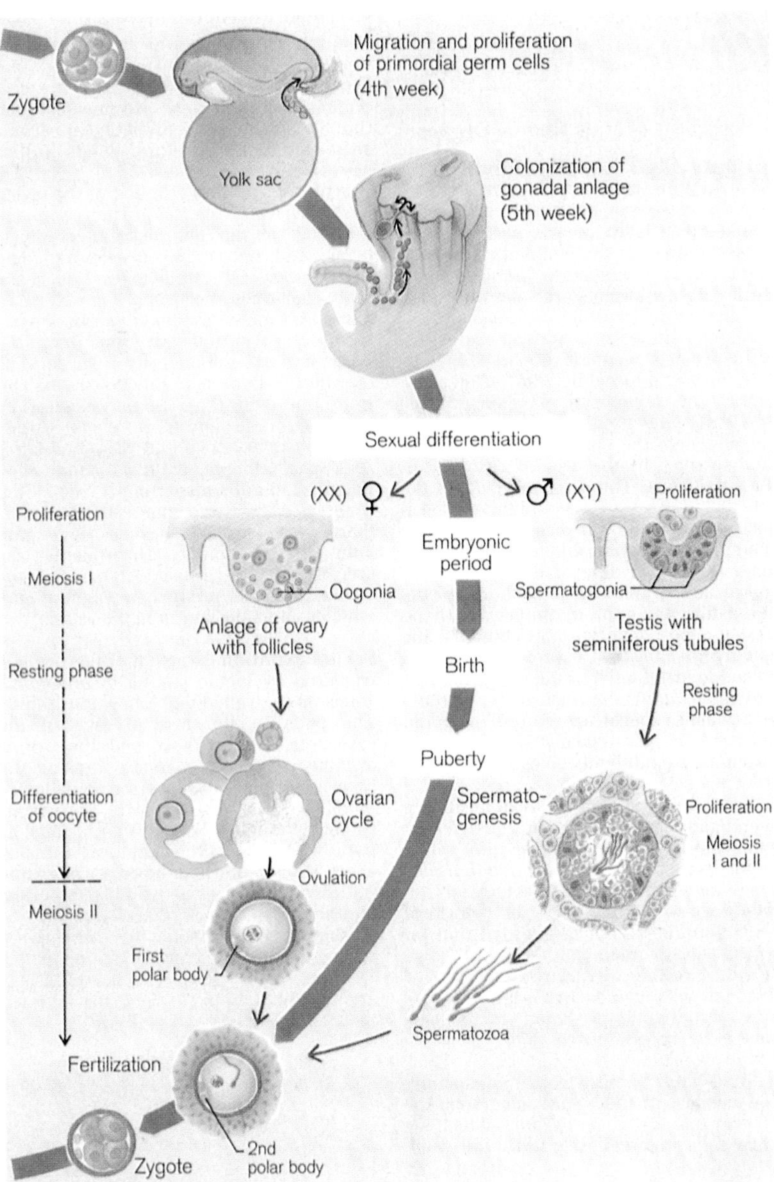

Human germ line

Plate 15.1

Overview: human germ line. Reproduced with permission from Thieme Medical Publishers Inc., New York, 1995, *Color Atlas of Embryology*, Ulrich Drews, Chapter 1: Reproduction

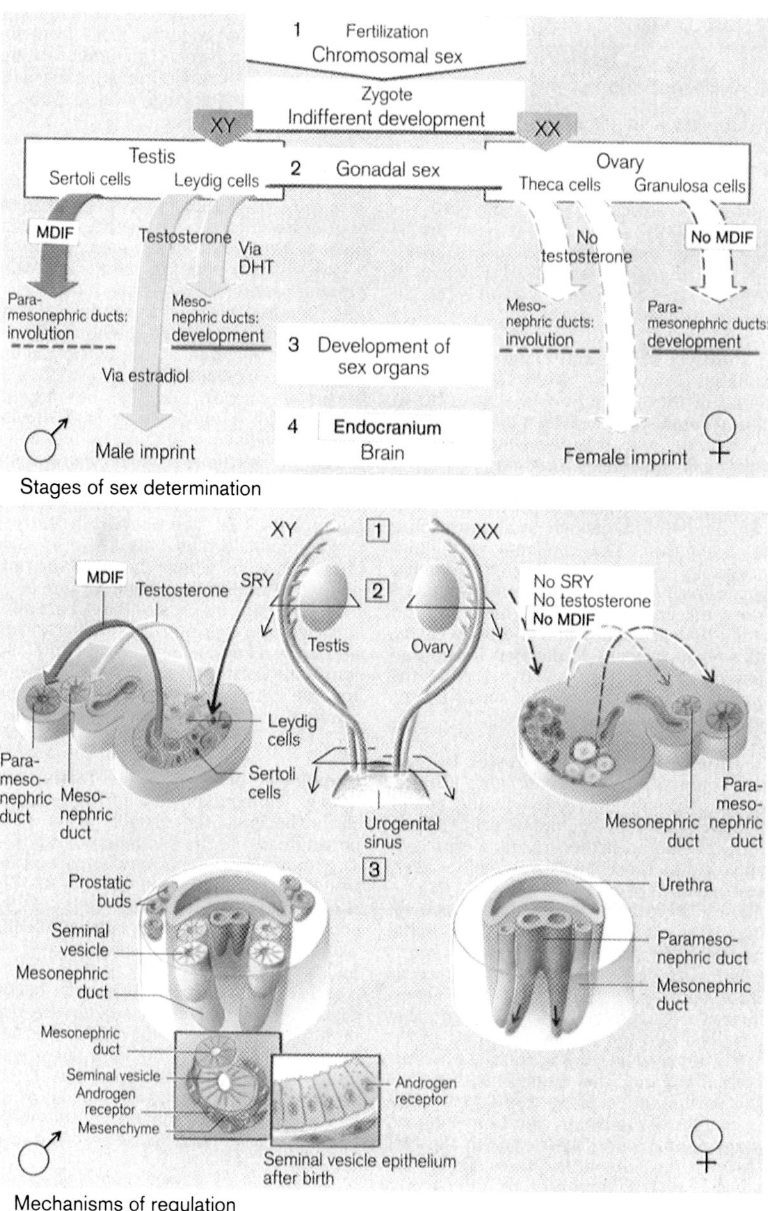

Plate 15.3

Sex determination. Reproduced with permission from Thieme Medical Publishers Inc., New York, 1995, *Color Atlas of Embryology,* Ulrich Drews, Chapter 8: Urogenital System.

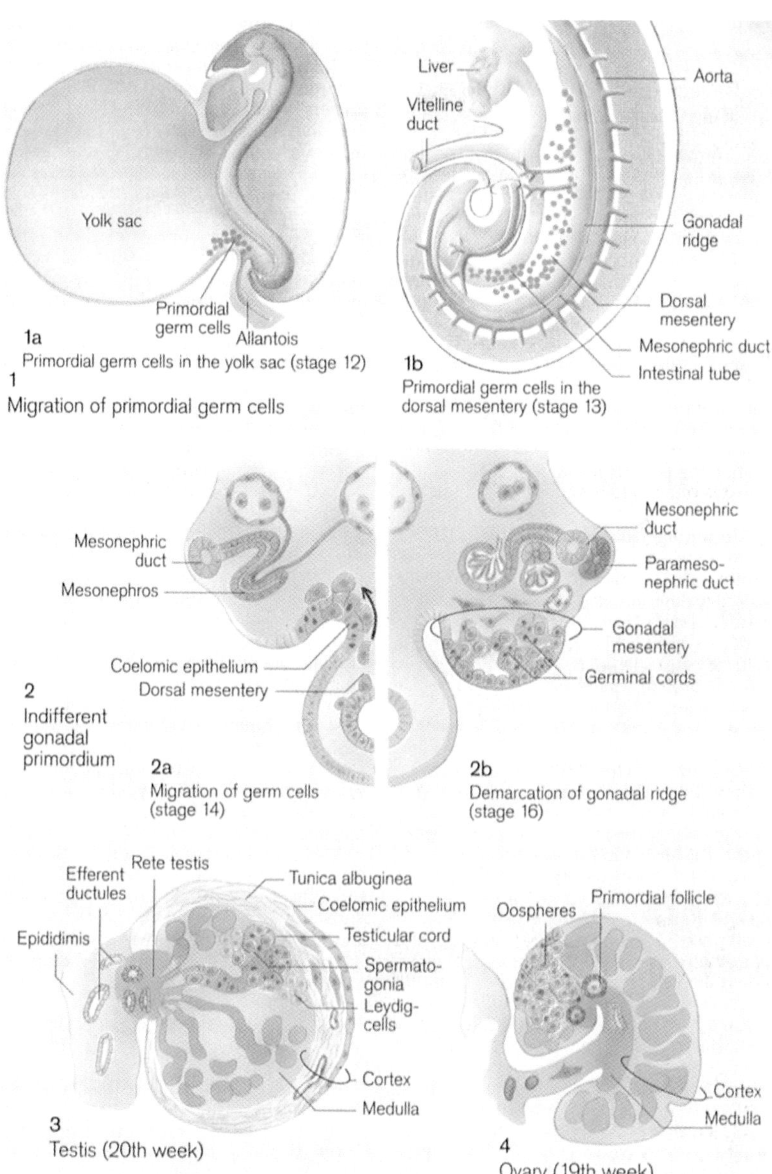

Plate 15.4

Differentiation of the gonads. Reproduced with permission from Thieme Medical Publishers Inc., New York, 1995, *Color Atlas of Embryology*, Ulrich Drews, Chapter 8: Urogenital System.

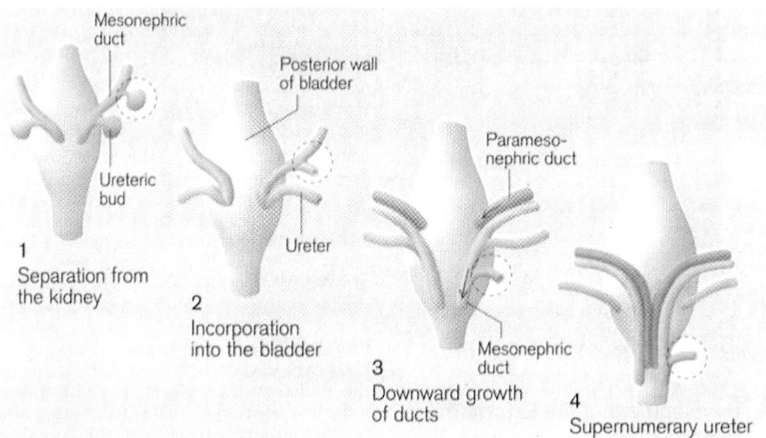

Separation of the ureter from the mesonephric duct

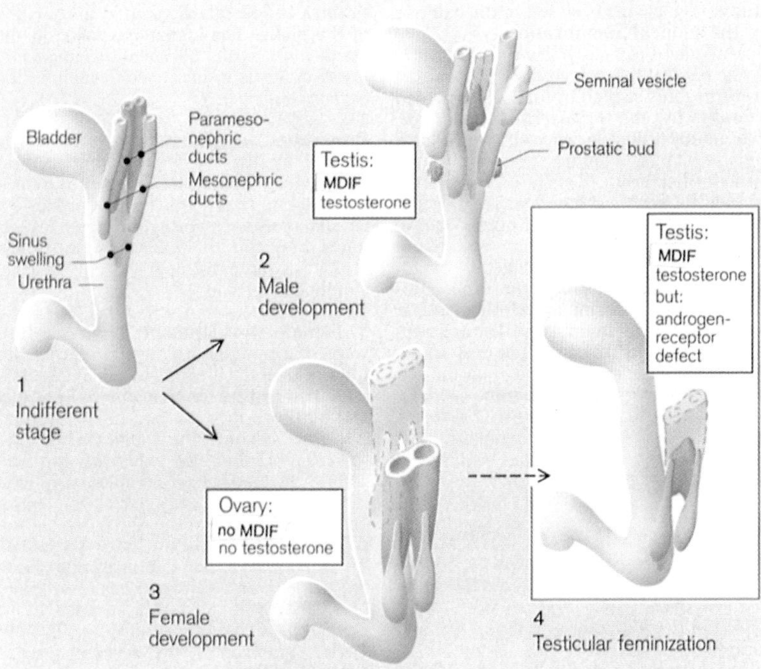

Downward growth of the vaginal anlage

Plate 15.5

Downward growth of the vagina. Reproduced with permission from Thieme Medical Publishers Inc., New York, 1995, *Color Atlas of Embryology*, Ulrich Drews, Chapter 8: Urogenital System.

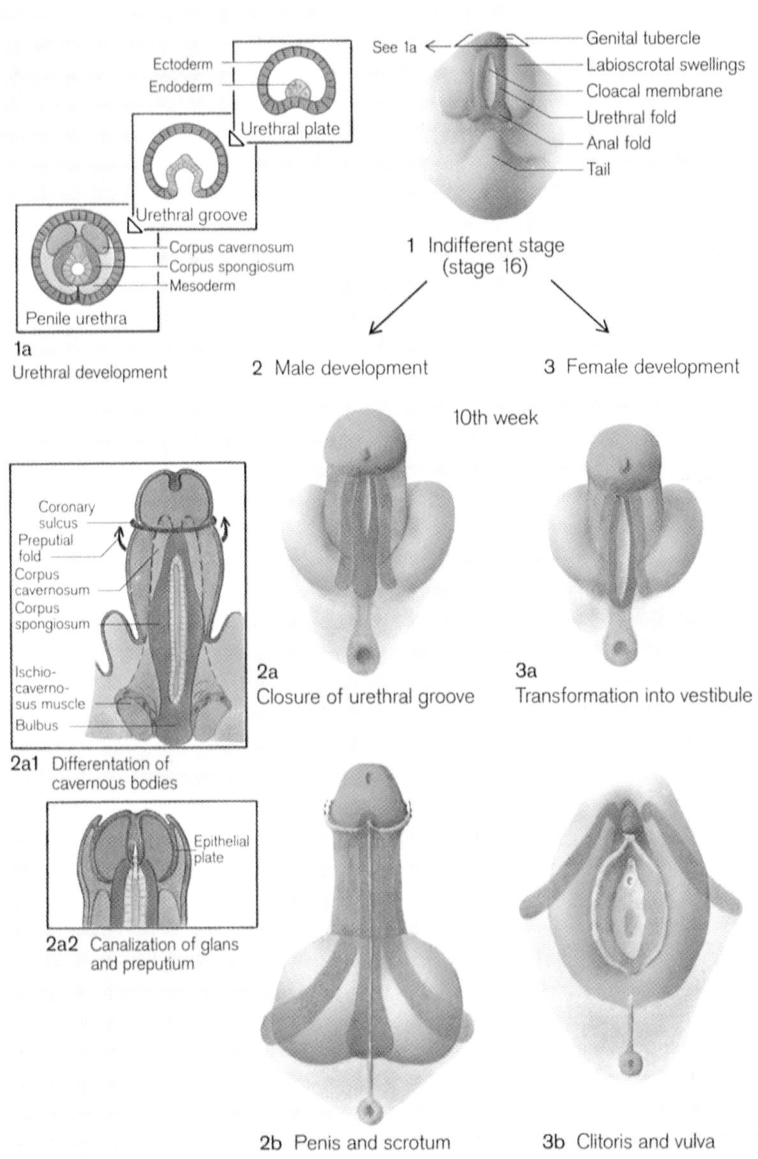

Development of the external genitalia

Plate 15.7

External genitalia. Reproduced with permission from Thieme Medical Publishers Inc., New York, 1995, *Color Atlas of Embryology*, Ulrich Drews, Chapter 8: Urogenital System.

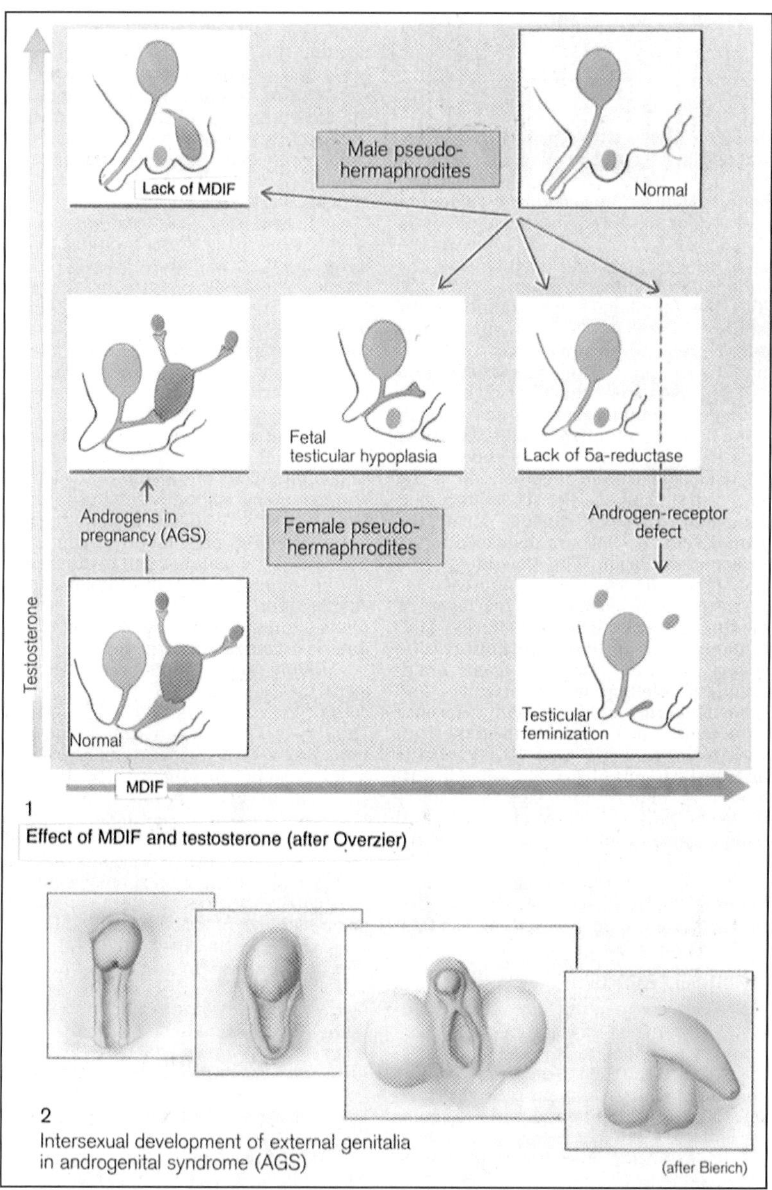

Disturbances in sexual development

Plate 15.9

Intersexuality. Reproduced with permission from Thieme Medical Publishers Inc., New York, 1995, *Color Atlas of Embryology,* Ulrich Drews, Chapter 8: Urogenital System.

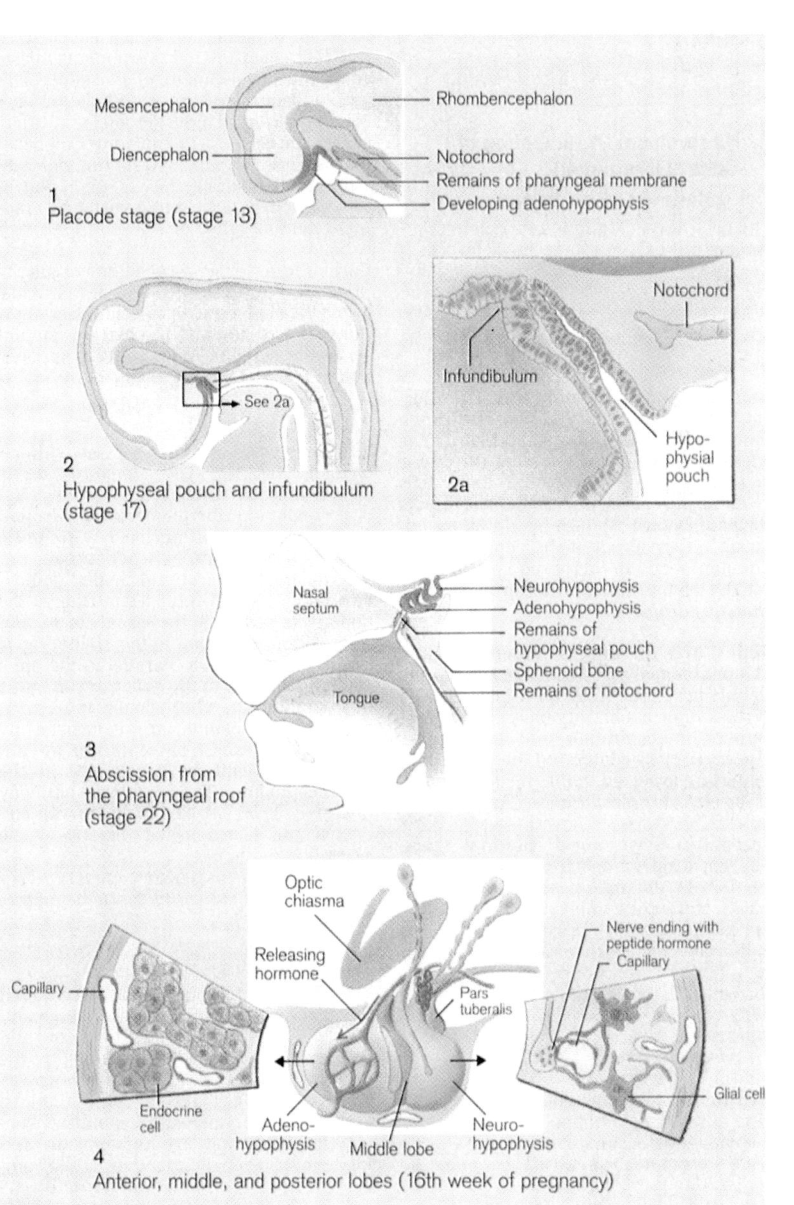

Development of pituitary gland

Plate 16.1

Pituitary gland. Reproduced with permission from Thieme Medical Publishers Inc., New York, 1995, *Color Atlas of Embryology*, Ulrich Drews, Chapter 4: Nervous System.

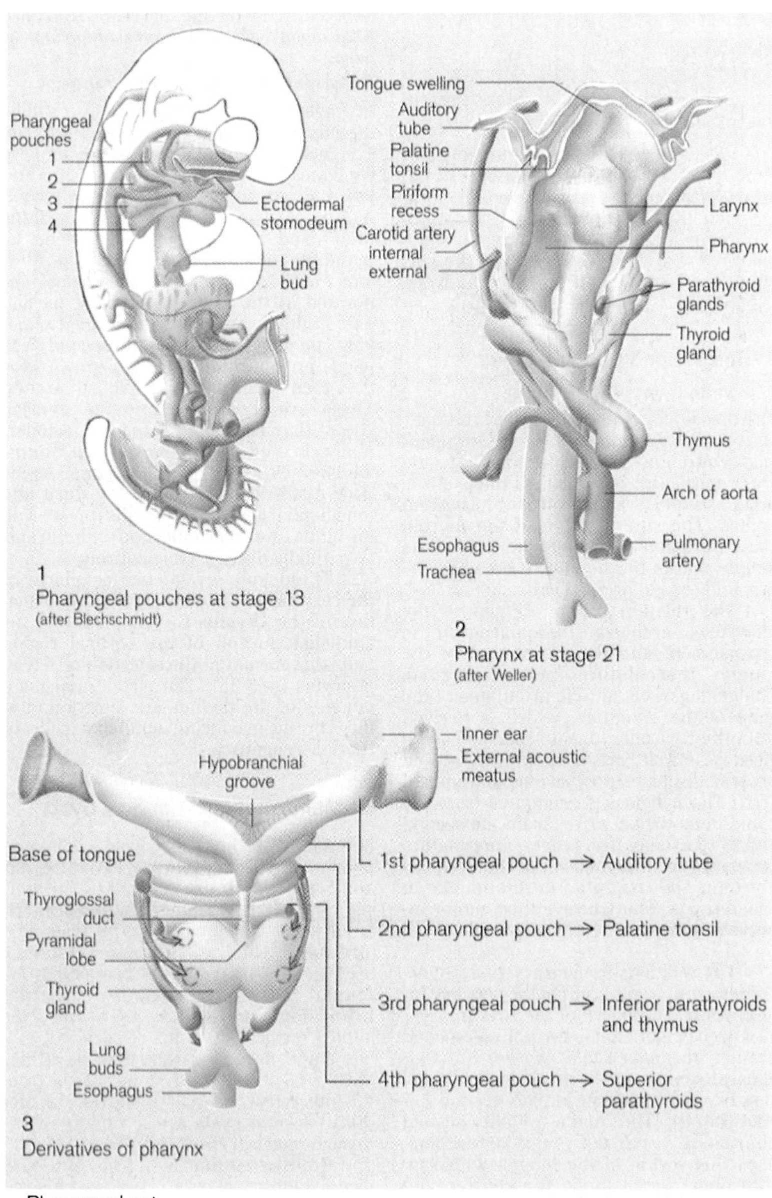

Plate 16.3

Derivatives of pharyngeal gut. Reproduced with permission from Thieme Medical Publishers Inc., New York, 1995, *Color Atlas of Embryology,* Ulrich Drews, Chapter 7: Gastrointestinal Tract.

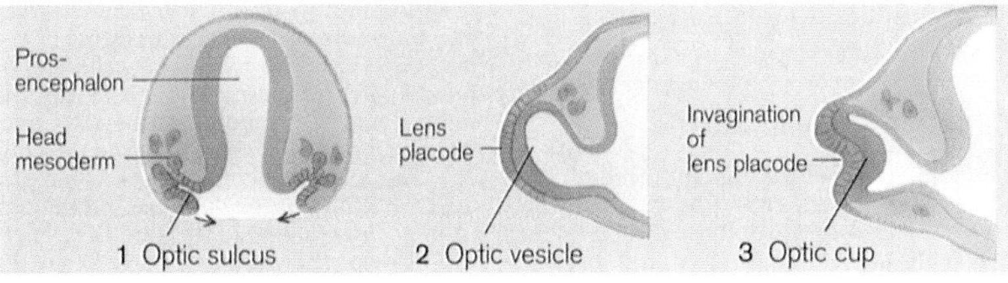

Development of the optic cup

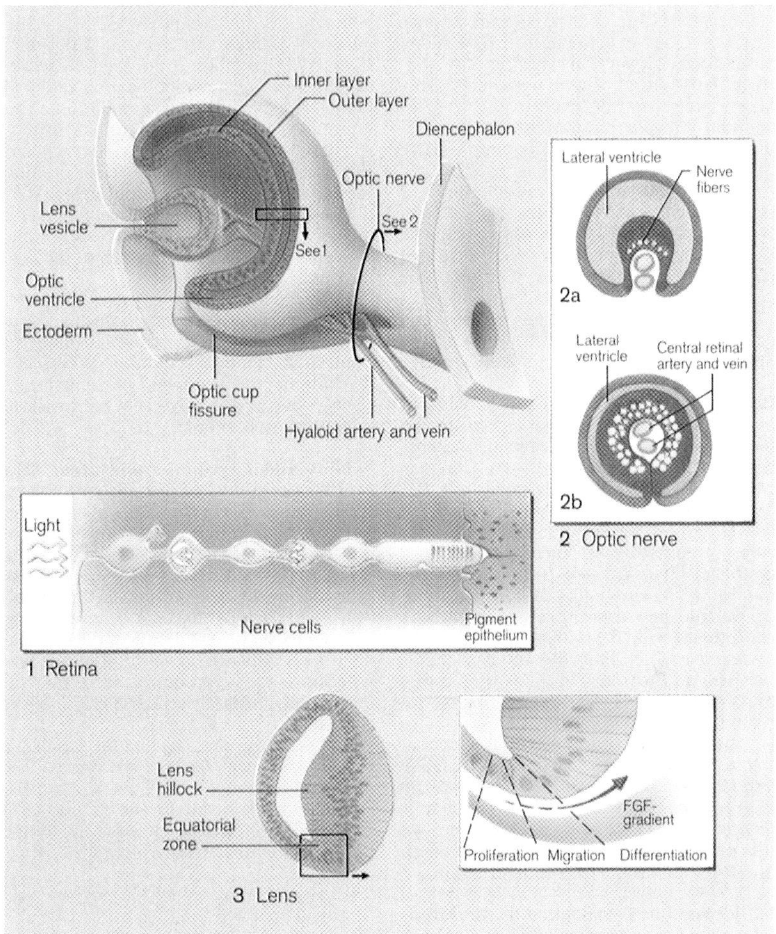

Optic cup and optic fissure

Plate 17.1

Optic cup. Reproduced with permission from Thieme Medical Publishers Inc., New York, 1995, *Color Atlas of Embryology*, Ulrich Drews, Chapter 5: Sense Organs.

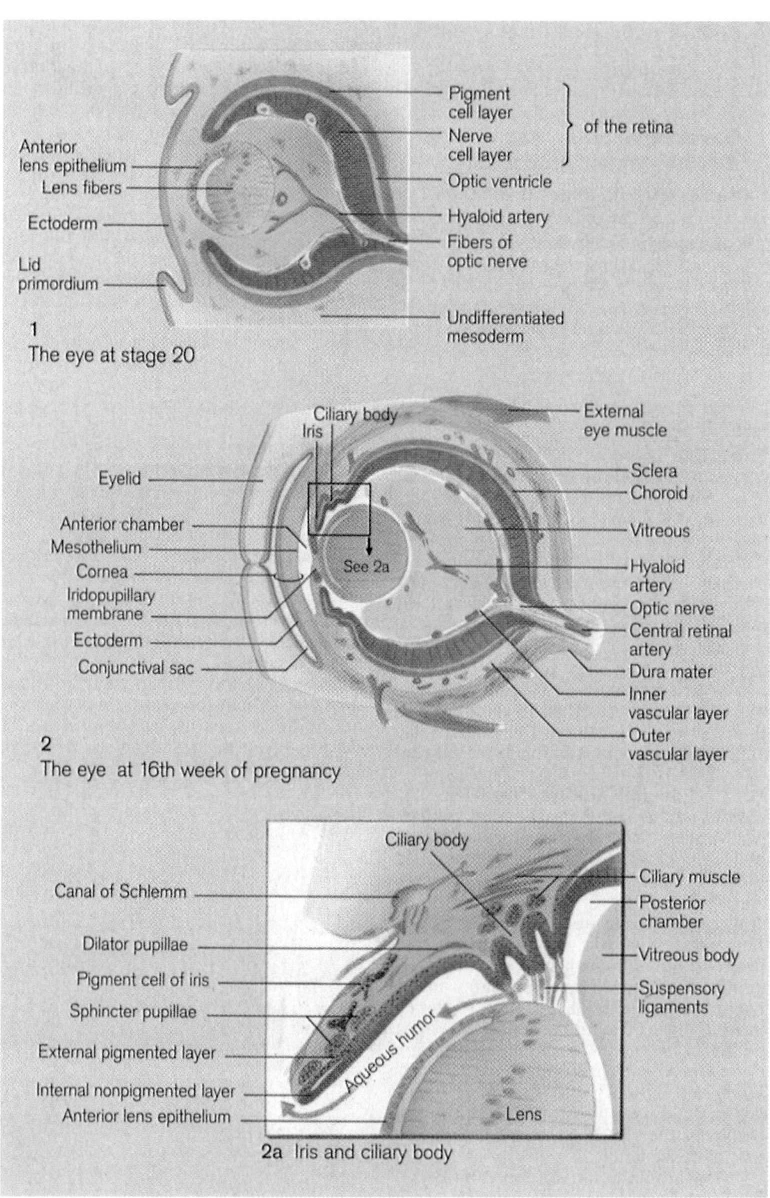

Accessory organs of the eye

Plate 17.2

Accessory organs of the eye. Reproduced with permission from Thieme Medical Publishers Inc., New York, 1995, *Color Atlas of Embryology,* Ulrich Drews, Chapter 5: Sense Organs.

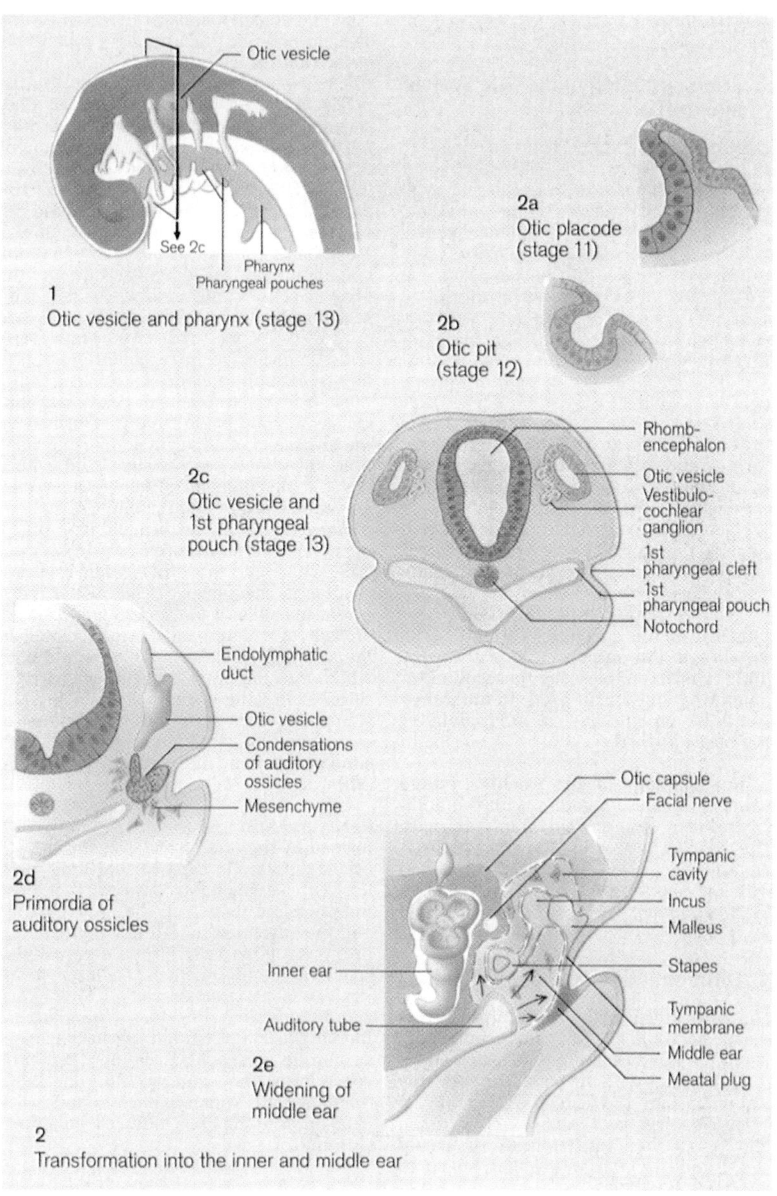

Development of the ear

Plate 17.3

Derivation of the ear. Reproduced with permission from Thieme Medical Publishers Inc., New York, 1995, *Color Atlas of Embryology*, Ulrich Drews, Chapter 5: Sense Organs.

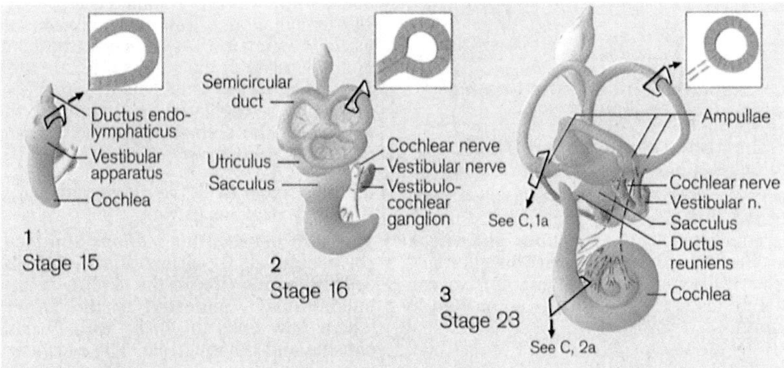

Transformation of the otic vesicle into the labyrinth

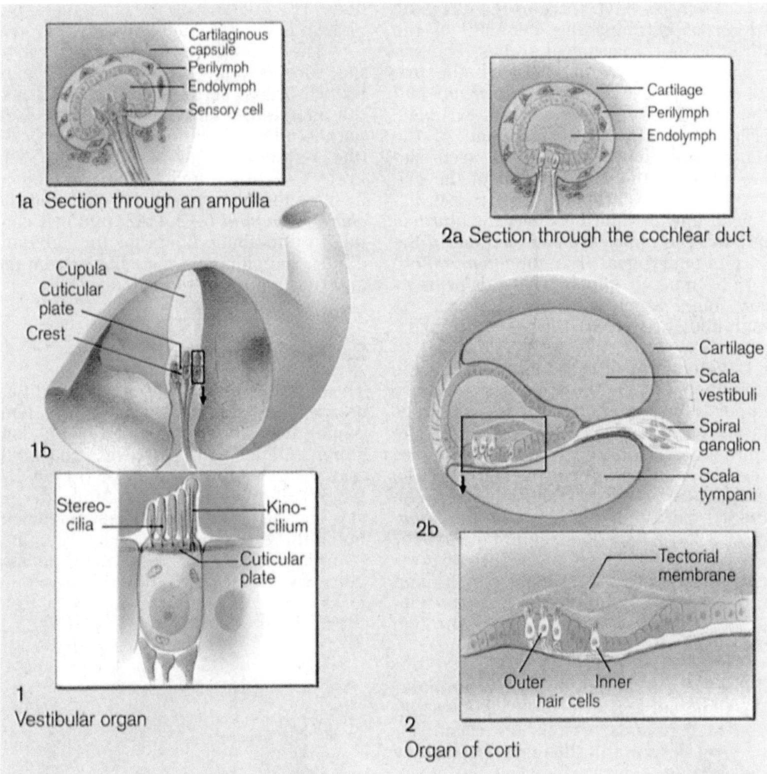

Differentiation of sensory cells

Plate 17.4

Inner ear. Reproduced with permission from Thieme Medical Publishers Inc., New York, 1995, *Color Atlas of Embryology,* Ulrich Drews, Chapter 5: Sense Organs.

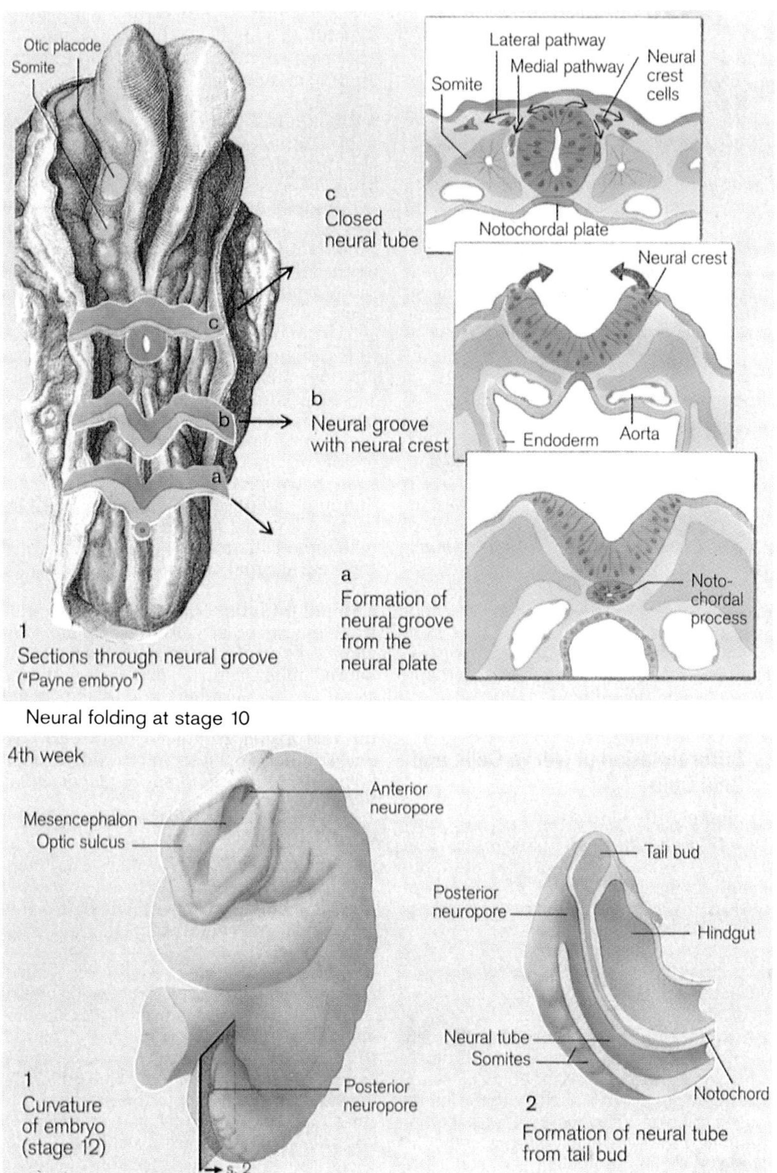

Plate 18.1

Neural tube. Reproduced with permission from Thieme Medical Publishers Inc., New York, 1995, *Color Atlas of Embryology,* Ulrich Drews, Chapter 4: Nervous System.

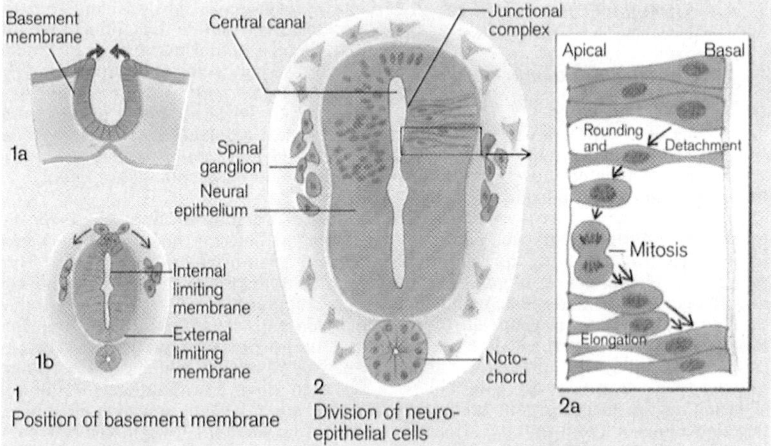

Proliferation of neuroepithelium

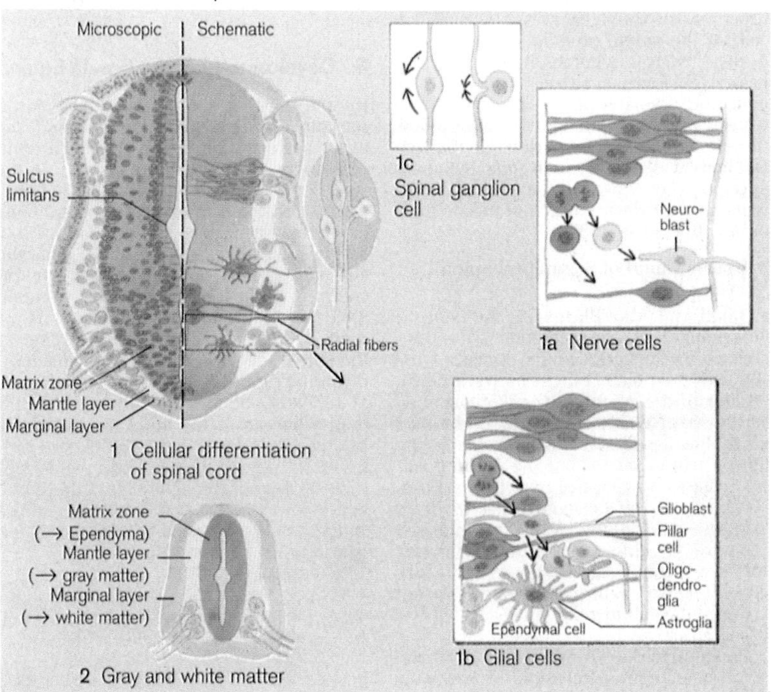

Differentiation of nerve cells and glial cells

Plate 18.2

Spinal cord. Reproduced with permission from Thieme Medical Publishers Inc., New York, 1995, *Color Atlas of Embryology,* Ulrich Drews, Chapter 4: Nervous System.

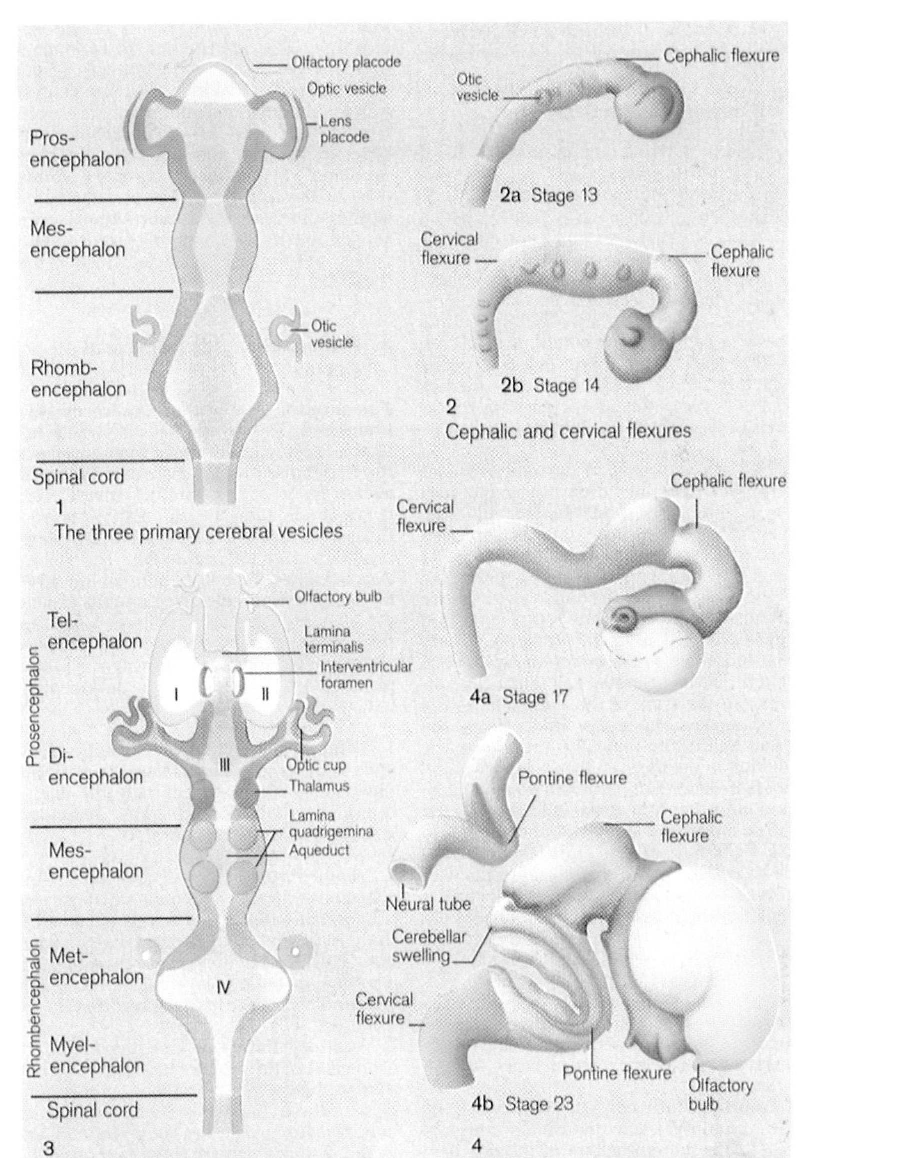

Plate 18.3

Brain vesicles. Reproduced with permission from Thieme Medical Publishers Inc., New York, 1995, *Color Atlas of Embryology,* Ulrich Drews, Chapter 4: Nervous System.

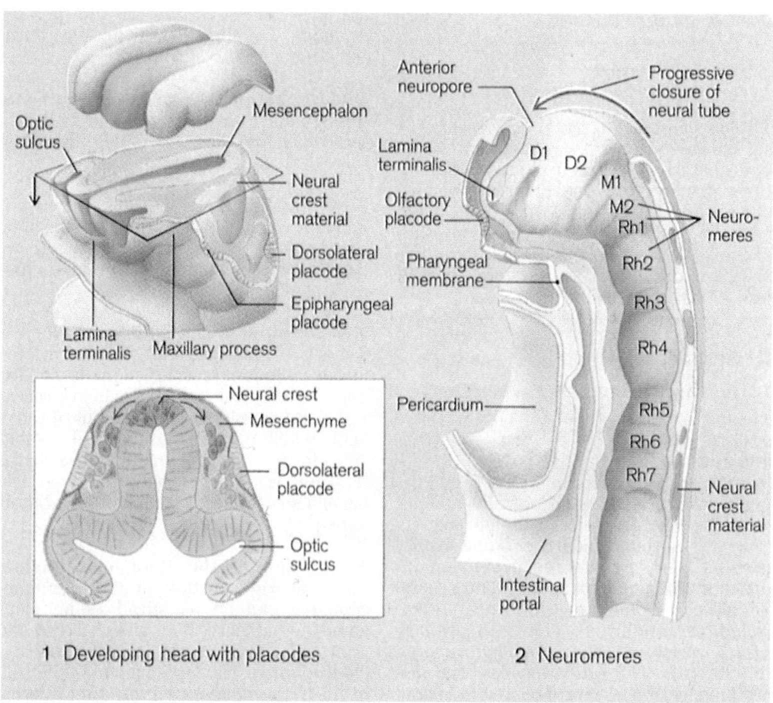

Neuromeres and placodes at stage 11

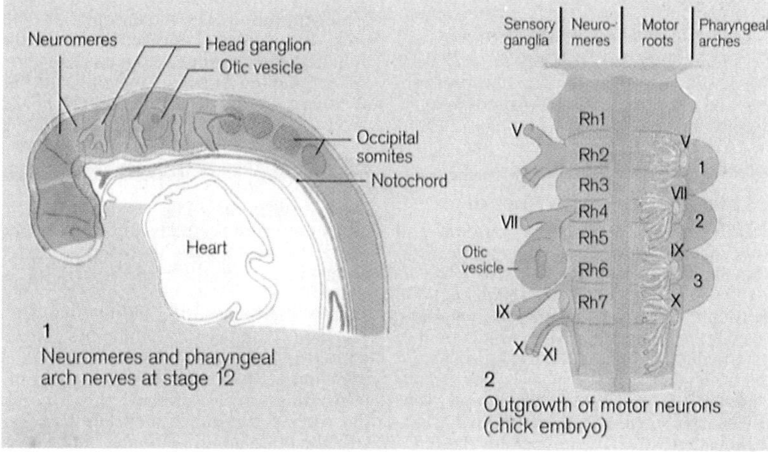

Segmental organization of pharyngeal arch nerves

Plate 18.4

Neuromeres and placodes. Reproduced with permission from Thieme Medical Publishers Inc., New York, 1995, *Color Atlas of Embryology,* Ulrich Drews, Chapter 4: Nervous System.

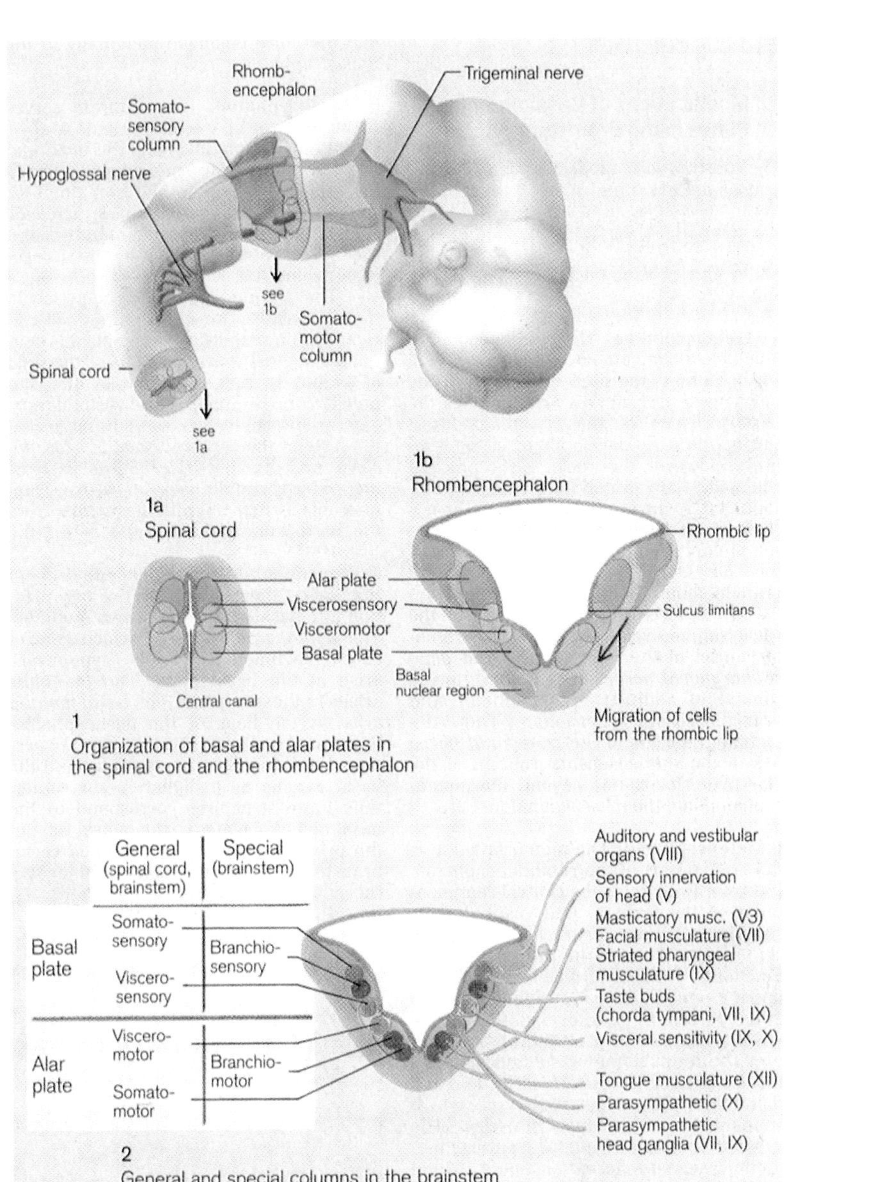

Basal and alar plates in the brainstem

Plate 18.5

Basal and alar plates. Reproduced with permission from Thieme Medical Publishers Inc., New York, 1995, *Color Atlas of Embryology,* Ulrich Drews, Chapter 4: Nervous System.

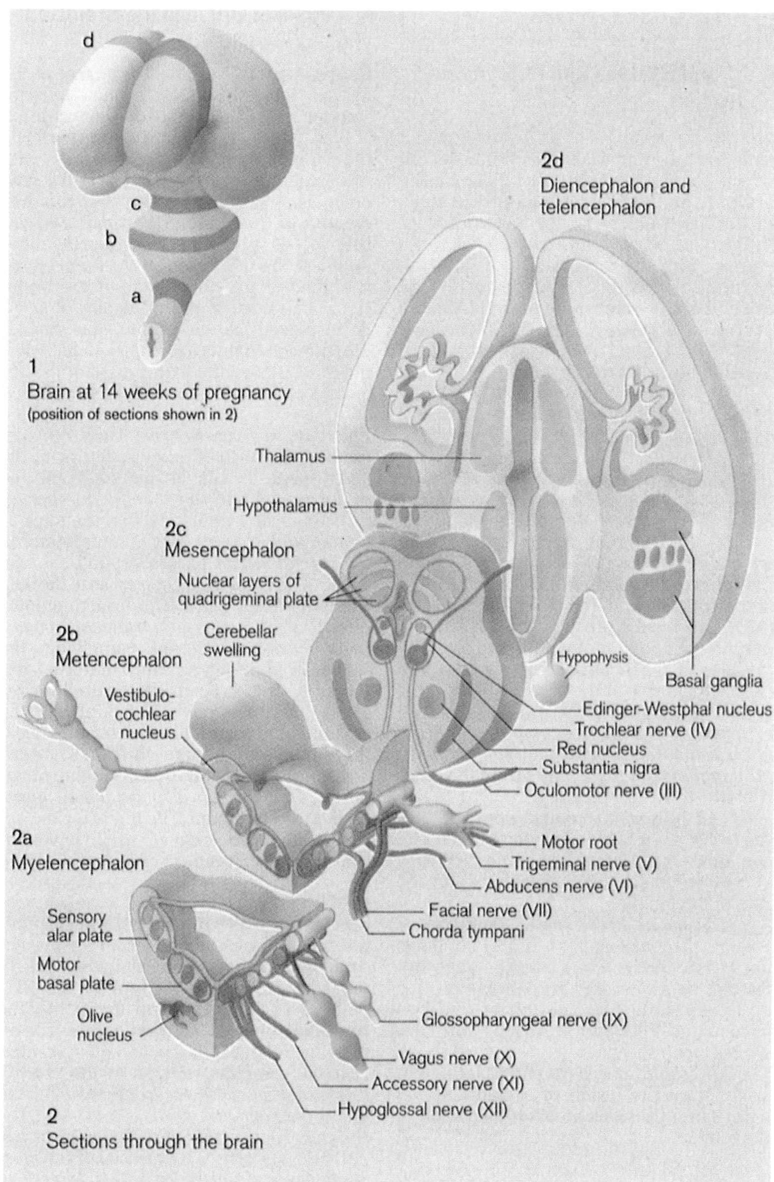

Modifications of basal and alar plates in the brainstem

Plate 18.6

Brain stem. Reproduced with permission from Thieme Medical Publishers Inc., New York, 1995, *Color Atlas of Embryology*, Ulrich Drews, Chapter 4: Nervous System.

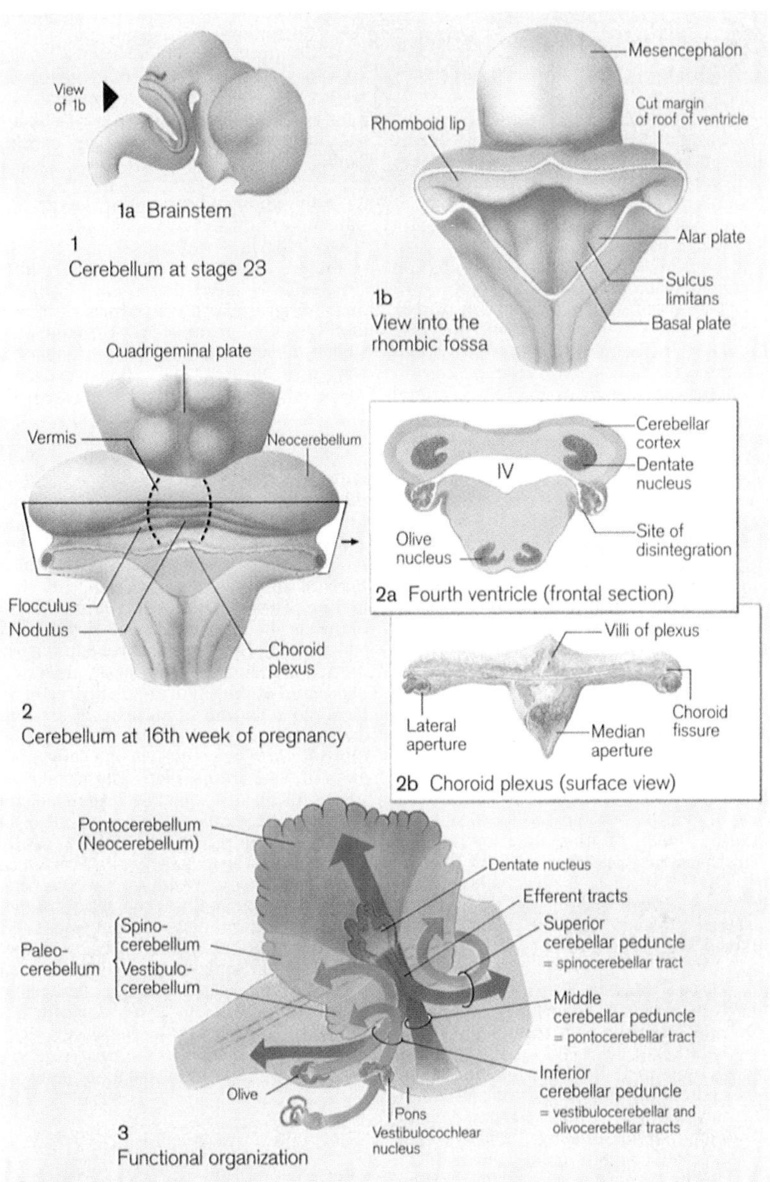

Development and organization of cerebellum

Plate 18.7

Cerebellum. Reproduced with permission from Thieme Medical Publishers Inc., New York, 1995, *Color Atlas of Embryology,* Ulrich Drews, Chapter 4: Nervous System.

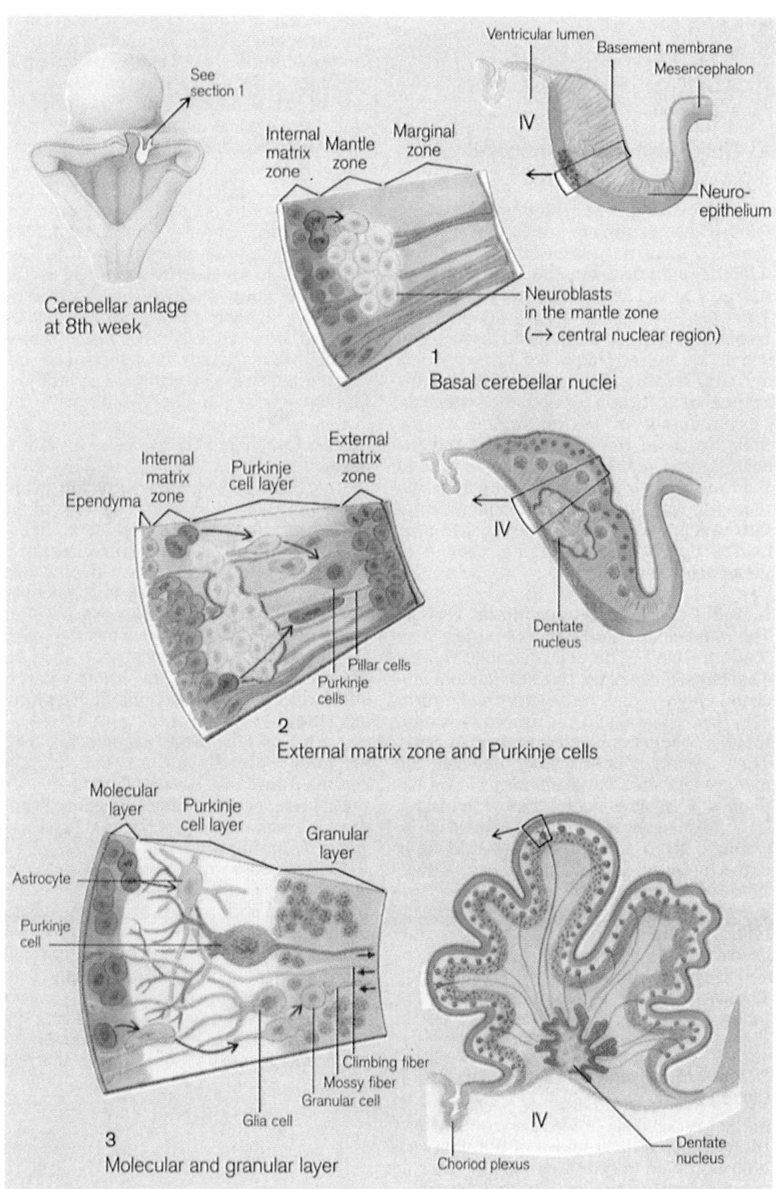

Histological development of cerebellar cortex

Plate 18.8

Cerebellar cortex. Reproduced with permission from Thieme Medical Publishers Inc., New York, 1995, *Color Atlas of Embryology*, Ulrich Drews, Chapter 4: Nervous System.

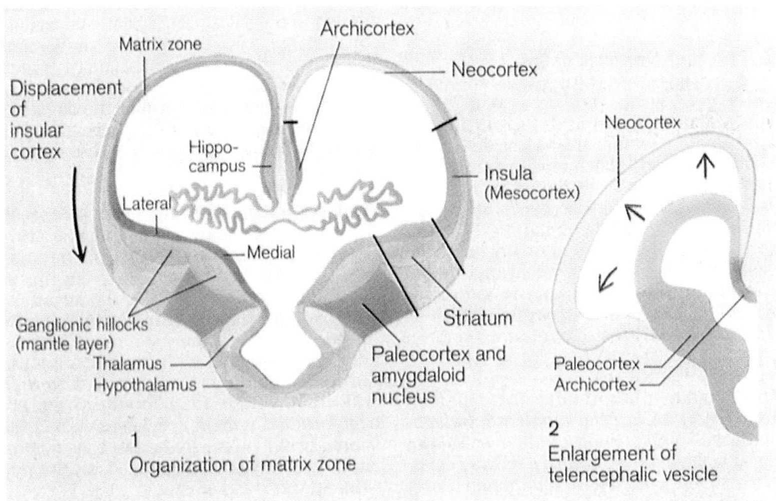

Plate 18.9

The paleo-, archi- and neocortex at the end of the embryonic period. Reproduced with permission from Thieme Medical Publishers Inc., New York, 1995, *Color Atlas of Embryology,* Ulrich Drews, Chapter 4: Nervous System.

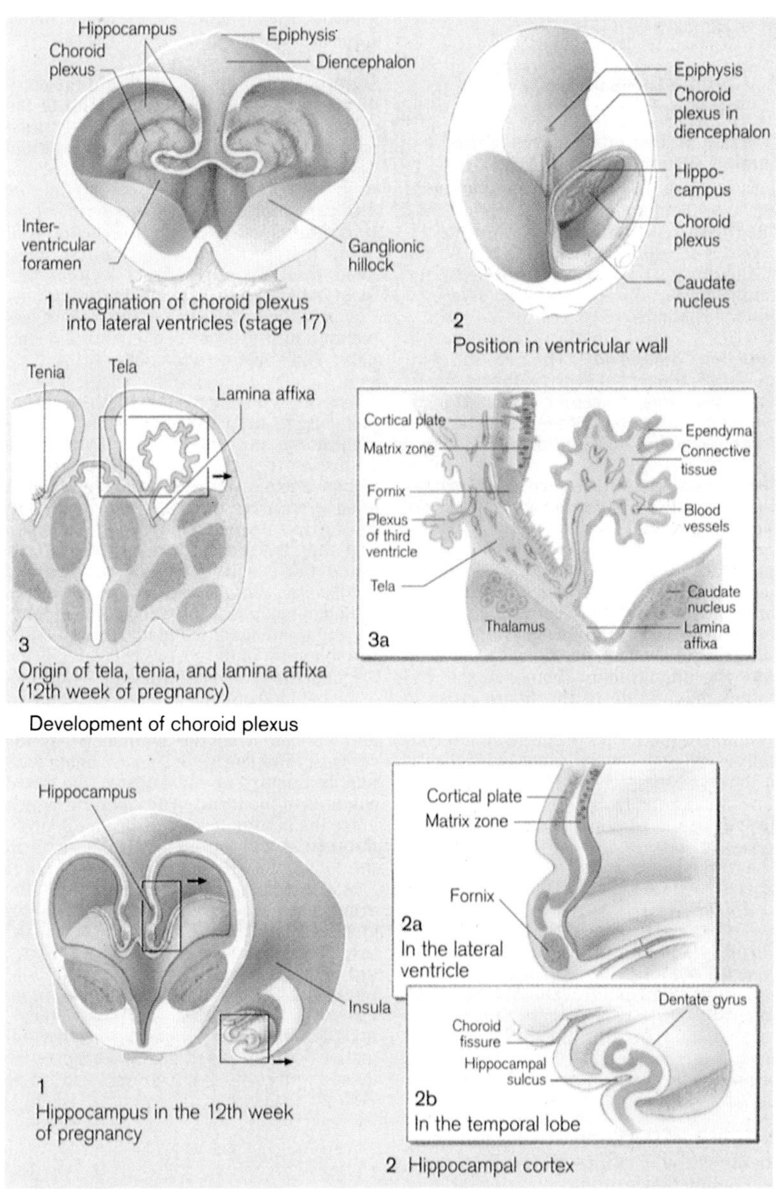

Plate 18.10

Choroid plexus and hippocampus. Reproduced with permission from Thieme Medical Publishers Inc., New York, 1995, *Color Atlas of Embryology*, Ulrich Drews, Chapter 4: Nervous System.

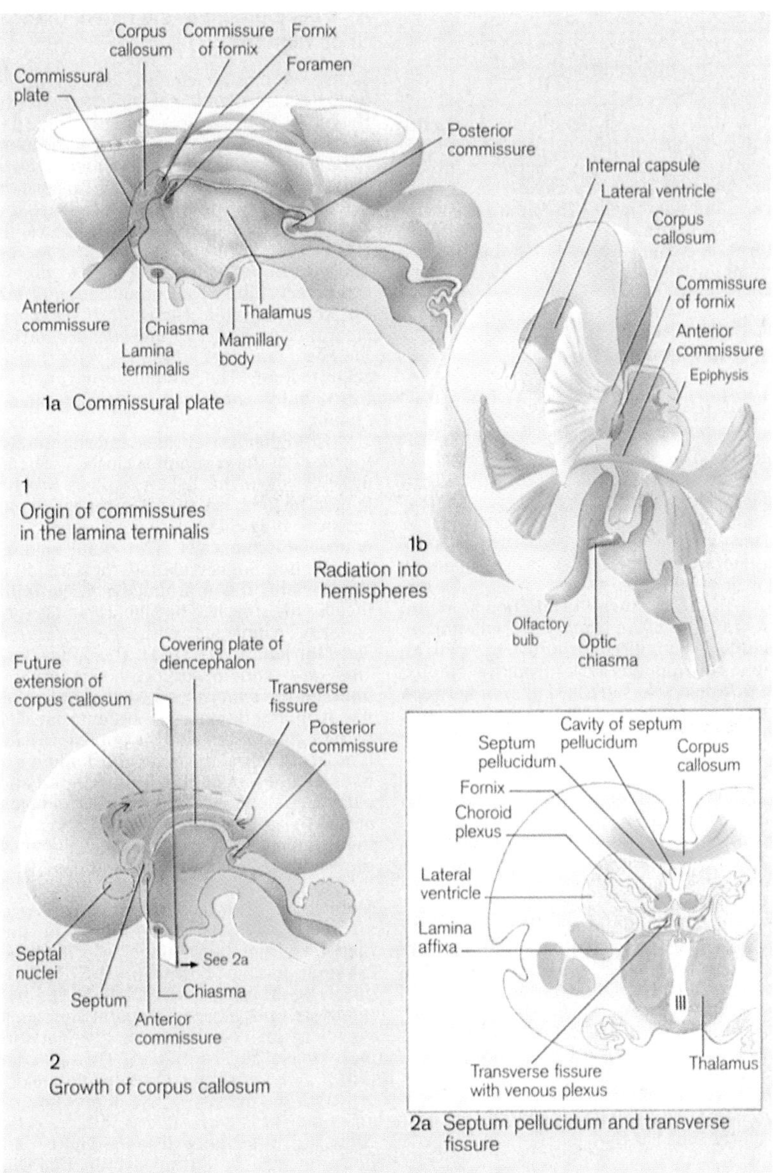

Commissures of telencephalon

Plate 18.11

Commissures. Reproduced with permission from Thieme Medical Publishers Inc., New York, 1995, *Color Atlas of Embryology*, Ulrich Drews, Chapter 4: Nervous System.

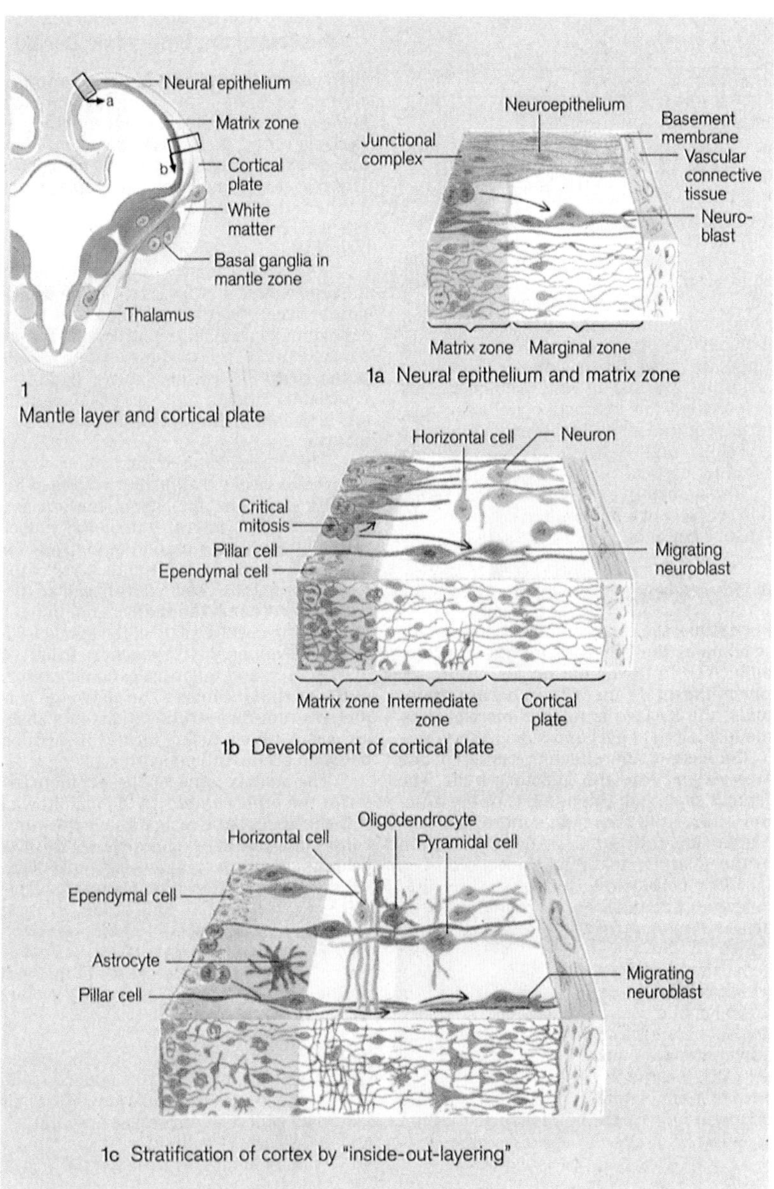

Development of cerebral cortex from neural epithelium

Plate 18.12

Cerebral cortex. Reproduced with permission from Thieme Medical Publishers Inc., New York, 1995, *Color Atlas of Embryology*, Ulrich Drews, Chapter 4: Nervous System.

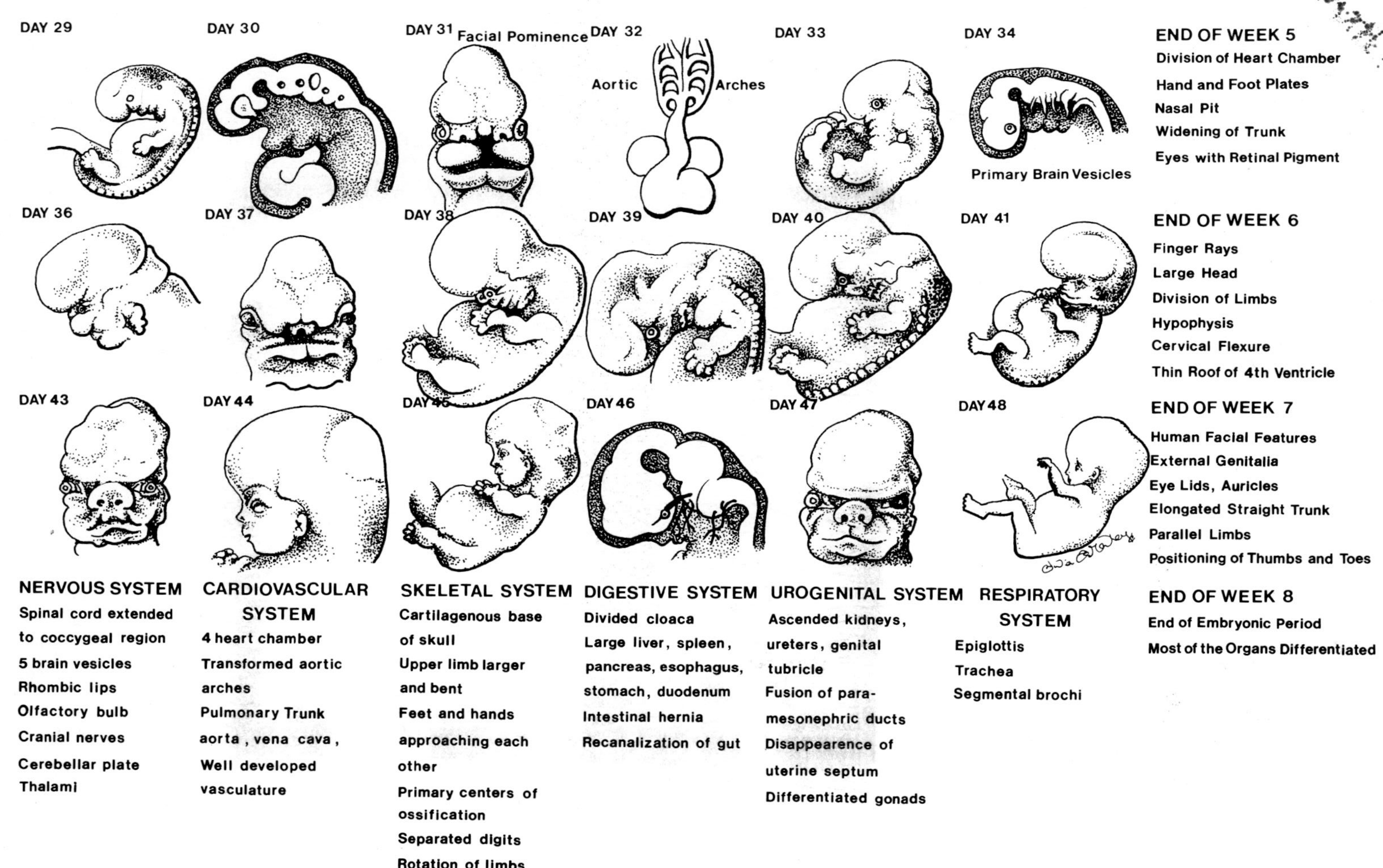
DAY 29
DAY 30
DAY 31
Facial Pominence
DAY 32
Aortic
Arches
DAY 33
DAY 34
Primary Brain Vesicles
END OF WEEK 5
Division of Heart Chamber
Hand and Foot Plates
Nasal Pit
Widening of Trunk
Eyes with Retinal Pigment
DAY 36
DAY 37
DAY 38
DAY 39
DAY 40
DAY 41
END OF WEEK 6
Finger Rays
Large Head
Division of Limbs
Hypophysis
Cervical Flexure
Thin Roof of 4th Ventricle
DAY 43
DAY 44
DAY 45
DAY 46
DAY 47
DAY 48
END OF WEEK 7
Human Facial Features
External Genitalia
Eye Lids, Auricles
Elongated Straight Trunk
Parallel Limbs
Positioning of Thumbs and Toes
NERVOUS SYSTEM
Spinal cord extended
to coccygeal region
5 brain vesicles
Rhombic lips
Olfactory bulb
Cranial nerves
Cerebellar plate
Thalami
CARDIOVASCULAR
SYSTEM
4 heart chamber
Transformed aortic
arches
Pulmonary Trunk
aorta , vena cava ,
Well developed
vasculature
SKELETAL SYSTEM
Cartilagenous base
of skull
Upper limb larger
and bent
Feet and hands
approaching each
other
Primary centers of
ossification
Separated digits
Rotation of limbs
DIGESTIVE SYSTEM
Divided cloaca
Large liver, spleen,
pancreas, esophagus,
stomach, duodenum
Intestinal hernia
Recanalization of gut
UROGENITAL SYSTEM
Ascended kidneys,
ureters, genital
tubricle
Fusion of para-
mesonephric ducts
Disappearence of
uterine septum
Differentiated gonads
RESPIRATORY
SYSTEM
Epiglottis
Trachea
Segmental brochi
END OF WEEK 8
End of Embryonic Period
Most of the Organs Differentiated